Advances in Comparative and Environmental Physiology 11

Volumes already published

Vol. 1: NaCl Transport in Epithelia
Edited By *R. Greger* (1988)

Vol. 2: From the Contents: Hibernation – Vapor Absorption
Nutrient Transport – Stable Pollutants (1988)

Vol. 3: Molecular and Cellular Basis of Social Behavior in Vertebrates
Edited by *J. Balthazart* (1989)

Vol. 4: Animal Adaptation to Cold
Edited by *L. C. H Wang* (1989)

Vol. 5: From the Contents: Osmoregulation – Chemoreception – Temperature – Spectroscopy – Metalloproteins Prostaglandins (Eicosanoids) – Maternal-Fetal Relationships (1989)

Vol. 6: Vertebrate Gas Exchange: From Environment to Cell
Edited by *R. G. Boutilier* (1990)

Vol. 7: From the Contents: Protein Turnover – Xenobiotic Metabolism Lipoproteins (1991)

Vol. 8: From the Contents: Dormancy in Aquatic Invertebrates-Root Effect – Red Blood Cell Functions in Fish Diving Mammals (1991)

Vol. 9: Volume and Osmolality Control in Animal Cells
Edited by *R. Gilles, E. K. Hoffmann,* and *L. Bolis* (1991)

Vol. 10: Comparative Aspects of Mechanoreceptor Systems
Edited by *F. Ito* (1992)

Vol. 11: Mechanics of Animal Locomotion
Edited by *R. McN. Alexander* (1992)

Volumes in preparation

Vol. 12: Muscle Contraction and Cell Motility: Molecular and Cellular Aspects
Edited by *H. Sugi* (1992)

Vol. 13: Blood and Tissue Oxygen Carriers
Edited by *Ch. P. Mangum* (1992)

Advances in

Comparative and Environmental Physiology 11

Mechanics of Animal Locomotion

Guest Editor: R. McN. Alexander

With Contributions by
R. McN. Alexander · M. B. Bennett · T. M. Casey · T. L. Daniel
M. E. DeMont · D. Grunbaum · A. I. Houston · C. Jordan
J. L. van Leeuwen · T. McGeer · G. R. Spedding

With 92 Figures

Springer-Verlag
Berlin Heidelberg New York
London Paris Tokyo
Hong Kong Barcelona
Budapest

Guest Editor:

Professor R. McN. Alexander
University of Leeds
Department of Pure and Applied Biology
Leeds LS2 9JT
United Kingdom

ISBN-13: 978-3-642-76695-4 e-ISBN-13: 978-3-642-76693-0
DOI: 10.1007/978-3-642-76693-0

Library of Congress Cataloging-in-Publication Data
Mechanics of animal locomotion / guest editor, R. McN. Alexander; with contributions by R. McN. Alexander ... [et al.]. p. cm. – (Advances in comparative and environmental physiology; 11) Includes bibliographical references and index.

1. Biomechanics. 2. Animal locomotion. 3. Physiology, Comparative. I. Alexander, R. McNeill. II. Series. QP303.M42 1991 591.1'852 – dc20 91-27959 CIP

Softcover reprint of the hardcover 1st edition 1992

Typesetting: Thomson Press (India) Ltd., New Delhi
31/3145 543210 – Printed on acid-free paper

Foreword to the Series

The aim of the series is to provide comprehensive, integrated reviews giving sound, critical and provocative summaries of our present knowledge in environmental and comparative physiology, from the molecular to the organismic level.

Living organisms have evolved a widespread range of basic solutions to cope with the different problems, both organismal and environmental, with which they are faced. A clear understanding of these solutions is of course of fundamental interest for all biologists, zoologically or medically oriented. They can be best comprehended in the framework of the environmental and/or comparative approaches. These approaches demand either wide surveys of animal forms or a knowledge of the specific adaptive features of the species considered. This diversity of requirements, both at the conceptual and technological level, together with the fact that physiology and biochemistry have long been mainly devoted to the service of medicine, can account for the fact that these approaches emerged only slow amongst the other new, more rapidly growing disciplines of the biological sciences.

The field has now gained the international status it deserves and the organization of a series devoted to it appeared timely to me in view of its actual rapid development and of the interest it arouses for a growing number of biologists, physiologists, and biochemists, independently of their basic, major orientation.

Liège, Belgium *Raymond Gilles*

List of Editors

Those interested in contributing to this series should contact Professor R. Gilles or one of the Series Editors. Review proposals should include a short outline – and a Table of Contents – briefly defining the aims and scope of the review.

Preface

Geoff Goldspink and I edited a book, *Mechanics and Energetics of Animal Locomotion*, which was published by Chapman and Hall in 1977. It dealt at an advanced level with all aspects of animal locomotion, emphasizing particularly the topics of then current research. Since then there have been more elementary books on the subject, including my own *Locomotion of Animals* (Blackie 1982), and specialized books on such topics as swimming and flight, but (despite very substantial progress in research) there has been no advanced book covering the whole range of animal locomotion. It seemed to me and to Professor Gilles (editor of this series) that a new book was needed.

Plainly, a book of this length cannot contain everything that is known about animal locomotion. We have not attempted to make it encyclopedic, but have tried to show where the study of animal locomotion stands now, and where it is going. Older books remain useful as sources of long-established information.

In selecting authors, I passed over some eminent authorities who have made enormous contributions to the field. They would have written excellent chapters, which would however almost inevitably been rather similar to the many reviews and book chapters that they had previously written. Instead, I have selected authors who have made important contributions to our knowledge but had not, in most cases, written previous wide-ranging reviews. The chapters that they have written seem to me to be fresh, original and enlightening, and I am confident that this book will be recognized as a major landmark in the study of locomotion.

It is an advanced book, designed primarily for research workers, graduate students and university teachers. Some of the material in it will inevitably be found difficult, especially for readers whose training has not included much physical mechanics. I have tried to make it a little easier for them, by including an introductory chapter that explains very simply most of the mechanics used later. If you find parts of the book difficult, please do not be discouraged: it is full of profound insights that will help us all to understand how animals move.

Leeds, U.K., February 1992 *R. McNeill Alexander*

Contents

Chapter 3

The Aerodynamics of Flight
G. R. Spedding

Chapter 4

Principles of Walking and Running
T. McGeer

Chapter 5

Empirical Studies of Walking and Running
M. B. Bennett

Chapter 9

Optimization of Locomotion

A. Houston

Chapter 1

Basic Mechanics

R.McN. Alexander

Contents

1 Introduction

Mechanics is one of the most basic and widely applicable branches of science, but its methods are not very familiar to some biologists. For that reason, it seems useful to start this book on the mechanics of animal movement with an elementary summary and explanation of some of the terms and concepts of mechanics. If you are a physical scientist or someone already experienced in biomechanics you will probably choose to ignore this chapter and move immediately to Chap. 2. If, at the other extreme, you are a biologist whose training in physical science was poor, you may wish to read this whole chapter before moving on. There will also be an intermediate group of readers who will at first simply note that this chapter is here, but will refer back to it whenever they need help.

A single chapter can give only a very brief outline of the branches of mechanics that are useful in the study of animal locomotion. More information can be found in Alexander (1983), Vogel (1981) and Wainwright et al. (1976), and in physics and engineering textbooks.

The basic units of mechanics are the metre (abbreviation m) for length, the kilogramme (kg) for mass and the second (s) for time. We will soon see how other

Department of Pure and Applied Biology, University of Leeds, Leeds LS2 9JT, UK units are derived from them. Prefixes are used to indicate fractions or multiples, both of the basic units and of the derived ones. For example, a millisecond (ms) is one thousandth of a second and a kilometre (km) is one thousand metres. Notice the anomaly that the kilogramme (one thousand grammes) rather than the gramme is regarded as the basic unit of mass in the internationally agreed (Système Internationale or S.I.) system of units. If you want to put numbers into an equation to calculate something, the first step is generally to convert any quantities expressed in milliseconds, kilometres etc. to the basic units. The most commonly used fractions and multiples are

milli- (m) 10^{-3}; micro- (μ) 10^{-6}; nano- (n) 10^{-9}
kilo- (k) 10^{3}; mega- (M) 10^{6}; giga- (G) 10^{9} .

2 Movements and Forces

The position of a point is often described by giving its coordinates in a Cartesian system, with X and Y axes to define position in a plane or X, Y and Z axes for three-dimensional space. Think of a point moving in just one dimension, along the X axis, on which it has position x at time t. Its velocity u is the gradient of a graph of x against t:

$$u = \frac{dx}{dt}. \tag{1}$$

It is measured in metres per second ($m\,s^{-1}$). Its acceleration a (rate of change of velocity, $m\,s^{-2}$) is the gradient of a graph of u against t.

$$a = \frac{du}{dt} = \frac{d^2x}{dt^2}. \tag{2}$$

The velocity of a body changes only if an unbalanced force acts on it. The force F required to give acceleration a to a body of mass m is given by Newton's second Law of Motion

$$F = ma = m\frac{du}{dt}. \tag{3}$$

If the mass and acceleration are in basic S.I. units (kg and $m\,s^{-2}$) this equation gives the force in newtons (N). If the body is rotating, different parts of it have different velocities and different accelerations. In that case the acceleration given by Eq. (3) is the acceleration of the body's centre of mass.

The momentum of a body is its mass multiplied by its velocity: it is mu. Notice that Eq. (3) can be written in a slightly different form,

$$F = \frac{d(mu)}{dt}, \tag{4}$$

which tells us that force equals rate of change of momentum. Also, integration of Eq. (3) gives

$$\int F \cdot dt = \int m \cdot du. \tag{5}$$

The expression on the left hand side is called impulse, so the equation tells us that the impulse applied to a body equals the resulting change of momentum.

It is often convenient to think in terms of impulse and momentum rather than force and velocity. For example, you would need a lot of information to calculate the force that acts briefly on a foot when it hits the ground, bringing it to rest, but to calculate the impulse you need only know the mass of the foot and its initial and final velocities. The momentum of a system of bodies remains constant unless an unbalanced external force acts on the system: this is the Principle of Conservation of Momentum. For example, when a startled fish pushes on the water with its tail, the velocities of both fish and water change, but the forward momentum given to the fish is matched by the backward momentum given to the water, so the total momentum of the system is unchanged.

Gravity gives a freely falling body a downward acceleration g (9.8 m s^{-2} anywhere animals are likely to live but less, for example, on the moon). That implies [by Eq. (3)] that gravity exerts a force mg on a body of mass m. This force is called the body's weight. Note that weight is not the same as mass: a giraffe with a mass of 1000 kg has a weight of 9800 N. Gravity acts on all parts of a body but it is generally convenient to think of the weight as a single force acting at the body's centre of mass (also called the centre of gravity).

Some quantities such as mass are fully described by a number and a unit (for example, the mass of my body is 70 kg). They are called scalars. Other quantities, called vectors, are not fully described until their direction is given as well as their size. They include velocity, acceleration, force, momentum and impulse: for example, a velocity of 3 m s^{-1} towards the north is quite different from 3 m s^{-1} eastwards or vertically upwards. It is often convenient to describe vectors in terms of components acting at right angles to each other. Thus the aerodynamic force on a wing is usually described by giving the component acting at right angles to its direction of motion through the air (the lift component) and the component that acts backwards along the direction of motion (the drag). A lift L and a drag D would have the same effect as a single resultant force $(L^2 + D^2)^{1/2}$ acting at an angle arctan (L/D) to the direction of motion (Fig. 1a). Conversely, a resultant force F acting at an angle θ to the

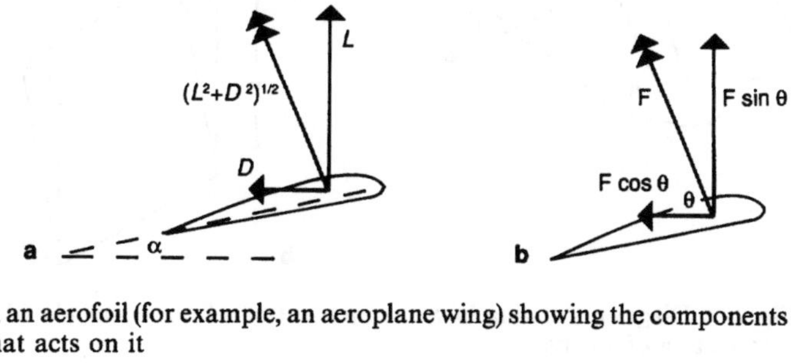

Fig. 1 a, b. Sections through an aerofoil (for example, an aeroplane wing) showing the components of the aerodynamic force that acts on it

direction of motion could be replaced by a lift component $F \sin \theta$ and a drag component $F \cos \theta$, which together would have the same effect (Fig. 1b). Velocities, accelerations etc. can be treated in the same way.

To describe the movement of a body, it is not enough to give its position, velocity and acceleration: it is also necessary to describe its rotation. This is done in terms of angles which are sometimes given in degrees but are properly given in the S.I. system in radians (rad). One radian is 57.3° and a complete revolution (360°) is 2π radians. The angular velocity is the rate of rotation and should be given in radians per second (rad s^{-1}). The angular acceleration is the rate of change of angular velocity and is given in rad s^{-2}.

If a body is moving in a plane (that is, in two dimensions) it can only rotate about axes perpendicular to the plane. Three-dimensional movement allows rotation about axes pointing in any direction, but any angular velocity or angular acceleration can be described as the resultant of components about three axes, parallel to the X, Y and Z axes of a Cartesian system.

The angular velocity of a body changes only if forces exert unbalanced moments about axes through its centre of mass. The moment of a force about an axis (also called the torque about that axis) is the force multiplied by the perpendicular distance of the axis, from the line of action of the force. For example, in Fig. 2, force F exerts a moment Fs about the axis perpendicular to the paper, through point **A**. If the force is expressed in newtons and the distance in metres, the moment is obtained in newton metres. Just as Eq. (3) relates linear acceleration to force, Eq. (6) (below) relates angular acceleration α to torque Q. The quantity I in the equation is the moment of inertia, which depends on the masses of all the parts of the body and on how far they are from the axis. Mechanics textbooks explain how moments of inertia can be calculated. In the S.I. system they are expressed in kg m^2.

$$Q = I\alpha = I\frac{d\omega}{dt} \tag{6}$$

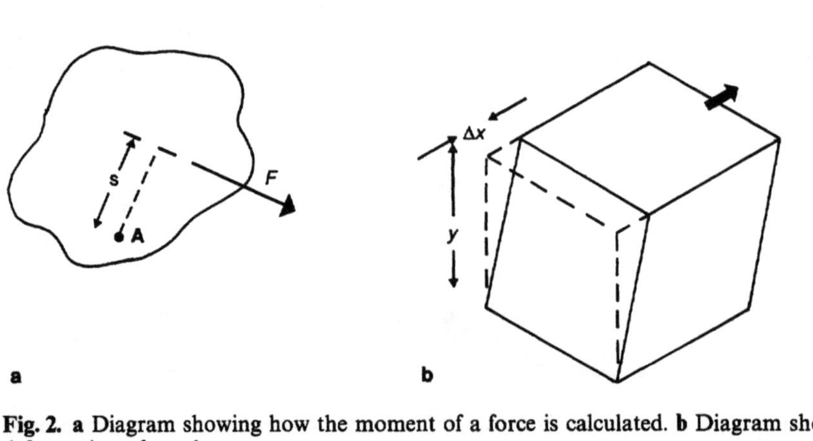

Fig. 2. a Diagram showing how the moment of a force is calculated. **b** Diagram showing shear deformation of a cube

(ω is angular velocity). This equation can also be written

$$Q = \frac{d(I\omega)}{dt}. \tag{7}$$

$I\omega$, moment of inertia multiplied by angular velocity, is called angular momentum. Just as the linear momentum of a system remains constant unless an unbalanced external force acts on it, so also the angular momentum about an axis remains constant unless an unbalanced torque acts. This is the Principle of Conservation of Angular Momentum.

Angular velocity, acceleration and momentum, and torque, are vectors: to specify them, you must give the direction of an axis, as well as the quantity. Also, a body does not have a unique moment of inertia, but has different moments of inertia about different axes. One good way of giving complete information about the moments of inertia of a body is to give the moments of inertia about three axes (mutually at right angles) through the centre of mass. Moments of inertia about axes, that do not pass through the centre of mass, are sometimes needed. For example, in calculations about legs swinging forward during walking, it is not the moment of inertia about the centre of mass that is wanted, but the moment of inertia about the hip joint. If a body of mass m has moment of inertia I_{cm} about an axis through its centre of mass, its moment of inertia about another axis, parallel to the first but at a distance s from it is

$$I = I_{cm} + ms^2. \tag{8}$$

There are many different ways of describing the position of a rigid body, moving in a plane. (You might, for example, regard the body of a beetle, running across the floor, as a rigid body moving in a plane). One way is to give the X and Y coordinates of a chosen point on the body and the angle θ of some line on the body to the X axis. (Thus you might give the distances of the beetle's snout from two of the walls of the room, and the angle of the beetle's long axis to one of the walls). For any such description, three quantities (distances or angles) must be stated, so three equations are needed to define the body's motion. For example, Newton's second Law of Motion [Eq. (3)] may be used to calculate the components of the body's linear acceleration, and Eq. (6) to calculate its angular acceleration

$$\left.\begin{aligned} \frac{d^2x}{dt^2} &= \sum \frac{F_x}{m} \\ \frac{d^2y}{dt^2} &= \sum \frac{F_y}{m} \\ \frac{d^2\theta}{dt^2} &= \sum \frac{Q}{I}. \end{aligned}\right\} \tag{9}$$

In these equations, x, y are the coordinates of the body's centre of mass at time t and θ is the angle of its long axis to the X axis. $\sum F_x$ means the sum of all the components parallel to the X axis, of the forces acting on the body (taking forces in the positive direction as positive, and in the opposite direction as negative) and $\sum F_y$ has the corresponding meaning for the Y axis. $\sum Q$ means the sum of the moments of all the forces about the centre of mass. (Anticlockwise moments are conventionally regarded as positive, and clockwise as negative).

For a rigid body that is free to move in three dimensions, six equations of motion are needed, giving, for example, three components of linear acceleration and three of angular acceleration. Conveniently, many problems can be treated as two-dimensional, although there is nothing to confine motion to a plane: for example, many mathematical models of human walking treat it as motion in a vertical plane. If a structure is composed of rigid segments jointed together, three equations of motion are needed for each segment (for motion in a plane) or six (for three-dimensional movement). The human leg is often treated as an assembly of two rigid segments, jointed together at the knee, or of three (with a second joint at the ankle). Three-dimensional analysis of the movement of multi-segment systems is formidably complicated, involving very large numbers of equations.

A body is in equilibrium when all forces and torques on it are balanced, so that it has no linear or angular acceleration. It may then be stationary, or if it is moving its linear and angular velocities remain constant. Equilibrium may be stable or unstable. If it is stable; the system tends to return to its initial condition after a disturbance. For example, the equilibrium of a mass suspended by a string is stable because it tends to swing back to its initial vertical position after being pushed to one side. Equilibrium is unstable if the system tends to move further from the equilibrium state after a disturbance. A pencil balanced on its point is unstable because it will fall over if pushed in any direction (and will in any case soon fall over because slight random disturbances are inevitable). The concept of stability applies to moving systems as well as to stationary ones. For example, the drag on a gliding bird or aircraft first decreases and then increases, as the speed of gliding increases. Below the minimum drag speed, gliding is unstable in that if speed increases for some reason, drag will fall and the speed will increase still more. Above the minimum drag speed, however, any increase in speed results in increased drag, which slows the bird or aircraft down again, so gliding is stable. The concept of stability is also applied to repetitive motion, for example by McGeer in this Volume (Chap. 4), in which he describes a model of walking that moves passively, without muscle action. He shows that if it is set moving in certain ways its cycle of movement is stable: it will recover from a small disturbance.

Stability is a technical term with a precise meaning but is often misused even by professional biomechanicists. Be suspicious of authors who write that a muscle or ligament stabilizes a joint: they usually mean something different from stability in the technical sense.

An object moving in a circle may have constant speed but its velocity is not constant – its direction is continually changing. If its constant speed is v and the radius of the circle is r, it has acceleration v^2/r towards the centre of the circle. If its mass is m, a centripetal force mv^2/r is needed, pulling towards the centre of the circle, to prevent it from flying off at a tangent. There may be no solid connection to the centre of the circle: the centripetal force on a circling bird is a component of the aerodynamic force on its wings.

3 Work and Energy

A force F does work W when it moves its point of application a distance s along its line of action

$$W = Fs. \tag{10}$$

If we use basic S.I. units (force in newtons, distance in metres) we obtain the work in joules. Notice the phrase "along the line of action": no work is associated with movements at right angles to it. For example, the centripetal force on a body moving in a circle always acts at right angles to the velocity and so does not work. Similarly, the engine of an aircraft does work against drag (which acts backwards along the direction of motion) but not against lift (at right angles to the direction of motion). Work is also done by torques. When a body acted on by a torque Q rotates through an angle θ about the axis of the torque, the work done is

$$W = Q\theta. \tag{11}$$

If we use basic S.I. units (newton metres for the torque, radians for the angle) we again get work in joules.

Work done on a rigid body gives it kinetic energy (energy of movement) or gravitational potential energy (energy due to height) or both. A body of mass m travelling with speed v and rotating with angular velocity ω about an axis through its centre of mass, about which its moment of inertia is I, has kinetic energy $\frac{1}{2}mv^2$ due to its linear movement, plus $\frac{1}{2}I\omega^2$ due to its rotation. If it is at a height h from the ground its gravitational potential energy relative to the ground is mgh, where g is the gravitational acceleration. (Notice that potential energy is always measured relative to some reference level). All these energies are measured in joules, the same units as work. The energy gained by a system when work is done on it, is equal to the work. The Principle of Conservation of Energy states that energy cannot be created or destroyed (except in atomic explosions etc., where there is some interconversion of mass and energy). However, work does not necessarily appear as mechanical energy. For example, when work is done against friction, that work is converted into heat energy.

Power is the rate of doing work. Notice that this technical meaning of the word is quite different from the meanings it is often given in common speech, for instance when non-scientists speak of a powerful grip. Power is measured in joules per second, also called watts. Work is force multiplied by distance, so power P is force multiplied by velocity (that is, by the component of the velocity in the direction of the force).

$$P = Fv.$$

Thus the power needed to propel an aircraft is its speed (relative to the air) multiplied by the drag.

Biologists studying animal movement often use the concept of negative work. When I accelerate my car, the engine does (positive) work, giving it kinetic energy. When I use the brakes they do negative work, degrading kinetic energy to heat and so slowing the car down. Positive work is done on a body when it moves in the direction of the force applied to it, and negative work when it moves in the opposite direction. The mechanical eneregy of an animal walking at constant speed on level ground is the same at corresponding stages of successive strides so its muscles do no net work (apart from the small quantities needed to overcome aerodynamic drag and joint friction). However, the body rises and falls in each stride, gaining and losing potential energy. It accelerates and decelerates, gaining and losing kinetic energy. Thus the muscles must do (positive) work at some stages of the stride and almost equal quantities of negative work at others.

The energy that muscles convert to work is chemical energy from foodstuffs. It may be supplied by aerobic metabolism (oxidation of foodstuffs) or by anaerobic processes that release energy without any immediate need for oxygen. Anaerobic metabolism releases far less energy than aerobic metabolism of the same food so, if the rest of the energy is not to be wasted, oxygen must be used afterwards to oxidise some of the products and supply the energy needed to convert the rest back to the original foodstuffs: the oxygen debt that has accumulated must be repaid. Anaerobic metabolism is not limited by the rate at which the animal can take in oxygen, so it can proceed very fast, but it cannot continue for long because the body can tolerate only limited concentrations of its products (often lactic acid). Anaerobic metabolism is used by vertebrates, cephalopods and crustaceans for short bursts of violent activity, but prolonged activity (and most of the activity of other animals) is powered by aerobic metabolism (Margaria 1976). For example, a human 100-metre sprint depends almost entirely on anaerobic metabolism and a 42-kilometre marathon almost entirely on aerobic metabolism. Whether carbohydrates, fats or proteins are being metabolized, aerobic metabolism using one cubic centimetre ($10^{-6}\,m^3$) of oxygen (measured at atmospheric pressure and 0 °C) releases about 20 joules chemical energy from food.

Only a fraction of this energy ever appears as work. For example, in classic experiments, Margaria (1976) measured the oxygen consumption of people walking up and down gradients. They used oxygen faster when walking up gradients than when walking at the same speed on the level. When walking up steep gradients, a 70-kg man used an extra 130 cm^3 oxygen, making 2600 joules chemical energy available, for every metre that he rose. The weight of his body was $70 \times 9.8 = 690$ N, so the potential energy gained with each metre of height was 690 J, only 0.27 of the extra metabolic energy consumption.

Efficiency is the ratio of useful work output to energy input, so could be reckoned as 0.27 in the case just cited. However, efficiency can be calculated in several different ways which may give very different values. We have just calculated net efficiency, work output divided by the *extra* input needed to obtain it. We might alternatively have calculated gross efficiency, which in this case would have been the potential energy gain divided by the *total* metabolic energy used in the climb, and would have been much smaller. Metabolic energy that does not appear as work is generally lost as heat. The Principle of Conservation of Energy tells us that it cannot simply disappear, and also that efficiency can never exceed one.

Net efficiencies of about 0.25 have been measured for various other human and animal activities, for example for budgerigars flying up a gradient (Tucker 1968).

However, efficiency is not always a useful concept in studies of animal locomotion. A muscle that is contracting isometrically (exerting tension without either lengthening or shortening) is doing no work but uses metabolic energy much faster than when resting, so its efficiency is zero. If it does positive work by shortening, it uses metabolic energy even faster (for the same force) but if it does negative work it uses energy less fast. The metabolic energy cost of running (in which muscles do almost equal quantities of positive and negative work) is more satisfactorily explained in terms of the cost of exerting force, than of the cost of doing work (Kram and Taylor 1990). In such circumstances, efficiency is a less useful concept than economy, which is the force exerted by unit length of muscle fibre, divided by the rate at which it is using metabolic energy.

Cost of transport (also called specific resistance) is a useful concept when comparisons are to be made between the energy costs of locomotion of different animals. It is the energy required to move unit weight of animal through unit distance. It is common to define it thus, in terms of weight rather than mass, for the following reason. Energy is measured in newton metres or joules and (weight × distance) is also measured in newton metres, so cost of transport defined in this way has no units. It is important to explain clearly how cost of transport is defined, whenever the concept is used. The energy cost referred to may be mechanical work (as in De Mont's account of soft-bodied crawlers, Chap. 6) or metabolic energy. In the latter case it may be the gross cost (the entire metabolic rate of the animal) or the net cost (in excess of what it would have used if it had remained stationary).

4 Stresses, Strains and Pressures

Pressure and stress are both force per unit area, and are measured in newtons per square metre, also called pascals (Pa). Pressure is a property of fluids, and exerts a force at right angles to any surface in the fluid. Stresses may act either in solids or in fluids. Two distinct kinds of stress are mentioned in this book, of which the more familiar is direct stress. When a human Achilles tendon of cross-sectional area $90\,\mathrm{mm}^2$ ($9 \times 10^{-5}\,\mathrm{m}^2$) transmits a force of 5000 N, a direct stress of $5000/9 \times 10^{-5} = 56 \times 10^6$ Pa acts on it. (These data are realistic for an adult man running: see Chap. 6). In this case the stress is tensile because the tendon is being stretched, but compressive stresses are also direct. Shear stresses act in the jam of a sandwich when the slices of bread are made to slide over each other. Similarly, shear stresses act on the air of the boundary layer, close to the surfaces of a flying bird or aircraft. In Chap. 3, Spedding emphasises the distinction between the pressure drag on a bird, due to pressure exerting forces at right angles to its surface, and friction drag due to shear stresses in the boundary layer.

The terms "stress" and "strain" are often confused in everyday speech, but stress is force per unit area and strain is the deformation that it causes. Tensile strain is fractional change in length: for example, the Achilles tendon may be stretched by about 5% of its initial length, in a running step, suffering a tensile strain of 0.05.

Compressive strain is negative tensile strain. Shear strain occurs when an initially rectangular block is deformed so that its sides become parallelograms: in Fig. 2b, the shear strain is $\Delta x/y$. Tensile and compressive strains are always accompanied by shear strains, and vice versa. For example, in Fig. 2b, tensile strain occurs along the diagonal of the sheared block.

Strains are accompanied by stresses because of two properties of matter, elasticity and viscosity. We think of elasticity as a property of solids and viscosity of liquids, but some solids are markedly viscoelastic, for example the mesogloea of sea anemones, which takes many hours to reach equilibrium under constant stress. In a purely elastic material the stress is a function of the strain and in a purely viscous one it is a function of the rate of strain. Thus the force needed to stretch an (elastic) steel spring depends on how far it is stretched, not on how fast; but the force needed to stir (viscous) oil depends on how fast the stirring stick is moved, not on how far. Ideally, the stress σ in an elastic material is proportional to the strain ε. For tensile stress and strain

$$\sigma = E\varepsilon, \tag{12}$$

where E is Young's modulus. Strain has no units (being a ratio of two lengths) so Young's modulus has the same units as stress, Pa. In Chap. 5, Bennett refers to the tangent Young's modulus of tendon. This is the slope of a graph of stress against strain, $d\sigma/d\varepsilon$. For materials like tendon, stress is not precisely proportional to strain, and the tangent modulus changes with the stress. An equation similar to (2) relates shear stress to shear strain and defines the shear modulus.

For an ideal viscous liquid, shear stress σ_s is proportional to the rate of change of shear strain ε_s

$$\sigma_s = \mu \frac{d\varepsilon_s}{dt}. \tag{13}$$

The constant of proportionality μ is called the viscosity, and has units Pa s. Its values for water and air (at 20 °C and atmospheric pressure) are 1.0×10^{-3} and 1.8×10^{-5} Pa s.

The density of a material is the mass of unit volume: 1000 kg m^{-3} for pure water, 1026 kg m^{-3} for sea water and 1.2 kg m^{-3} for air at 20 °C and atmospheric pressure. A body of volume V, density ρ_b has mass $\rho_b V$, weight $\rho_b g V$. Archimedes' Principle tells us that if it is immersed in a fluid of density ρ, an upward force acts on it equal to the weight of fluid that would occupy the same volume, ρV. Thus the net downward force (weight minus upthrust) is $(\rho_b - \rho) g V$. Animal tissues are generally denser than either fresh water or sea water, so swimming animals that have no special buoyancy organs must generate upward hydrodynamic forces to prevent themselves from sinking: for example, tunnies (*Thunnus*) use their pectoral fins like aeroplane wings to produce lift. Many other animals have buoyancy organs (for example, the gas-filled swimbladders of most teleost fishes) which match their density to that of the water ($\rho_b = \rho$).

The pressure at depth h in a liquid of density ρ is $\rho g h$.

The volume V of n moles of a gas depends both on the pressure p and the temperature T. The ideal gas equation tells us that

$$pV = nRT, \qquad (14)$$

where R is the universal gas constant, $8.3\ \mathrm{J\,mol^{-1}\,K^{-1}}$. It does not hold exactly for real gases, but the deviations become important only at very high pressures and very low temperatures.

5 Fluid Dynamics

Fluid motion is described by a set of three equations, known as the Navier–Stokes equations. (There is one equation each for the X, Y and Z direction of three-dimensional space). These equations are complicated, and cumbersome to use (but are used by Daniel and his co-authors in Chap. 2). Roughly speaking, they say that the force available to accelerate a particle of fluid is the vector sum of the pressure difference acting across it, its weight and the viscous force due to shear gradients in the surrounding fluid.

Bernoulli's equation is very much simpler but applies only in restricted circumstances: viscous forces must be small enough to be ignored, and the flow must be steady. In steady flow, the velocity at any point in space is constant, but the velocity of a particle of fluid may change as it travels from one point to another. The paths taken by particles of fluid are called streamlines. Bernoulli's equation says that for all points on any particular streamline

$$\tfrac{1}{2}\rho v^2 + \rho g h + p = \text{constant}, \qquad (15)$$

where ρ is the density of the fluid, v is its velocity, h its height, p its pressure and g the gravitational acceleration. Notice that $\frac{1}{2}\rho v^2$ and $\rho g h$ are the kinetic energy and the gravitational potential energy of unit volume of fluid. Bernoulli's equation shows how the pressure differences along a streamline account for the changes of kinetic and potential energy of the moving fluid. The term $\frac{1}{2}\rho v^2$ is called the dynamic pressure.

Fluids exert forces on objects that move through them, or over which they flow. It does not matter whether it is the object or the fluid that is moving; the forces are the same for the same *relative* velocities. The forces could in principle be calculated using the Navier–Stokes equations, but in practice this is often formidably difficult. Instead, they are often calculated from empirically determined drag coefficients and lift coefficients.

We have already encountered drag and lift (Fig. 1). The drag D on a body moving at speed v through fluid of density ρ is

$$D = \tfrac{1}{2}\rho v^2 A C_D \qquad (16)$$

(the dynamic pressure multiplied by area A and drag coefficient C_D). Confusingly, different areas may be used in the equation with correspondingly different drag coefficients. Commonly, the frontal area is used – that is the area of a full-scale front view, seen from the direction of flow. (This is usually about equal to the maximal cross-sectional area). Sometimes, however the total surface area ("wetted area") is used. The reason is that drag has two components due to different physical effects. Friction drag is due to shear stresses in the boundary layer and pressure drag is due to pressure differences between the front and rear of the body, which occur because the body drags a wake of moving fluid along behind it. In comparisons between different bodies, friction drag is often more or less proportional to wetted area and pressure drag to frontal area. Streamlined (torpedo-shaped) bodies are designed to develop as little wake as possible, and the drag on them is mainly friction drag. Spheres and other shapes that form large wakes suffer mainly pressure drag.

Drag coefficients are dimensionless, so have no units. Their values depend on the shapes of the bodies, on the smoothness of their surfaces and on their orientation. (Much more drag acts on a knife blade moved broadside-on through water than when it moves edge-on at the same speed). Drag coefficients also depend on a quantity called the Reynolds number, which will be explained soon.

Aerofoils and hydrofoils are designed to give as much lift and as little drag as possible. The lift L and drag D are given by the equations

$$L = \tfrac{1}{2}\rho v^2 A_P C_L, \tag{17}$$

$$D = \tfrac{1}{2}\rho v^2 A_P C_D, \tag{18}$$

Confusingly, the area A_P conventionally used here is neither the frontal area nor the wetted area but the plan area, the area of a full-scale plan of the aerofoil or hydrofoil including (in the case of aeroplane and animal wings) a band across the fuselage (Fig. 3a). The lift and drag coefficients, C_L and C_D, depend on the size and shape of the aerofoil or hydrofoil, on the angle of attack (α, Fig. 1a) and on the Reynolds number (which will be defined below). As the angle of attack increases, the lift coefficient increases up to a certain limit at which stalling occurs and the coefficient falls steeply.

The extra lift that results from an increase an angle of attack is necessarily accompanied by increased drag. The reason is that upward lift is obtained by driving air downwards (Fig. 1b). The lift equals the rate at which momentum is given to the air [Eq. (4)], and if the air is given momentum (mv) it must also be given kinetic energy ($\tfrac{1}{2}mv^2$). Work must be done, to supply this kinetic energy: it is done against

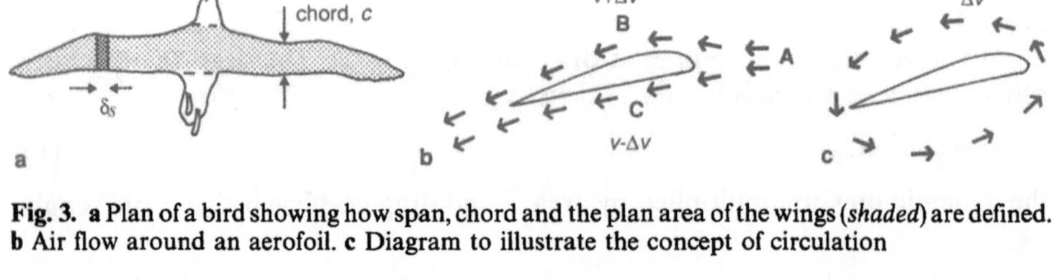

Fig. 3. a Plan of a bird showing how span, chord and the plan area of the wings (*shaded*) are defined. **b** Air flow around an aerofoil. **c** Diagram to illustrate the concept of circulation

additional drag, known as induced drag. The total drag on an aircraft is the sum of the drag on the fuselage (parasite drag), the drag that would act on the wings in the absence of lift (profile drag) and the additional drag associated with the lift (induced drag). In Chap. 3, Spedding derives an equation for induced drag. Here we need only note that induced drag is inversely proportional to aspect ratio.

$$\text{aspect ratio} = \frac{\text{wing span}}{\text{mean chord}} = \frac{(\text{wing span})^2}{\text{plan area}} \qquad (19)$$

(Fig. 3a). In other words, long narrow wings suffer less induced drag than short broad ones of the same area.

The lift on an aerofoil can be explained in two different but equivalent ways. First, air is given downward momentum, so there must be an upward force on the aerofoil (Fig. 3b). Alternatively, air that flows just over the aerofoil and air that flows just under it start off in adjacent streamlines (at **A**, Fig. 1b) where they must have the same pressure. As it passes over the aerofoil, however, the air at **B** is travelling faster than the air at **C** so, by Bernoulli's Principle [Eq. (15)] its pressure must be less. This pressure difference across the aerofoil accounts for the lift. If the air travels over the wing with speed $v + \Delta v$ and under it with speed $v - \Delta v$ the pressure difference Δp [from Eq. (15)] is

$$\Delta p = \tfrac{1}{2}\rho((v + \Delta v)^2 - (v - \Delta v)^2 \simeq 2\rho v \cdot \Delta v. \qquad (20)$$

The lift can be calculated from the pressure difference. For example, the narrow slice of wing shown darker than the rest in Fig. 3a has area $c \cdot \delta s$, so the lift on it is

$$\begin{aligned} \delta L &= \Delta p \cdot c \cdot \delta s \\ &= \rho v \cdot (2\Delta v \cdot c)\delta s. \end{aligned} \qquad (21)$$

The velocities ($v + \Delta v$ over the aerofoil and $v - \Delta v$ under it) are the same as if a uniform velocity v were combined with an anticlockwise circulation of air at speed Δv (Fig. 3c). The stronger the circulation, the greater the lift. "Circulation" is a technical term which is defined formally by Spedding in Chap. 3. In the simple case just defined (assuming constant speed over each surface of the aerofoil) it is simply Δv multiplied by the perimeter of the aerofoil, $2c$: it is the quantity in parentheses in [Eq. (21)]. Both in this case and more generally, lift per unit wing span dL/ds is related to circulation Γ by the equation

$$\frac{dL}{ds} = \rho v \Gamma. \qquad (22)$$

6 Dynamic Similarity

Zoologists often want to make comparisons between the shapes of animals of different sizes. The simplest possible relationship is geometric similarity. Two shapes are geometrically similar to each other if they could be made identical by a uniform change of the scale of length. For example, a triangle of sides 3, 4 and 5 cm is geometrically similar to one of 6, 8 and 10 cm because all the linear dimensions of one are double the corresponding dimensions of the other. If all the lengths in one of two geometrically similar animals are λ times those of the other, all areas have the ratio λ^2 and all volumes λ^3. If the animals are made of materials of the same densities, corresponding parts have masses proportional to their volumes. Hence lengths are proportional to (body mass)$^{0.33}$ and areas to (body mass)$^{0.67}$.

The concept of dynamic similarity applies to motion rather than to shape. Two motions are dynamically similar if they could be made identical by uniform changes in the scales of length, time and force; two pendulums swinging through the same range of angles will serve as an example. If gravitational forces are important (as in the cases of swinging pendulums and running animals) dynamic similarity is possible only if the motions have equal Froude numbers

$$\text{Froude number} = \frac{v^2}{gl}, \tag{23}$$

where g is the gravitational acceleration, and v and l are a speed and a length characteristic of the motion: in comparisons between running animals, running speed and leg length are usually used. Ideally, it does not matter how the speed and length are defined, provided they are defined in the same way for all the animals being compared. In practice, animals are never precisely geometrically similar and do not move in precisely dynamically similar fashion, so it is sensible to use a speed and a length that have obvious relevance to the motion. It would be possible to use jaw length as the length in the Froude number, but it would not be sensible to do so. Froude numbers are dimensionless, so they have no units.

Reynolds numbers are another very important class of dimensionless numbers. Just as dynamic similarity requires equal Froude numbers, if gravitational forces are important; it requires equal Reynolds numbers, if viscous forces are important. The forces on bodies moving through fluids depend on the pattern of flow around them, and viscous forces are important. We cannot assume that lift and drag coefficients will be the same, even for geometrically similar bodies, unless the patterns of flow around them are dynamically similar – which requires equal Reynolds numbers.

$$\text{Reynolds number} = \frac{lv\rho}{\mu}, \tag{24}$$

where l and v are a characteristic length and speed, as before, ρ is the density of the fluid and μ is its viscosity. The ratio ρ/μ is $1.0 \times 10^6\,\mathrm{s\,m^{-2}}$ for water at 20 °C; and $7 \times 10^4\,\mathrm{s\,m^{-2}}$ for air at 20 °C and atmospheric pressure.

7 Vibrations

Vibrations occur, about positions of stable equilibrium. For example, a pendulum will swing to either side of the vertical and a mass suspended from a spring will bounce up and down, above and below its equilibrium position. The frequency f of the motion (the number of cycles in unit time) is measured in the S.I. system in herz (Hz, cycles per second). It is the reciprocal of the period T (seconds). For a pendulum of length l

$$f = 1/T = (1/2\pi)(g/l)^{1/2} \tag{25}$$

and for a mass m hanging from a spring of stiffness K

$$f = 1/T = (1/2\pi)(K/m)^{1/2}. \tag{26}$$

(Stiffness means the force needed to stretch or compress the spring through unit distance. It is expressed in newtons per metre).

Angular frequency means 2π times the frequency. The rationale for this is that a revolution is 2π radians, so f revolutions per second is $2\pi f$ radians per second.

Fish tails and bird wings are hydrofoils or aerofoils that make repetitive movements, beating from side to side or up and down, and changing their angle of attack, in each cycle of movement. The motions of different tails or wings cannot be dynamically similar, unless each travels the same number of chord lengths in each cycle. If the speed is v and the frequency f, the distance travelled in a cycle is v/f. Thus dynamic similarity requires equal values of v/fc. This of course implies equal values of $2\pi fc/v$, a dimensionless number that is known as the reduced frequency. As for Froude numbers and Reynolds numbers, different lengths can be used to define reduced frequencies. Spedding explains this in Chap. 3.

Flow around beating tails and wings is unsteady: the velocity at any particular point (relative to the tail or wing) changes in the course of each cycle. Lift and drag coefficients measured for steady motion give reasonably accurate results in calculations for beating tails and wings, only if the reduced frequency as defined above is less than about 0.5 (that is, if a distance of at least 12 chord lengths is travelled in each cycle).

References

Alexander RMcN (1983) Animal Mechanics 2nd edn. Blackwell, Oxford

Kram R, Taylor CR (1990) Energetics of running: a new perspective. Nature (Lond) 346: 265–267

Margaria R (1976) Biomechanics and energetics of muscular exercise. Clarendon, Oxford

Tucker VA (1968) Respiratory exchange and evaporative water loss in the flying budgerigar. J Exp Biol 48: 67–87

Vogel S (1981) Life in moving fluids. Willard Grant, Boston

Wainwright SA, Biggs WD, Currey JD, Gosline JM (1976) Mechanical design in organisms. Arnold, London

Chapter 2

Hydromechanics of Swimming

T. Daniel, C. Jordan, and *D. Grunbaum*

Contents

1 Introduction: the Physics and Problems of Aquatic Locomotion

Swimming animals span a tremendous range of sizes and shapes, from among the tiniest of living organisms to the world's largest extant animals. Along with this impressive range of sizes we find a myriad of modes of propulsion ranging from the

Department of Zoology, NJ-15, University of Washington, Seattle, WA 98195, USA gentle rowing motions of cilia to the powerful oscillations of whale flukes. In the face of such diversity, we are challenged with quantifying the ways in which animals generate force and their concomitant energy expenditure. In doing so, we seek an understanding not just of the mechanisms of movement but, more importantly, of the consequences of variation in size and shape to aspects of an animal's biology that are related to movement.

This chapter is meant neither to be exhaustive in its review of the literature nor all inclusive in its survey of aquatic locomotor schemes. Instead, we illustrate some of the basic principles of fluid mechanics to show the kinds of problems that have been addressed and the kinds of dilemmas we face in future analyses of animal locomotion. We begin with a very brief review of the physical problems faced by swimming animals and some of the mechanisms they employ in propelling themselves. This simple review sets the stage for why we divide our studies of animal locomotion into analyses of the slow and small, or the big and fast. We then examine some particular applications of the physical principles where we have gained a great deal of insight into the mechanics and biology of movement in fluids. We conclude this chapter with some case studies that suggest new directions for research in aquatic locomotion.

1.1 The Basic Problem of Propulsion

All aquatic organisms move actively by transferring momentum from moving portions of their body to the fluid surrounding them. The rate at which the animal transfers momentum determines the amount of thrust it generates. The mechanisms underlying that exchange of momentum and the effectiveness of exchange (how much of the animal's momentum is manifest in the fluid) depend upon size and speed of the moving portions of the animal and the viscosity and density of the fluid surrounding them. To understand these basic processes and to determine how size, shape, and behavior all interact to determine thrust, swimming speed, and energy expenditure, we examine below some very basic relations for moving fluids. We restrict our attention to only water, a fluid with a dynamic viscosity (μ) of about $10^{-3}\,\mathrm{kg\,m^{-1}\,s^{-1}}$ and a density (ρ) of about $1000\,\mathrm{kg\,m^{-3}}$.

The most general description of the relationship between force and flow in a moving fluid is given by the Navier–Stokes equations, an imposing set of partial differential equations that are based on conservation of mass and momentum. In words, they are not so bad. They state that for any element of fluid about a moving animal:

$$\text{mass} \times \text{acceleration (or rate of accumulation of momentum)} =$$
$$\left\{\begin{matrix}\text{spatial gradient in}\\ \text{pressure stress}\end{matrix}\right\} + \left\{\begin{matrix}\text{spatial gradient in}\\ \text{viscous stress}\end{matrix}\right\} + \left\{\begin{matrix}\text{external}\\ \text{forces}\end{matrix}\right\} \tag{1}$$

and in mathematical symbols for two dimensions (see for example Batchelor 1967):

$$\rho\frac{Du}{Dt} = \frac{\partial P}{\partial x} + \mu\left(\frac{\partial^2 u}{\partial x^2} + \frac{\partial u^2}{\partial y^2}\right) + \rho g, \tag{2}$$

where P is the local pressure, u the local velocity, and g the value for gravitational acceleration in the x-direction.

The substantial derivative in Eq. (2), follows changes in momentum in the frame of the moving fluid element with both temporal and spatial changes:

$$\frac{D}{Dt} = \left(\frac{\partial}{\partial t} + \frac{\partial u}{\partial x}\right). \tag{3}$$

Thus if we monitor a small element of fluid near an animal, the above equation shows that the change in momentum of that element is determined by any differences in the flux of momentum due to pressure and viscous stresses and by the net forces acting on the body.

The good news is that with these equations, under appropriate boundary conditions, we can determine thrust and energy requirements for aquatic locomotion. The bad news is that for most biological applications, no solutions exist to the full form of these equations! This dilemma arises because the full form of these equations is nonlinear, and precludes the development of analytic solutions. Today, there are powerful computational schemes that are developed to solve these equations for specific geometries. These schemes are often too restrictive and too embroiled in mathematical detail to be useful to biologists. Instead, two important simplifications to the equations have been made for which analytic solutions exist. One simplification is that viscous stresses in the fluid are negligible; the other is that inertial stresses are negligible. These two extremes are indexed by a dimensionless parameter, the Reynolds number ($Re = Ul/v$ where U is the body speed; l, its length; v, the kinematic viscosity of water), which represents the ratio of inertial to viscous stresses. Thus the nearly noninertial extreme corresponds to tiny Reynolds numbers ($\ll 1$) and results in simpler equations, called Stokes equations, which state that the gradients in pressure stresses are precisely balanced by viscous stresses. In one dimension, this is simply

$$\frac{\partial P}{\partial x} = \mu\left(\frac{\partial^2 u}{\partial x^2} + \frac{\partial^2 u}{\partial y^2}\right). \tag{4}$$

In this "Stokesian" regime, the absence of inertia tells us that gliding is not possible, flows are entirely reversible, and all of the relevant forces result from fluid shearing (velocity gradients near boundaries). Reversible flows, in the absence of inertia, tell us that for slow, tiny organisms, thrust may only be generated by viscous shearing with some asymmetric motion, such as in cilia, or with continuous motions, such as in planar and helical waves on flagella.

The nearly nonviscous regime (inviscid or potential flow) yields another simple set of equations, called Euler's equations, which state that accelerations of the fluid element yield pressure gradients in the fluid. Again, for uni-directional flow, and in the absence of gravitational forces this equations is

$$-\frac{\partial P}{\partial x} = \rho\frac{Du}{Dt}. \tag{5}$$

This Eulerian realm of fluid motion tells us that thrust forces (pressures integrated over the surface of moving portions of the body) arise, not from steady viscous shearing, but from accelerational components of the flow.

These simplified equations have been used to solve a variety of problems about animal propulsion. As we examine in more detail further on in this chapter, their solutions give insight into the consequences of size and shape to some measure of performance, such as speed, acceleration, efficiency, or energy requirements for movement. Unfortunately, the Reynolds number restrictions to these equations often precludes their application to organisms commonly encountered in the plankton or in streams. Such animals fall in a nasty size and speed range for which neither inertial nor viscous forces are negligible. In this size range, theoreticians waver and experimentalists rule. We rely on experimentally defined coefficients of force. For forces parallel to the direction of motion, we define a drag coefficient by the amount of force measured in that direction:

$$C_d = \frac{\text{Measured drag force}}{\left(\rho \dfrac{SU^2}{2}\right)}. \tag{6}$$

A similar definition exists for lift forces; that is those forces that are perpendicular to the direction of fluids motion:

$$C_1 = \frac{\text{Measured lift force}}{\left(\rho \dfrac{SU^2}{2}\right)}. \tag{7}$$

These coefficients are about all we have to rely on in attempting predictions of thrust for moving appendages. A particularly elegant example was developed by Nachtigall (1960) and Nachtigall and Bilo (1975) to describe thrust generation by the rowing hind-limbs of dytiscid beetles. As we shall see, more recent studies have successfully used such coefficient approaches to examine a range of propulsion problems. The advantage of such approaches lies in their ease of use; their disadvantages lie in their lack of generality.

1.2 Moving Parts and the Divisions of Locomotion

Armed with this primer to fluid motion we can now attempt to physically characterize the spectrum of structures and gizmos that animals use as propulsors. The two obvious divisions we have encountered so far are the high and low Reynolds number swimmers. The higher ones such as whales rely either on fluid accelerations, pressure drag, or lift in generating force. The lack of any apparent wheels in this range of swimming animals imposes a requirement for undulating, oscillating, or reciprocating thrust-producing devices. Thus we are often concerned with the unsteady (time-varying) component of the forces generated by such structures. In some cases, a simple quasi-steady analysis works well. Such an analysis examines only those forces associated with velocity per se and neglects accelerational forces.

In many cases, however, unsteady forces must be incorporated in our analysis. Classic examples of these are the accelerational forces generated by body undulations of eels (Lighthill 1971, 1975), pectoral fin rowing by angelfish (Blake 1981a, b), or the lift forces generated by fins, flukes, and flippers (Daniel 1988; Hui 1988; Yates 1986).

At the other extreme, we find ciliary and flagellar modes of propulsion. While undulations apparently work quite well for low Reynolds number creatures, the physical rules that govern thrust are quite different. The overwhelming prevalence of reversible viscous forces prevents such schemes as jets or lifting surfaces from working in this range.

Alternative divisions of aquatic locomotor schemes have been developed. Most of these allude to particular animals (mostly fish) and to the assumed mechanism of thrust production. For example, Breder (1926) developed a classic scheme that corresponds to the amount of a fish's body involved in thrust production. Thus, whole body undulatory swimming was termed "anguilliform" locomotion after the genus *Anguilla*. Similarly, he identified carangiform, subcarangiform, and tuniform modes of locomotion, among others (for reviews see Lighthill 1975; Webb 1984). While these serve some useful purpose in generating a visual picture of the animal, they say little about the relevant mechanisms. Instead, the divisions of accelerational-, lift-, drag-, and shear-based modes identify both the relevant mechanisms and the relevant size range of organisms. These latter divisions were used to examine a whole variety of fish swimming problems and to illuminate the relevant morphological features that strongly affect thrust production (Blake 1983; Webb 1984).

In the following we examine the physical, morphological, and behavioral implications of a variety of swimming animals following the divisions according to Reynolds number and mechanism.

2 The Realm of Low Reynolds Numbers: Viscosity Waves the Rules

In the world of ciliary and flagellar propulsors, viscous shearing is the sole mechanism available for generating thrust. Recall from the Introduction that this can be expressed mathematically by the Stokes equations. As mentioned above, reversibility and the lack of inertia are the fundamental physical problems faced by such creatures. A dramatic demonstration of the lack of inertia is seen in Berg's (1983) calculation for the motion of a bacterium when its flagellum suddenly stops: the cell coasts at most one tenth of an angstrom, or the diameter of a hydrogen atom!

Reversibility in the flows at these low Reynolds numbers is guaranteed by the linear nature of the Stokes equations. For example, if a known force is applied to the fluid such that when an element from point A moves to point B and is then followed by an application of the exact same force, but with opposite sign, the element will move back to exactly point A. This example is nothing more than a demonstration that symmetrically reciprocating structures generate no net force at low Reynolds numbers because the forward progress resulting from a power stroke would be exactly canceled by the rearward force from a recovery stroke.

Therefore, at low Reynolds numbers, there must be some asymmetry between power and recovery strokes in the shape reciprocating structures for any thrust to arise. Merely changing the speed between power and recovery strokes cannot result in net thrust. Thus a low Reynolds number creature having a fast power stroke and a slow, but geometrically similar, recovery stroke makes no forward progress.

How then do animals, plants, protists, and bacteria manage to swim in this realm of elegant symmetric flows? Two primary mechanisms for generating thrust predominate, corresponding to the use of cilia or one or more flagella. There are alternatives but they boil down to variations on the theme of a flagellum, or cilia. For example, helically shaped bacteria like Spirochaetes and *Spirilla* swim by "corkscrewing" through a fluid. In this case the flagellum (internal or external) does not generate thrust but merely counteracts the moments set up by the spinning body (Brennen and Winet 1977).

2.1 Swimming with an Undulating Tail: Sperm and Protozoan Flagella

The mechanics of undulatory swimming at low Reynolds numbers has intrigued biologists and mathematicians for several decades, beginning with the work of Hancock (1953) and culminating in recent computational methods for examining morphological and kinematic determinants of thrust and energy requirements (Taylor 1951; Hancock 1953; Gray and Hancock 1955; Holwill 1966; Wu 1977; for nice reviews see Yates 1986 and Lighthill 1976).

With the exception of the bacterial rotor, all cilia and flagella are cellular projections consisting of a $9+2$ arrangement of microtubules linked by several classes of microtubular-associated proteins. The present view is that the progressive activation of a molecular motor, dynein, results in adjacent microtubular sliding (Warner and Satir 1974). The result of such motion is a bending wave that propagates along the flagellum from the axoneme towards the tip. We do not know how much of the organelle is actively involved in bending, and how much is passive. Nevertheless, this mechanism can be used to generate both planar and helical waves simply by varying the axis of the bending moments along the length of the flagellum. Despite such variation in wave kinematics, flagella are morphologically similar across a wide taxonomic range; flagella are generally 0.2 μm in diameter, though they range from 10 to 1000 μm in length, and the number of waves on a flagellum is often around 1.3, but can be as high as 4 to 5 (Brennen and Winet 1977).

Prokaryotic flagella are morphologically and mechanically different from eukaryotic cilia and flagella. In general, they are smaller (0.02 μm in diameter and up to 20 μm long) and less flexible structures (Holwill 1977). Prokaryotic flagella function not by bending, but by rotating. In this case, only the very base of the organelle is active, while the rest is passively turned to generate thrust. Continuous rotation is driven by a proton gradient across the plasma membrane (de Pamphilis and Adler 1971).

Regardless of the how ciliary or flagellar motion is generated, thrust arises from the same basic mechanism: viscous shearing. At low Reynolds numbers the force generated by a moving object is proportional to the speed of the object, its size, and the dynamic viscosity of fluid surrounding it. The constant of proportionality, or coefficient of drag (C_d), is a function of the object's shape and orientation to the flow. Thus we can determine the fluid force required to move an object from the following simple expression:

$$F = \mu C_d U_a, \tag{8}$$

where U is the fluid velocity relative to the object and a is its characteristic length. For cylindrical and spherical creatures, a is the radius. The value of C_d is known explicitly for only a few very simple geometries, and must be measured for all others. Fortunately, the geometry of a cilium or a flagellum can be approximated by a long thin cylinder, for which force coefficients can be calculated. It is then straightforward to represent the force generated by a waving flagellum as the sum of the forces produced by many small, straight, cylindrical segments (Fig. 1). This method, entitled "resistive force theory" (Gray and Hancock 1955; Hancock 1953), has been quite successful in predicting thrust for sea urchin spermatozoa. By this scheme, each segment encounters fluid motion at some angle (θ), and we express the sectional thrust as the resolution of the forces produced by fluid motion normal and tangential to that segment:

$$F = \mu\{\cos(\theta)C_{dt}U_t + \sin(\theta)C_{dn}U_n\}, \tag{9}$$

where the subscripts t and n correspond to tangential and normal flows. The ratio of is C_{dt} to C_{dn} is 1.0 for a sphere, but approaches 0.5 for a cylinder of infinite length (Cox 1970). As a result, an inclined cylinder moving through a fluid generates a force perpendicular to the direction of motion (Fig. 1). This slewing force arises from the fact that C_{dn} is always greater than C_{dt}.

The beauty of resistive force theory lies in its simplicity and ease of application. However, the theory ignores any upstream or downstream effects of adjacent sections or any hydrodynamic interactions between the propulsor and the nearby surfaces of the organism. This limitation is a problem when the theory is applied to motions of very large amplitude, swimming near a boundary, or when there are closely packed propulsors, as is the case with cilia.

Such awkward flow situations have been addressed with various improvements to resistive force theory. These are solutions to the Stokes equations, giving predictions of fluid motion everywhere around a moving body. For example, slender body theory, as developed by Lighthill and others (Lighthill 1975, 1976; Johnson 1977, 1980; Johnson and Brokaw 1979), is a solution to the Stokes equations for a cylinder moving with arbitrary amplitude at Re = 0. The solution is developed in the form of a line of distributed singularities – point forces and point sources and sinks of momentum. It is possible to use this technique because the Stokes equations are linear. Thus, if a combination of these singularities, along with the appropriate boundary conditions (fluid is attached to the surface of the creature), is used to generate the flows produced by an undulating cylinder, then the total thrust is known explicitly, and there is no need to approximate the situation with

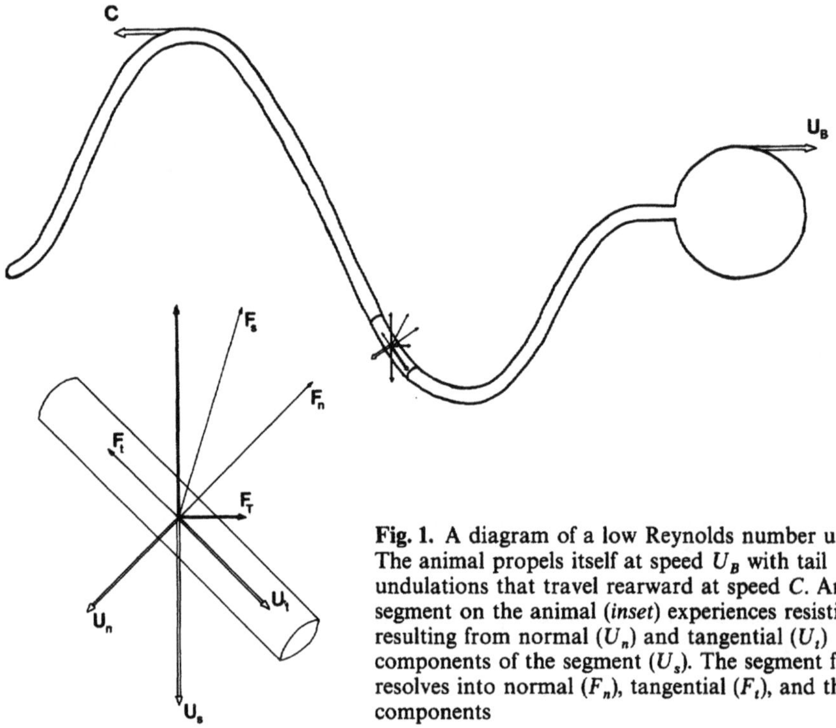

Fig. 1. A diagram of a low Reynolds number undulator. The animal propels itself at speed U_B with tail undulations that travel rearward at speed C. Any segment on the animal (*inset*) experiences resistive forces resulting from normal (U_n) and tangential (U_t) components of the segment (U_s). The segment force (F_s) resolves into normal (F_n), tangential (F_t), and thrust (F_T) components

force coefficients. A drawback to this method is that it can be computationally intensive for arbitrary motions and geometries. In fact, Hancock (1953) originally used a distribution of singularities to estimate the force coefficients that he then used in the resistive force theory. A direct comparison of slender body theory and resistive force theory for large amplitude planar waves shows that a 35% increase in the resistive force coefficients results in very good agreement between the two theories predictions for distribution of forces and energy expenditure (Johnson and Brokaw 1979). Johnson and Brokaw conclude that resistive force theory is an appropriate approximation unless the cell body is large, or has a large lateral amplitude. This is because resistive force theory neglects any hydrodynamic interactions between the head and tail.

To apply either form of analysis, it is convenient to examine the "no thrust condition", or the case in which the thrust produced just balances the drag on the head. The simplest case is that of a spherical head of radius r, whose drag (D) is given by the Stokes solution,

$$D = -6\pi\mu rU. \tag{10}$$

The total rate of work done by the flagellum is given by the limiting sum of the sectional rate of work $[\mu(C_{dn}U_n^2 + C_{dt}U_t^2)]$, plus the rate of work done in moving the head ($6\pi\mu rU^2$). We can, therefore, define an efficiency of propulsion (η) as the ratio of the rate of useful work done to the total rate of working. Lighthill (1975) had found that the maximum efficiency for headless flagellar propulsion with planar waves is given by:

$$\eta = \left[1 - \left(\frac{C_{dt}}{C_{dn}}\right)^{1/2}\right]^2. \tag{11}$$

The efficiency is largely independent of the size of the cell body; this is due to the observation that the drag on the head is small compared to that of the tail – at most one tenth – and fairly constant over the size and shape range of flagellated cells (Lighthill 1975). Thus, when the ratio of drag coefficients is unity efficiency is zero (thrust is also zero in this case), and as this ratio approaches the limiting value of 0.5, efficiency reaches a maximum of 0.085. Lighthill (1976), however, argues that a helical wave will have a higher power requirement, but coupled with the production of thrust along its entire length, may have a higher efficiency.

The power requirements of prokaryotic flagella involves another term, the energy required to rotate the body and thus conserve angular momentum. In this case efficiency is a function of the size of the cell body. Chwang and Wu (1971) estimate for organisms with spherical heads of radius 10 to 20 times the radius of the flagella that propulsive efficiency reaches a maximum of 0.28. This optimum arises because, as the head to tail ratio approaches 1, the forward speed becomes zero as the angular velocity of the head and flagellum become equal and opposite. In fact, for organisms with heads of radius less than five times that of the flagellum, a planar wave is the only effective mode of swimming. In the other extreme, as the head becomes much larger than the flagellum, forward speed declines as the drag on the head increases.

2.2 Swimming with Cilia: the Same Rules Applied to a Field of Low Re Structures

Ciliates differ from flagellated organisms by being larger (on the order of 25 to 1000 μm) and having hundreds to thousands of short, 15-μm-long, simple cilia (Sleigh and Blake 1977). Unlike the sinusoidal or helical waves of flagella, cilia typically beat in an asymmetrical rowing fashion, stiff and straight during their power stroke, and sliding tangentially during their recovery stroke (Fig. 2a). We know from above that this asymmetry is required in the Stokesian realm, and that the resistance of the recovery stroke is roughly half that of the power stroke.

In addition, ciliated surfaces exhibit a coordinated beating, or metachrony, that can be in any direction relative to the direction of the effective stroke (Fig. 2b). For example, a symplectic wave moves in the same direction as the power stroke of the cilium; antiplectic waves move in the opposite direction. Some examples of ciliary and wave parameters for two well-studied ciliates, *Opalina* and *Paramecium*, are summarized in Table 1.

These two organisms represent some of the kinematic diversity that exists among ciliates. Intriguingly, however, large ciliates swim at lower specific speeds, measured in body lengths per second (Sleigh and Blake 1977). Specific speed also apparently decreases with increasing numbers of cilia, or increasing values for the product of

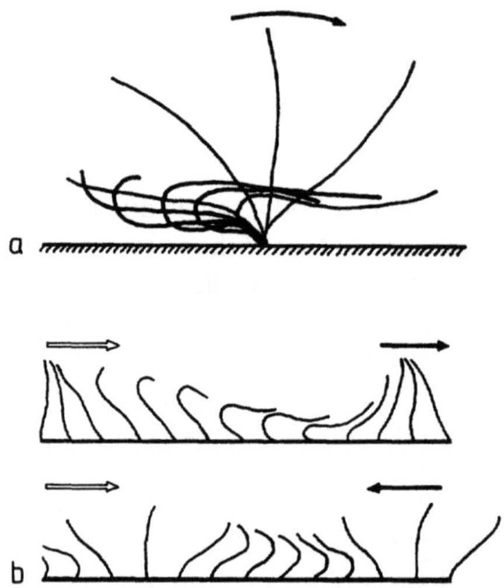

Fig. 2 a, b. Motions of cilia. **a** The power stroke in which the cilium moves through an arc in a straight configuration is followed by a recovery stroke in which the cilium experiences mainly tangential flows. **b** Symplectic (*upper panel*) and antiplectic (*lower panel*) waves of ciliary beating. The *open arrow* denotes the direction of the wave and the *filled arrow* denotes the direction of the power stroke (after Blake and Sleigh 1975)

Table 1. Parameters for cilium geometry and wave characteristics. (After Sleigh and Blake 1977; Brennen and Winet 1977)

Parameter	Opalina	Paramecium
Cell size (μm)	400	200
Ciliary spacing (μm)	3 × 0.3	3 × 3
Beat frequency (Hz)	5	30
Metachronal wavelength (μm)	40	10
Metachronal wavespeed (μm/s)	150	200
Wave form	Symplectic	Dexioplectic
Swimming speed (μm/s)	50	1500

cilium length and cilium number. The results of these observations is that absolute swimming speed remains constant across a broad size range.

How can we apply what we have learned from analyses of swimming with a single flagellum to cells with fields of closely packed propulsors? What is it about ciliated surfaces that allows the organisms to be so much bigger than flagellated cells, but at the same time appears to set an upper bound on size by constraining absolute speed? These questions surround an important difference between the above approaches and applications to ciliates: closely packed cilia do not act independently. There are hydrodynamic interactions between neighboring cilia and between cilia and the cell surface beneath them. Thus their contribution to the total thrust cannot be determined by summing the forces generated by a number of independently beating cilia.

Three models – the envelope, sublayer, and traction layer models – of ciliary propulsion have been developed to address these questions. The models predict the velocity distributions around ciliated organisms, the forces they generate, and how the morphology and behavior of these animals affect swimming performance. The envelope model (Brennen 1975) works well for densely packed arrays of cilia by approximating the field or ciliary tips as a continuous surface deforming in a manner equivalent to the metachronal waves. This model generates reasonable predictions for swimming speeds for densely, and uniformly ciliated symplectic swimmers.

The sublayer model (Blake 1972) examines the flow within the ciliary layer and is especially applicable to nonsymplectic metachrony, or widely separated cilia. In this model the force on every cilium is found by using a distribution of singularities as we have seen for the slender body theory. A system of image singularities, sprinkled beneath the cell surface, accounts for the interesting hydrodynamic interactions between the surface of the body and the ciliary field. This system of equations, integrated over the body, gives the average swimming speed and mean velocity profiles within the ciliary field. Unfortunately, such models provide only time-averaged velocity profiles and neglect specific interactions between neighboring cilia.

To address these problems, Keller et al. (1975) developed the so-called traction layer model. They consider the entire ciliary sublayer as a continuous force distribution. This approach implicitly includes interaction terms, and preserves the oscillatory component of the sublayer flow. Interestingly, both ciliary layer flow models predict a region of reversed flow near the cell wall, a result supported by the observations of Cheung and Winet (1975).

What then can these analyses tell us about the morphology and kinematics of ciliated structures? Scaling arguments imply that there is a limit to the length of an individual cilium, as well as to the size of a ciliated cell. The rates of working and bending moments of individual cilia increase with l^3 (Sleigh and Blake 1977), where l is the length of the cilium. The flexural stiffness increases as l^2 (Baba 1972), and the velocity profile of the ciliary layer, only as l^1. This means that if we were to lengthen a cilium, the work done on the cilium by the fluid, and the energy required to affect a power stroke increase faster than the cilium's ability to resist bending, and much faster than the resultant fluid velocity at the cilium's tip. Clearly these structural and energetic limits will restrict the maximal length of cilia. However, the length of an individual flagellum is up to 50-fold greater. This difference is due to the fact that a flagellum is passing bending waves along its length rather than bending as a whole, so the above scaling arguments would not apply. The only options then for larger body size are more than one or two flagella, or many cilia. The first option is only possible if speed is of no concern since flagellates swim roughly ten times slower than comparably sized ciliates (Sleigh and Blake 1977). In order to have effective ciliary locomotion, the cell needs the number (n) of cilia to be much larger than the ratio of body length to the cilium length ($n \gg L/l$) (Sleigh and Blake 1977). Thus one should not expect to see ciliated organisms smaller than 20 μm in length. The upper size limit to body size is set by the efficiency of propulsion.

By defining hydrodynamic efficiency (η) as the rate of moving an inert cell to the rate of working for all the cilia on the cell we obtain,

$$\eta = \frac{K\mu L U^2}{nP}. \tag{12}$$

where U is the velocity of propulsion and proportional to ωl, P is the rate of working for an individual cilium and proportional to $\omega^2 l^3 \mu$, ω is the angular frequency, and

K is a nondimensional shape constant. Substituting in all of the above proportionalities we find that

$$\eta \propto \frac{L}{nl}. \tag{13}$$

Because ciliary density is independent of body size, and fairly constant, n is proportional to L^2, thus η declines as L^{-1}. This implies that the larger the organism the lower the hydrodynamic efficiency of ciliary propulsion.

As a final note, we might add that ctenophores propel themselves with cilia and are up to two orders of magnitude larger than the largest ciliate. They do so with compound cilia that reach 1 mm in length and contain some 10^5 cilia, thereby satisfying any structural constraints on the individual cilium. Moreover, many of the analyses above were developed for an entirely viscous fluid, but the size and speed of ctenophores places them in a realm where inertial stresses are of comparable magnitude to viscous stresses.

2.3 Swimming with Bristled Oars

Copepods and crustacean larvae inhabit the upper end of the low Reynolds number regime. Unlike the examples above, locomotion for these creatures is often intimately connected to their feeding biology. Using intricate rowing motions of appendages made from modified mouth parts or antennae, these small crustaceans both propel themselves and capture minute food particles (Koehl and Strickler 1981; Koehl 1981). While traditional views maintained that the bristled appendages of these creatures, covered with numerous tiny setae, acted as sieves, filtering food particles out of their fluid environment (Kerfoot 1980), flows at low Reynolds numbers suggests otherwise. Indeed, in a recent computational study of flow around bristled appendages moving at low Reynolds numbers, Cheer and Koehl (1988) showed that there is negligible flow between the setae. Such appendages act, therefore, not as rakes, scrubbing food out of the water, but more like paddles, spooning food particles towards the mouth. Thus, with little investment of skeletal material, creatures swimming with oars can use brush-like devices to create, in hydrodynamic sense, a paddle.

3 The Realm of High Reynolds Numbers I: Inertia Rules the Waves

We have seen a variety of devices that rely on shear to propel creatures at low Reynolds numbers. Our understanding of how these work relies on both classical solutions to the Stokes equations as well as more recent numerical schemes. Some common themes are: (1) reversibility requires shape change for net thrust to arise, (2) forces are directly proportional to the speed of a propulsor, and (3) hydrodynamic interactions may significantly affect the flows generated by arrays of structures moving at low Reynolds numbers. In the section below, the first two themes no longer apply, and the latter has only recently been explored. In this new regime of high Reynolds numbers, we change the rules that govern motion and explore, again, how size, shape, and kinematics affect performance.

3.1 Swimming with an Undulating Body

Euler's equations tells us that for animals whose Reynolds numbers are quite large, one viable mechanism for thrust production arises from acceleration of fluid elements relative to the body. As an element of the body accelerates laterally at some angle with respect to the fluid, there is a reaction in the fluid to that acceleration. This reactive force has a component in the direction of thrust (Fig. 3). Indeed, this is apparently the scheme employed by animals such as eels, larger fish, and to a lesser extent, smaller undulators such as leeches. The classic theory of Lighthill (1971) shows that we can estimate the thrust generated by such creatures from analytical solutions to Euler's equations. The goal of such solutions is to determine the average force generated by a body with lateral undulations. Most such analyses begin with an estimate of the instantaneous sectional force generated by lateral accelerations. This force is written formally as the local rate of change of fluid momentum:

$$F_{\text{sectional}} = \frac{D(mw)}{Dt} = \frac{\partial(mw)}{\partial t} + \frac{U\partial(mw)}{\partial x}, \tag{14}$$

where w is the lateral velocity of the body section and m is its virtual mass. The latter quantity is approximated by

$$m = \rho\pi b^2, \tag{15}$$

where b is the depth of the section. This term, representing a reaction in the fluid to acceleration, acts as if the accelerating section has some increase in its mass. This increase depends upon the shape of that section, with deep sections disturbing proportionately more fluid around them as they accelerate (Lighthill 1975; Daniel 1984).

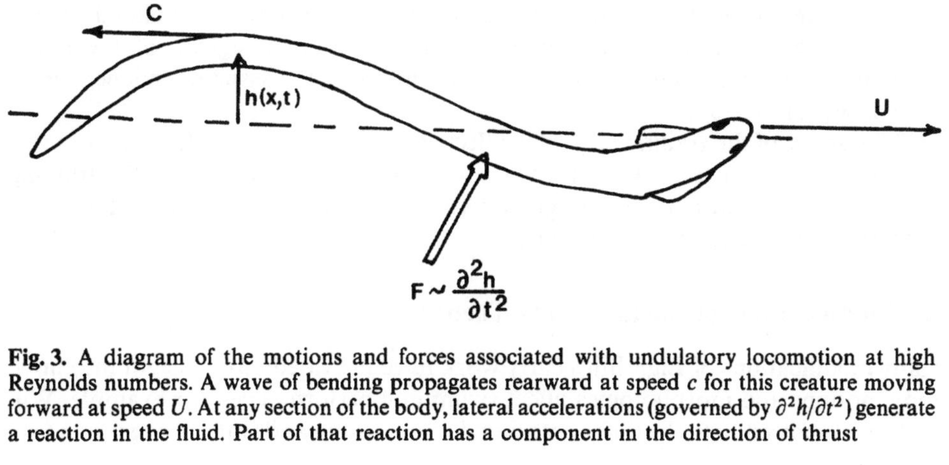

Fig. 3. A diagram of the motions and forces associated with undulatory locomotion at high Reynolds numbers. A wave of bending propagates rearward at speed c for this creature moving forward at speed U. At any section of the body, lateral accelerations (governed by $\partial^2 h/\partial t^2$) generate a reaction in the fluid. Part of that reaction has a component in the direction of thrust

Momentum imparted by lateral accelerations of body sections is transferred down the body in the direction of the rearward propagating wave. Without detailing the derivation of Lighthill's (1971) slender body theory, we find that the total average thrust is

$$T = m\left(\frac{V^2}{2} - \frac{V\partial h}{\partial t}\right), \tag{16}$$

where T, the average thrust, is evaluated at the trailing edge ($x = L$), V is the velocity of the trailing edge following its motion ($V = Dh/Dt = \partial h/\partial t + U\partial h/\partial x$) and $\partial h/\partial t$ is the lateral velocity of the trailing edge. We use only the time-averaged value of the entire quantity in the brackets. The hydrodynamic efficiency (η: the ratio of the rate at which useful work is done to the total rate of mechanical power input) is

$$\eta = \frac{1 - U}{c}, \tag{17}$$

where U is the forward speed of the animal and c is the speed of the rearward propagating wave (Fig. 3).

These two equations tell us that the average thrust generated by a generic undulator can be estimated by the rate at which momentum is shed from the rear end of the animal. This makes sense if we consider the analogy of a jet engine in which fluid approaching the turbine is accelerated and then ejected at some higher speed. The thrust generated by this scheme is simply the rate at which momentum is ejected by the turbine. So, too, is thrust generated by creatures such as squid, salps, and jellyfish, who eject boluses of momentum in the form of discrete pulsatile jets (Bone and Trueman 1983; DeMont and Gosline 1988; Daniel 1985; Packard 1969; and especially DeMont, this Vol.). Similarly, for swimming fish and eels, we can estimate the average thrust from the rate at which the animal ejects momentum from its trailing edge.

With the above analogies in mind, we can use Lighthill's equations to examine how the motion and shape of the trailing section of the animal affects thrust. For example, ever-larger lateral accelerations yield greater thrust. One must be careful not to carry this to extremes. As the amplitude of the lateral oscillations becomes very large, many of the assumptions underlying the theory are violated. Most notably, exceedingly large amplitudes will affect flows in regions of the body upstream from the trailing edge. Under such conditions, the assumed independence of each section no longer occurs.

The shape of the trailing edge also affects thrust production. We note from the spatial derivative in Eq. (14) that if the body depth increases towards the trailing edge, thrust rises. Moreover, the greater the lengthwise rate at which the depth increases, the greater will be the thrust.

3.2 Modifications of Slender Body Theory

Many modifications of Lighthill's early work have provided more insight into how body motions and morphologies affect swimming performance. For example, Wu (1971 a–c) showed how the flows associated with leading and trailing edges of a fish may significantly affect thrust. Most interesting is Wu's treatment of a phenomenon called "leading edge suction". As its name suggests, this force arises from the unique flows that occur about the leading edge of oscillated bodies or appendages. Large changes in momentum are required for a fluid to negotiate the often sharp turns about such shapes. These changes in momentum result in regions of very low local pressure which, when integrated over the tiny area of a leading edge give a suction force that acts perpendicular to that edge (Batchelor 1967). The magnitude of that force is

$$T_{\text{suction}} = \frac{1}{2} V^2 \frac{\partial m}{\partial x}, \tag{18}$$

where $V = Dw/Dt$ (for a nice review see Yates 1983). Here again, we note the importance of the body shape: the value of $\partial m/\partial x$ (gradient of the body depth) at the leading edge determines the magnitude of this suction force. As with the above analyses, one must be careful not to carry this to an extreme. Ever-steeper spatial gradients of body depth may lead to morphologies for which the slender body assumptions no longer apply. Below we indicate new engineering methods that can be used to avoid this problem. In practice, however, leading edge suction is small, contributing only about 10% of the total thrust for a swimming fish such as a salmon (Yates 1983).

Despite the low magnitude of the leading edge suction force, it has some intriguing implications. For example, heterocercal tails found in many elasmobranchs have sharp leading edges oriented at an oblique angle with respect to the long axis of the body. A common interpretation for this unique morphology is that the force such tails generate has components pointed in both anterior and dorsal directions (Alexander 1965; Thomson 1976). The dorsal component of the force is thought to be important in preventing fore-aft pitching of the animal, a phenomenon which arises from an awkward placement of the center of gravity relative to the center of fluid forces acting on the animal. What mechanism gives rise to that upward component? One common answer is that it arises solely from the tilt of the lift force generated by the tail (Alexander 1965, 1982). This tilt requires a slight lateral bend to the tail. An alternative mechanism that can yield sufficient upward force is the leading edge suction term. Acting perpendicular to the line of leading edge, it will always have a dorsal component, propping the back end of the animal up (Fig. 4).

The trailing edges of fish bodies, fins, flukes, and flippers are also important regions that determine flows and forces generated by motions. As we noted above, the rate at which momentum is ejected from the trailing edge of a body determines part of the total thrust. But at the trailing edge of a body or an appendage, fluid tends to continue tangentially along its path, separating from the surface of the animal. This result, often referred to as the Kutta condition, yields a structure called a vortex wake. In the next section, we examine in more detail the physics of such vortical flows. Here we simply indicate that the presence of vortex shedding from the trailing edge makes any variation in sectional added-mass unimportant, as such reaction forces are computed only for unseparated flows, that is flows for which no wakes form. The consequences of vortex shedding to fish swimming have been examined in considerable detail by Wu (1971 a–c) and recently reviewed by Yates

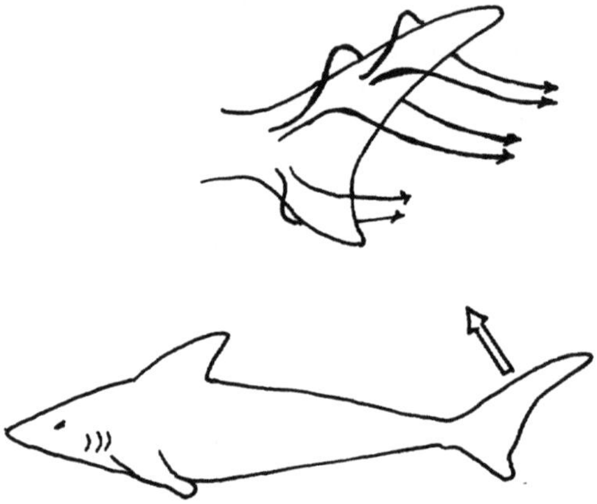

Fig. 4. A diagram of the flows and forces associated with leading edge suction acting on the heterocercal tail of some elasmobranch. This suction force arises from the large pressure associated with flows around sharp leading edges

(1983, 1988). An intriguing twist to the consequence of vortex shedding is that caudal fins of many fish lie in the vortex wake generated by dorsal fins. The energy lost in shedding vorticity from the dorsal fin can be recovered by the caudal fin and used to create more lift than if the caudal fin were oscillated in an undisturbed flow (Yates 1983).

In addition to fin-fin interaction, fin-body interactions have been a focus of recent studies (Lighthill and Blake 1990). In such cases, the slender body theory outlined above has been modified to examine how the peculiar flows that occur between a fin and the body of a fish yield greater thrust.

Our discussions so far have dealt largely with isolated organisms moving along straight paths at constant velocity. Yet there are many important situations in which such a view is too constraining. For example, attack of prey or excape prediation both include significant turns and accelerations. In addition, many aquatic organisms form schools or aggregations in which significant hydrodynamic interactions may arise. Both of these situations were examined by Weihs (turning and acceleration: 1973a; schooling: 1973b). In his study of turning, Weihs modified slender body theory to account for body accelerations and changes in direction. He found that the body trajectory could be predicted from the kinematics of the body bends, typically "C"-shaped bend followed by an "S"-shaped wave. Moreover, Weihs (1973a) showed that these kinematics can greatly increase the hydrodynamic efficiency of attack. Numerous additional papers have examined how body morphology and kinematics affect performance during such accelerational maneuvers (e.g., Webb 1978; Harper and Blake 1989), with the general conclusion that long, flexible bodies are more favorable.

Hydrodynamic interactions between individuals, or between an individual and some nearby surface, are also important determinants of swimming performance. Weihs' analysis of schooling fish suggests that, under appropriate spacing, individuals may experience some augmentation in the thrust they generate. Other hydrodynamic interactions may not be as favorable. For example, Webb et al. (1990) have shown that acceleration performance of fish is greatly reduced near the free surface of the water column. This result has interesting implications to the spatial distribution of various sized fish: their results indicate that larger fish experience greater decrements at the same water depth than do smaller fish. Thus in a predator-prey interaction, smaller fish should show greater survivorship nearer the surface.

3.3 Some New Tools and Directions

While the present theories generally indicate that ever-larger amplitudes of motion yield ever-greater thrust forces, with no consequence to efficiency, we cannot apply them to extremely large amplitude motions and cannot, therefore, determine physical or physiological limits to such amplitudes. Good news, however, is on the horizon with the advent of new computational methods for solving the basic flow equations. This area of "computational fluid dynamics" has only recently been applied to studies of swimming and flying. One such example is the development of numerical methods to evaluate the forces on swimming fish without the underlying assumption of low amplitude (Schultz, Chow and Webb, pers. commun.). Another method, described below, uses vortical flow theories to develop numerical schemes that estimate forces generated by undulated fins or flippers (Grunbaum and Daniel, in prep.). The power of such methods lies in their ability to deal with arbitrary motions, including large amplitudes, whole body accelerations, and complex geometries. Their disadvantage lies in the cumbersome mathematics and programming required for solutions to specific problems.

4 The Realm of High Reynolds Numbers II: Thrust from Lift and the Nature of Hydrofoils

The previous section alluded to wakes and vorticity as important components in our analyses of thrust. These phenomena become extremely important in studies of aquatic flight. This mode of locomotion is seen in animals such as penguins, sea turtles, whales, and tunnies, who propel themselves by producing thrust with a mechanism that is substantially different from the reactive force described above. With appropriate motions of appendages such as wings, flukes or fins these animals generate forces that are *perpendicular* to the direction of fluid motion along an appendage. These lift forces depend upon the shape and size of the appendage as well as its motion. As with the examples above, we seek an understanding of how such parameters affect the various measures of performance. We delve into a bit more detail than would seem necessary to explain several physical phenomena. Our reasons for this digression are twofold: one is that new analytical and computational methods analyzing swimming and flying rely on an understanding of vortical flows for predicting thrust and energy requirements, the other is that many previous descriptions of lift based solely on Bernoulli's principle give only a partial view of the important determinants of lift performance. Accordingly, we return once again to Euler's equations for exploring the physics of underwater flight.

4.1 The Basics: Vorticity, Circulation, and Lift

Recall that Euler's equations say that the pressure gradients about a body are determined by accelerations and by a peculiar velocity gradient. We follow Childress (1981) to expand this gradient for a simple two-dimensional flow, monitoring only the vertical pressure differences (Fig. 5).

$$-\frac{\nabla P}{\rho}=\frac{\partial \mathbf{u}}{\partial t}+\mathbf{u}\cdot\nabla\mathbf{u} \tag{19a}$$

$$=\frac{\partial \mathbf{u}}{\partial t}+\frac{\nabla(\mathbf{u}\cdot\mathbf{u})}{2}+\mathbf{u}\mathbf{x}\{\nabla\mathbf{x}\mathbf{u}\}, \tag{19b}$$

where the symbol ∇ is the differential vector operator $[\partial/\partial x, \partial/\partial y]$. Hidden in this last term is a parameter called vorticity ($\omega=\nabla\mathbf{x}\mathbf{u}$), an often poorly understood physical quantity that represents the tendency in a fluid to form a spin, or to create circulation. Vorticity is a flow vector that is orthogonal to the two flow directions (recall the right hand rule for the cross-product of two vectors and generates a force). This force has a direction that depends upon the spatial gradient of flow around a point and is in the plane of fluid motion. To see this more clearly, we expand the last term in the above equations for the simple, steady, two-dimensional flow shown in Fig. 5:

$$\frac{\partial P}{\partial y}=\frac{\rho u \partial u}{\partial y}. \tag{20}$$

Thus there is a force associated with this flow that is directed upwards from the lamina of reference. Without even mentioning Bernoulli's principle, we can account for a force that is perpendicular to the direction of motion. Bernoulli's equation, derived under the assumption that vorticity is zero, could not completely account for lift in this sense.

We note that vorticity yields lift. The next logical step is to determine the magnitude of that force and how it contributes to thrust. Clearly, the greater the vorticity, the greater the area over which vorticity is distributed, or the greater the

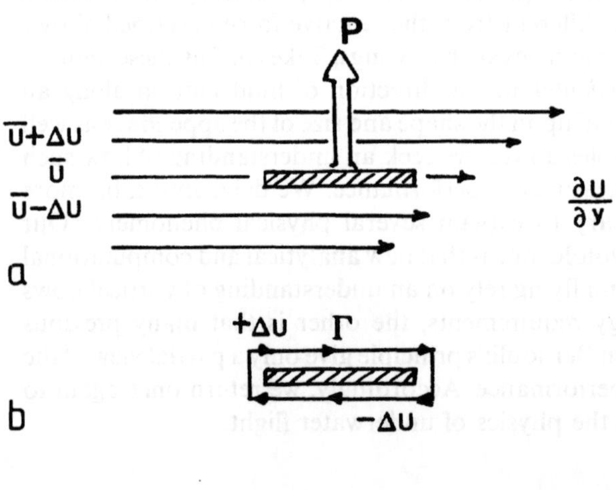

Fig. 5 a, b. Flows and forces associated with vortical flow. **a** A simple flow around a thick flat plate shows the relevant velocity gradients and the direction of the resulting pressure term. **b** The same figure redrawn in the frame of reference of fluid moving at speed *u*

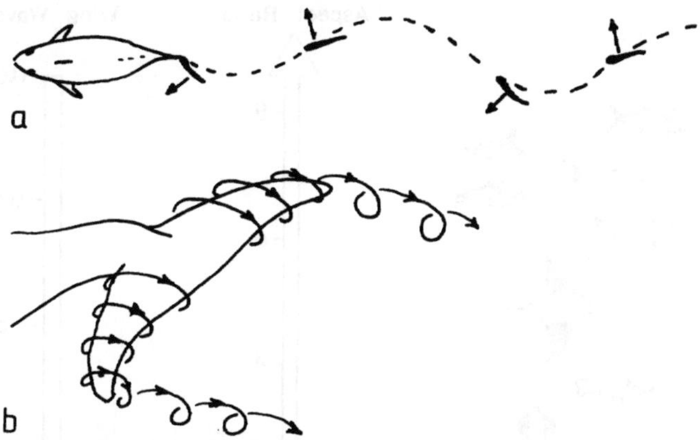

Fig. 6 a, b. Some lift-based devices used in aquatic propulsion. **a** the resolution of the lift vector is shown for various positions of the caudal fin of a tuna. **b** the flows and tip-vortex associated with heaving motions of a whale fluke

density of the fluid, the greater will be the total lift. We integrate the vorticity over the surface to obtain a term that is quite closely related to lift: circulation (Γ)

$$\Gamma = \int \mathbf{w}\, ds \tag{21}$$

$$\text{Lift} = \rho U \Gamma. \tag{22}$$

To see circulation we replot our two-dimensional flow over a surface in the frame of the mean relative velocity (Fig. 5b). In this frame, the fluid appears to spin about the surface. The rate of apparent spin depends upon the difference in velocity (the vertical gradient) above and below the surface and, therefore, is a measure of the vorticity distribution. The total lift, in turn, is directly proportional to the total circulation. Therefore any mechanisms or structures that augment and maintain circulation on a wing will have profound effects for the total lift and thrust generated by that wing.

To seen how lift can produce thrust we examine the forces produced by the caudal fin of a tuna (Fig. 6a). This appendage, shaped like a crescent, is oscillated from side to side with sharp changes in its angle of attack (the angle of the fin with respect to oncoming fluid). At each instant, some amount of lift is developed with some component of that force directed in the forward direction. Thrust from lift, therefore, relies on a mean forward tilt of the lift vector for one complete tail beat cycle. Armed now with our understanding of the origins of lift and thrust from lift, we can proceed to examine how appendage morphology and motion influence these forces.

An interesting example of the relationship between fin geometry, circulation, and thrust is seen in our illustration of a whale fluke (Fig. 6b). Here, circulation is developed by the powerful heaving and pitching motions of the tail. On the downstroke, flow over the fluke moves faster over the top than the bottom. This velocity gradient generates some distribution of vorticity over the fluke which, when integrated over the length of the fluke, yields some sectional circulation. Because the fluke has ends to it (it is finite in span), the circulation tends to roll off the tips yielding a "tip vortex". The very ends of the fluke contribute, therefore, little to the

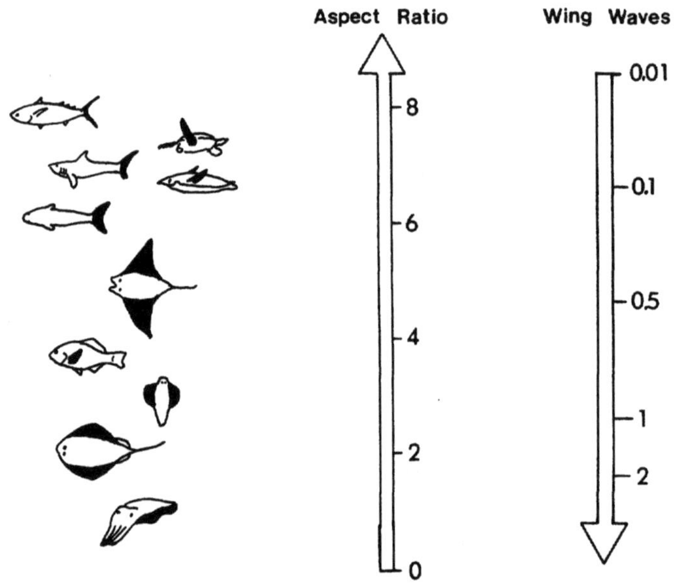

Fig. 7. A rough estimate of the relative aspect ratios for the wings of various flyers. Also shown is an estimate of the number of propulsive, retrograde, waves that propagate along the wing

overall forces that are generated. Flukes of ever greater span, however, have proportionately less of their surface associated with this loss of circulation and produce, therefore, proportionately more force. Thus, from this very simple verbal description of lift, we see one important morphological consequence of fin geometry: greater spans yield better lift performance.

The "aspect ratio" of the fluke is an additional morphological parameter that comes into play in our analysis of thrust production. While the aspect ratio has several definitions, we refer to the one used most commonly as the square of the span divided by the total area of the appendage. Thus fins or flukes of high aspect ratio are rather long and thin, whereas ones of low aspect ratio are short and stubby. From our above argument it seems that, for a given amount of wing area, higher aspect ratio wings yield greater thrust. Indeed, this is the view maintained by researchers of animal flight (e.g., Tucker 1977; Oehme 1977). Unfortunately, there are many cases in which animals produce thrust with "wings" of very modest, or remarkably low, aspect ratio (Fig. 7). Can we determine whether such forms are indeed "poor" swimmers or whether additional physical factors offset the effects of low aspect ratio?

One apparent correlate of low aspect ratio wings is that they deform extensively. This deformation is commonly the result of active waves of bending that travel along the wing. For a continuum of aquatic flyers we note (crudely) an inverse correlation between wing deformation (measured by the number of propulsive waves present along that structure) and the aspect ratio. To understand this phenomenon we must examine more details about the physics of aquatic flight.

4.2 Unsteady Aquatic Flight

Our previous descriptions of lift have been from the standpoint of steady-state fluid motion. Thus, when we examined the lift and thrust generated by an oscillating fluke or fin, we assumed that we could account for the relevant forces solely from the instantaneous velocity distribution about the wing. For wings that heave and pitch, the circulation changes sense between upstrokes and downstrokes. We assume that the circulation developed by that motion rises and falls in precise accord with changes in velocity. Unfortunately, circulation takes some time to develop. Accordingly, for motions that change sufficiently rapidly, the circulation may not be as great as we would assume from a steady or quasi-steady theory. Indeed, for wings that not only heave and pitch, but deform, the variations in local velocity could be large.

To determine the relative importance of oscillatory motion we use a dimensionless term called the reduced frequency parameter (σ, see also the Spedding Chap. 3, this Vol.):

$$\sigma = \frac{2\pi f l}{U}, \tag{23}$$

where f is the frequency of oscillatory motion and U is the forward speed of the body or appendage. In a sense, the reduced frequency parameter is a measure of the ratio of oscillatory to forward motion. If σ is much less than 1, we can treat the motion as essentially steady. The flows and forces for such situations can be estimated from simple steady theories. If, however, that parameter is much greater than 1, we must incorporate the time lags in the development of circulation. A graph of the amount of lift in phase with wing motion versus the amount lagging the motion shows how the time-dependence in growth of circulation affects thrust production (Fig. 8). While the mathematical representation of this graph is beyond the scope of this chapter, we note several important features: (1) at reduced frequencies below 0.1, almost all of the lift is in phase with the motion and its magnitude is unaffected by

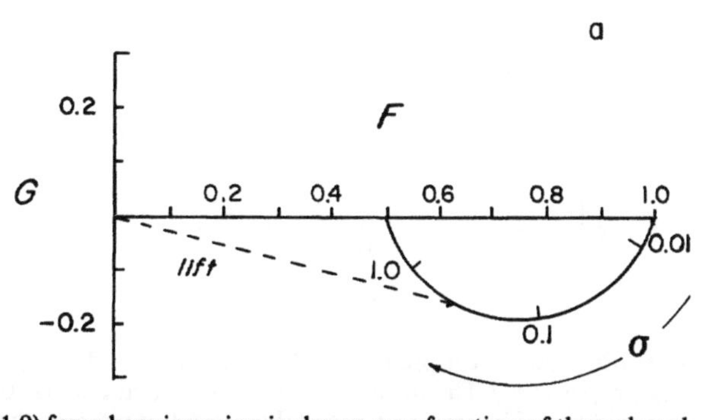

Fig. 8. The theoretical lift (scaled to 1.0) for a heaving wing is shown as a function of the reduced frequency parameter. The *angled line* gives the magnitude of the total lift. The component of that vector along the abcissa (F) gives the proportion of lift that is in phase with wing motion and the ordinate (G) gives its out-of-phase component. Even at moderate reduced frequencies, unsteady effects play an important role. (After Lighthill 1975)

oscillations; (2) at reduced frequencies of about 1.0, circulation is no longer fully developed and the lag between wing motion and lift is substantial; (3) at very high reduced frequencies the lift falls to about half of its steady value.

For some aquatic organisms, the reduced frequency parameter may reach 1.0 (Daniel and Webb 1987) or higher (Yates 1983; Magnuson and Prescott 1966). Thus we cannot neglect the time-dependent nature of circulation development in our analyses of thrust from lift.

While it appears that oscillation is "bad" for lift, a nice consequence of such motion is that it yields higher relative speeds. Thus while the peak value of lift for an oscillating fin will not be as high as steady-state theory would predict from the peak velocity, increases in instantaneous velocity can yield increases in lift and, therefore, thrust. Moreover, as accelerational flows arise in oscillatory wing motions, unsteady forces may contribute to thrust. This complex set of trade-offs between circulation development, acceleration forces, and lift was examined in detail by Wu (1971 a–c) and later applied to aquatic flight by rays and skates (Daniel 1988). Two intriguing results emerge from these analyses: (1) for wing sections of varying lengths, the thrust

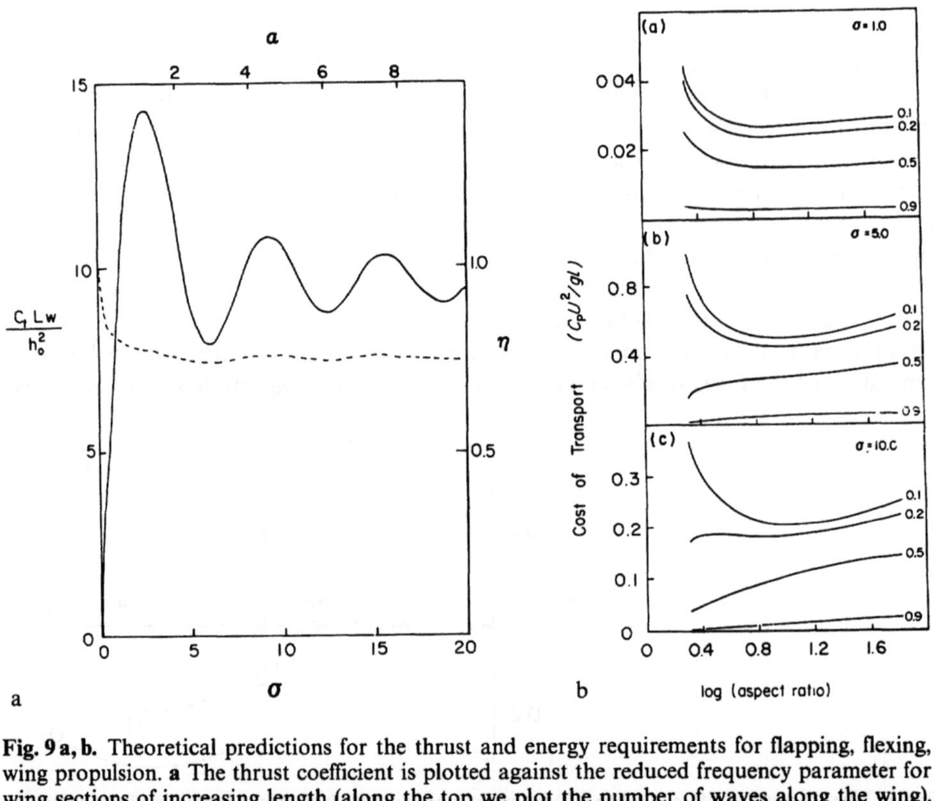

Fig. 9 a, b. Theoretical predictions for the thrust and energy requirements for flapping, flexing, wing propulsion. **a** The thrust coefficient is plotted against the reduced frequency parameter for wing sections of increasing length (along the top we plot the number of waves along the wing). **b** The predicted cost of transport as function of reduced frequency, relative wave speed (U/c, ranging from 0.1 to 0.9), and aspect ratio. The reduced frequencies are (a), 1; (b), 5 and (c) 10. Note that very flexible wings (those with low values for U/c) show lower costs of transport at lower aspect ratios. (After Daniel 1988)

generated by this unsteady mechanism has a complex dependence on the number of propulsive waves and the reduced frequency parameter, showing a distinct optimum for wing kinematics (Fig. 9a), and (2) when integrated over the surface of a wing whose flapping amplitude increases spanwise, there is a unique aspect ratio that minimizes the specific energy requirements for flight, as measured by the predicted cost of transport (Fig. 9b). Indeed, for animal appendages with propulsive waves traveling in the antero-posterior direction, ever lower aspect ratio wings yield lower energy requirements for locomotion, an observation that partially explains the apparent diversity in wing aspect ratio for aquatic flyers. All of the above results follow from the trade-offs between accelerational forces and circulation development in unsteady flow.

4.3 New Directions for Aquatic Flight

Analyses of the two major mechanisms of propulsion at high Reynolds numbers have relied on solutions to various forms of Euler's equations with the underlying assumption that the oncoming flows encountered by a segment of a body or an appendage are not influenced by the motion of that segment. The underlying equations and boundary conditions are thus said to be "linear" and can, therefore, be solved analytically. Unfortunately, biology does not care whether mathematicians can easily analyze the problems it poses. Indeed, in many examples of aquatic locomotion, the amplitudes of appendage or body motion may become so large that linear theories should not be applied. What can we do to analyze such problems, and what factors might limit the amplitude? Linear theory gives no insight into these questions as it always predicts increases in thrust with increases in amplitude, without any decrements in the efficiency of locomotion.

Computational schemes that are used to solve nonlinear flow problems are being developed to answer such questions. One such application is used to examine the forces and flows associated with flapping fin propulsion (Grunbaum and Daniel, in prep.). The spirit of the problem is to solve for the distribution of vorticity that is generated by a heaving, flexing, pitching wing section that generates some leading edge suction and satisfies the Kutta condition. The vorticity is then used to compute lift and thrust forces as well as the energy dissipated in the wake of the wing. Moreover, predictions of the flow generated by such complex motions arise from these methods (Fig. 10), showing the structure of complex time-dependent wakes. One preliminary result from this type of analysis is that the coefficient of thrust divided by the square of the heaving amplitude is not, as linear theory predicts, a constant, but falls with increasing amplitude. We have, therefore, a first glimpse into the flows generated by such motions and the potential limits to their amplitudes.

Numerical schemes, such as the one above, bridge analyses of swimming and flying, predicting wake structures that were suggested by Rayner (197a, b) and Ellington (1984) for flight, and provide new tools to analyze the mechanics of aquatic locomotion. Moreover, such methods allow integration of other factors, such as muscle mechanics, that are important determinants of locomotion performance. We anticipate increased use of such methods in biological fluid dynamics.

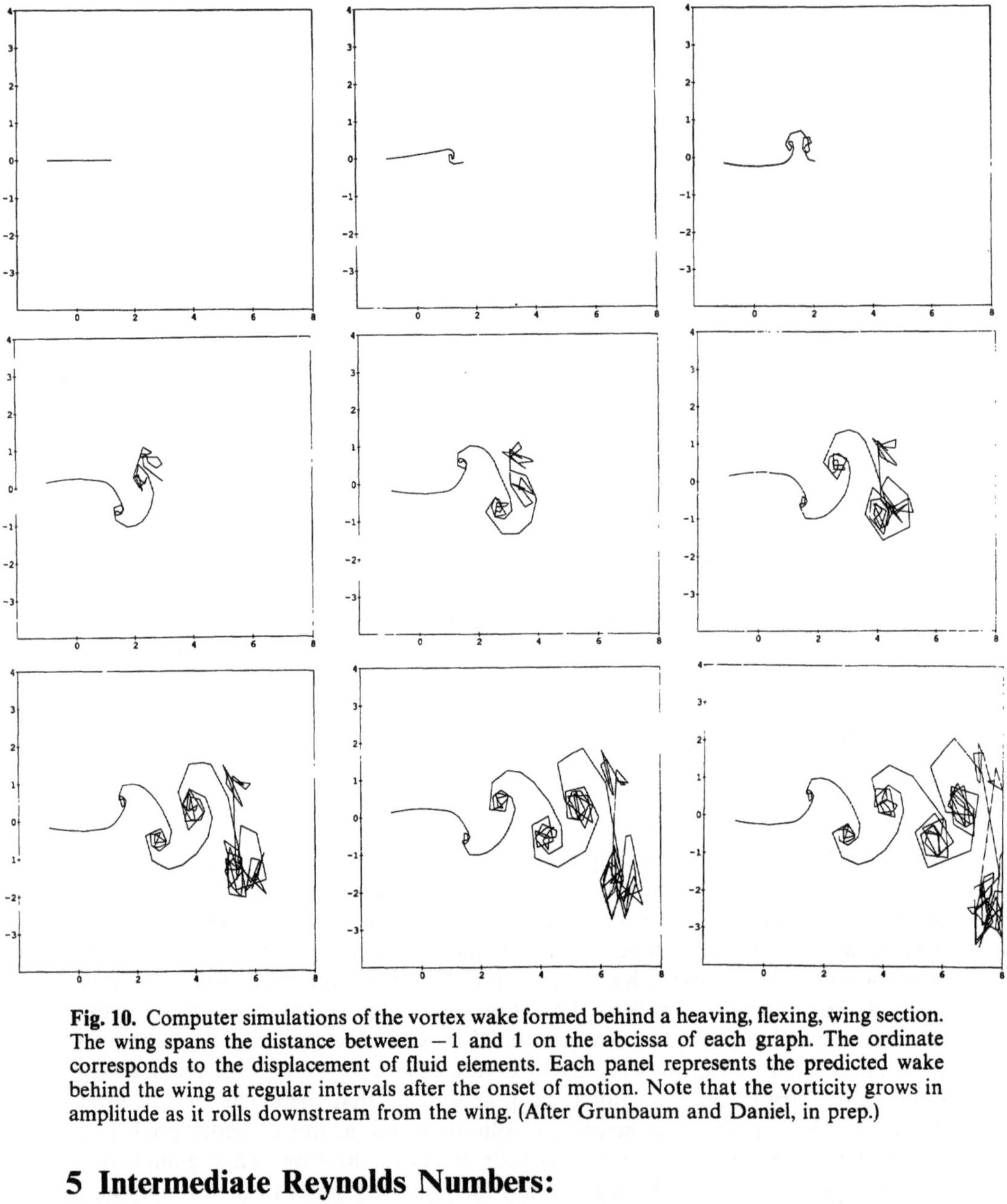

Fig. 10. Computer simulations of the vortex wake formed behind a heaving, flexing, wing section. The wing spans the distance between −1 and 1 on the abcissa of each graph. The ordinate corresponds to the displacement of fluid elements. Each panel represents the predicted wake behind the wing at regular intervals after the onset of motion. Note that the vorticity grows in amplitude as it rolls downstream from the wing. (After Grunbaum and Daniel, in prep.)

5 Intermediate Reynolds Numbers: Biology Loves this Region and Theoreticians Are Challenged by It

When the Reynolds number for bodies or appendages lies somewhere between 1 and 1000, the convenient simplifications of the Navier–Stokes equations no longer apply. This awkward region of Reynolds numbers occurs quite commonly in biology, including much of the planktonic community, many small fish, and the aquatic larval stages of countless creatures. To analyze the mechanics of locomotion in this realm, we must resort to an approach quite different from those above. Thus, rather than solving the equations of motion, we rely on empirical coefficients of force that were defined by Eqs. (6) and (7). These coefficients give fairly good estimates for the kinds of drag forces produced by rowing appendages and undulating bodies. We emphasize, however, that they should not be taken too seriously for motions with large changes in velocity, as drag, wakes, and the peculiar flows that occur at intermediate Reynolds numbers all take time to develop. Below we explore two examples of intermediate Reynolds number swimming: rowers and undulators.

5.1 Rowers: Pectoral Fins, Frogs, and Dyticid Beetles

Dyticid beetles, labriform fishes, the larvae of many crustaceans, and a host of other creatures propel themselves with rowing motions of some oar-like appendage. The basic theme for appendage motion is quite simple: a power stroke, in which an appendage is flung rearwards, generates greater force than a recovery stroke, in which the appendage returns to its starting position. The difference in force generated by power and recovery strokes is almost always the result of some shape change in the appendage, with setae or webbing offering greater resistance to motion during the power stroke. Since most animals propel themselves with paired appendages that move in synchrony, any lateral forces cancel.

The most extensively studied rowers are aquatic beetles (Nachtigall 1960) and labriform fishes (Blake 1981a, b). A principle goal of such studies was to illuminate how the size and shape of the body and propulsive appendages affect net thrust, body translations, and energy requirements for locomotion. Using a quasi-steady analysis, in which drag forces generated by power and recovery strokes counteract the drag on the body of water beetles, Nachtigall (1960) and Nachtigall and Bilo (1975) computed the overall efficiency of swimming to be about 0.45, a factor that arises from energy wasted in recovery strokes and in generating lateral forces as the appendages swing through their arc.

Because oscillatory motions always involve accelerations, the possibility arises that added-mass forces may be involved in thrust production. For aquatic beetles, hindlimb Reynolds numbers are probably sufficiently small to permit a quasi-steady analysis. However, for larger appendages, such as pectoral fins of labriform fishes (Blake 1981a, b), and the hindlimbs of frogs (Gal and Blake 1981a, b), added-mass forces may be important determinants of thrust. Indeed, for both organisms, added-mass forces are nearly one decade higher than drag forces, suggesting that much of the mechanics are dominated by accelerational rather than drag-based mechanisms.

An intriguing consequence of the role of such unsteady forces is that thrust can be generated at peculiar times during the cycle of a limb beat. An analysis of just added-mass forces for entirely symmetric rowing shows that, even without any shape change between power and recovery strokes, net thrust can be generated (Daniel 1984). This result follows from several features about unsteady rowing. First, many rowers oscillate their oars through an angle of about $2\pi/3$ radians about a stoke axis that is directed posteriorly (Nachtigall 1960; Blake 1981a, b; Daniel 1984; Gal and Blake 1988b). At the onset of the power stroke, the limb is accelerated from a position in which it may generate a net forward reaction in the fluid. At the end of the power stroke, the limbs are nearly parallel to the direction of motion and no net force can be generated, as lateral components cancel and forward components are negligible. At the end of the recovery stroke, as the limbs decelerate, a *forward* reaction in the fluid yields additional thrust. Thrust is, therefore, generated by this scheme at the onset of the power stroke and at the end of the recovery stroke. The latter point would not be predicted from a quasi-steady analysis.

Gal and Blake's (1988a, b) analysis of frog swimming showed one additional, and somewhat disturbing, point that may plague analyses of rowing. Even though both unsteady and quasi-steady forces were used to predict thrust their summed values were not large enough to account for the total thrust generated, as measured by body accelerations. Either the estimates of the drag and added-mass coefficients were too small or another mechanism may come into play. The authors favor the latter case, suggesting that, at the end of the kick, the hind limbs come together and generate a jet of fluid with rearward momentum, yielding more thrust than would be predicted from a model that neglects the hydrodynamic interactions between appendages.

A recent theoretical analysis of rowing with just one appendage during the rapid tail flexion of shrimp includes this hydrodynamic interaction (Daniel and Meyhöfer 1989). In addition, since only one appendage is responsible for thrust, lateral forces do not cancel and significant rotational motions arise. Predictions for instantaneous forces, and the moments they generate about the animal's center of gravity, are used to estimate both translational as well as rotational motions of the entire animal. The forces arise from a combination of drag and added-mass forces, with the latter dominating instantaneous thrust. At the end of the tail flip, as the abdomen is pressed against the cephalothorax, an additional force arises. This "squeeze" force is quite similar to the fling mechanism proposed by Weis-Fogh (1973) and elaborated by Ellington (1984) for hovering insects and to the jet reaction mechanism for squid, medusae, salps, and siphonopohores (Daniel 1984, 1985; DeMont and Gosline 1988; Bone and Trueman 1982, 1983; O'Dor 1988). The magnitude of this force is greater than any instantaneous drag and approximately equal to the peak force generated by the acceleration reaction. Such hydrodynamic interactions between appendages, or between an appendage and the body of an animal, are likely to contribute significantly to the total thrust generated by swimming animals. This observation is born out in a more recent study of locomotion in triggerfish in which fin-body interaction yields a threefold increase in the thrust that would be generated by a fin undulating in the absence of a body (Lighthill and Blake 1990).

As mentioned above, significant rotations result when only one oar is present. Thus, by relaxing the common assumption of rectilinear locomotion, Daniel and Meyhöfer (1989) could examine how body size and shape affect performance in this rarely used, but selectively important, mode of locomotion. Their results suggest that, while greater tail lengths yield ever more force, the moments generated by tail flips become so large that nearly pure rotational, rather than translational, motion results. There is a unique ratio of tail length to body length that maximizes translation. This ratio depends upon the size of the animal. Ever-larger tail flippers should have smaller tails to affect better translation.

5.2 Chaetognaths Swim with Undulations and Analyses Require Both High and Low Re Approaches

As we have seen from both the high and low Re realms, undulatory locomotion is a common mode of swimming. Yet, while the kinematics are quite similar, the mechanisms for generating thrust are quite different. These two realms represent the extremes in size and speed, which correspond to the simplifying cases of Stokesian and Eulerian flows. What can we say about all the undulators whose Reynolds numbers fall somewhere between those of a sperm whale and a whale sperm? As above, we must resort to experimentally determined force coefficients and steady-state approximations in estimates of thrust. However, the picture is not entirely bleak, as new computational methods being developed (Fauci and Peskin 1988; Fauci 1990) address such problems.

Chaetognaths are undulators that lend themselves to studies that combine high and low Re approximations. These animals are raptorial planktonic predators that swim with rapid dorso-ventral undulations. Their size and speed puts them well within the intermediate Re range, but their geometry is reminiscent of both low-Re resistive force swimmers, as well as high-Re slender body swimmers, with a long slender body equipped with fins. To determine the fluid forces and the relative

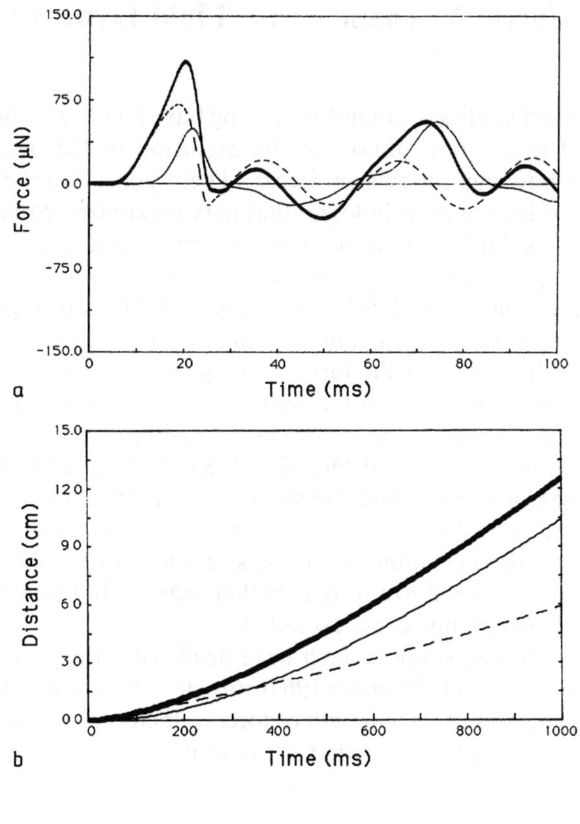

Fig. 11 a, b. Theoretical predictions of the forces and movements of chaetognaths swimming at intermediate Reynolds numbers. (After Jordan 1991). **a** Predictions of the instantaneous forces by component: total force (*bold line*), resistive component (*solid line*), and reactive component (*broken line*). Note the decay in the reactive component with time. **b** Predictions of the distance traveled from rest based on a purely resistive model (*solid line*), a purely reactive model (*dashed line*), and a combination of the two (*bold line*). Reactive theory works best for the first 200 ms of motion contribution of viscous and inertial stresses generated by an undulating body at this scale, Jordan (1991) used both the high Reynolds numbers approximations of Lighthill (1971) and the low Reynolds number theories of Gray and Hancock (1955). His results show nice agreement between measured and predicted body trajectories for animals starting from rest.

An interesting result of this study was the predictions of the relative magnitude of inertial and viscous forces generated by an accelerating animal. At the moment of start-up the inertial stresses are very large, but they decay rapidly, and resistive stresses predominate as the motion approaches steady state (Fig. 11a). If this model is compared to predictions from resistive force theory and inertial slender body theory we note some nonintuitive results: for a given set of body kinematics, the first 200 ms of an intermediate Re swimming event can be approximated by the purely inertial theory, but after 200 ms the resistive theory becomes a much better approximation (Fig. 11b). Thus, even at a Re of 100, where the inertial stresses are 100-fold more than the viscous stresses, thrust is dominated by the resistive component of the flows after the initial start-up of motion.

6 Performance, Limitations, and Scaling in Aquatic Locomotion: The Need for Integrating Muscle Mechanics with Fluid Dynamics

Most studies of animal swimming have focused on body kinematics and predictions of flow forces, almost to the exclusion of the mechanics of muscles involved in producing propulsive forces. Their goals often revolve around predictions of motions or morphologies that may maximize propulsive forces or minimize energy costs. Thus such studies may predict certain suites of characters (fin size and shape, amplitude of body undulations, etc.) which appear favorable yet may not be physiologically feasible. For example, the stress generated by propulsive devices may exceed the maximum isometric stress that muscles may generate (Daniel and Webb 1987), or the force-velocity relationship for contracting muscle (implicit in many studies of locomotion) may not correspond to the relationship between force and velocity for an appendage moving in a fluid. In two of the above examples (shrimp: Daniel and Meyhöfer 1989; chaetognaths: Jordan 1991) unique limits to the size of animals and the speed with which they move body parts arise by simply imposing the constraint that body motions and sizes may only increase to the point that the maximum isometric stress is realized. Beyond such points we must slow appendages down or reduce their size so that nowhere during a swimming cycle is the maximum stress exceeded.

Those studies which have dealt with muscle mechanics often emphasize fiber trajectories (Alexander 1969), muscle architecture (Wainwright 1983) and such basic processes as force-length or force-velocity relationships (Johnston 1983). Their goal is to hypothesize what structural, kinematic, or physiological features of the propulsor is allowable and how the design of the muscle system may reflect the functional needs of the moving animal.

While each approach to animal locomotion – fluid dynamics versus muscle mechanics – can be used to suggest limits to performance and morphology, the extent of integration is, at present, limited. Consider, for example, the following scenario: the force generated by a rowing appendage, such as on a beetle, depends approximately on the square of the speed of the appendage with respect to the fluid around it, but there is an inverse relationship between the force and speed of muscle contraction. Since the speed of the appendage is logically correlated with the speed of muscle contraction, we have a dilemma: both the equations of motion for fluid flow and the equations that describe muscle contractility must be satisfied. To our knowledge, this problem has not been addressed.

Two exciting offshoots arise from this combined approach. First, the method predicts both the kinematics of the propulsor and the kinematics of the body. Past approaches have relied on prescribed propulsor kinematics to predict body motions. A second consequence of combining muscle mechanics and appendage hydrodynamics to some measure of performance is that both morphological and physiological parameters can be examined. Past approaches examine the two rather independently.

A simple model for the power stroke of a hypothetical rowing creature shows how one combines muscle mechanics with fluid dynamics (Fig. 12). Letting the oars swing through an arc of 90 degrees, beginning with an angle of 45 degrees, we can write the instantaneous total force and its forward component (thrust) using the drag and accelerational forces acting on some paddle of radius, r, at an outboard distance, l_0:

$$\mathrm{Force}_{\mathrm{hydrodynamic}} = F_{\mathrm{drag}} + F_{\mathrm{accelerational}} = \frac{\rho(\pi r^2)C_d\left(\frac{l_0 d\theta}{dt}\right)^2}{2} + \frac{m l_0^2 d^2\theta}{dt^2} \tag{24}$$

$$\mathrm{Thrust} = \mathrm{Force\ hydrodynamic}\,\sin(\theta). \tag{25}$$

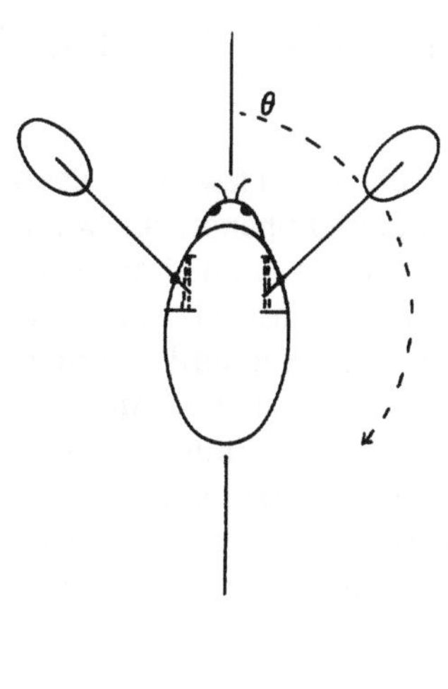

Fig. 12. Diagram of our hypothetical rowing creature. It swings its oars through an arc of 90 degrees. The oars are 1 cm long paddle is 0.4 cm in diameter, the muscle is attached to an apodeme that is 1 mm long. The muscles are each 2 mm long. These parameters were used in the equations described in the text

The added-mass term in this equation (m) corresponds to the reaction for a circular disc and is equal to πr^3 (Brennen, 1982). For simplicity, take C_d to be constant.

The relationship between force and velocity for contracting muscle is given by Hill's equation (Hill 1938; for a review see McMahon 1984) as

$$\text{Force}_{\text{muscle}} = \frac{(bT_0 - av)}{(v + b)}, \tag{26}$$

where v is the shortening velocity, T_0 is the maximum isometric force, and a and b are parameters that are specific to the particular muscle of interest.

If these oars are driven by a simple arrangement of internal muscles of length l_m, hooked on to an apodeme of length l_a, then the velocity in Eq. (24) can be written as $v = l_a d\theta/dt$.

Setting the moments generated by contracting muscle equal to the moments required to overcome drag, we arrive at a differential equation that predicts both the forces as well as the kinematics of the propulsor:

$$F_{\text{hydrodynamic}} l_0 = F_{\text{muscle}} l_a$$

$$\left\{\rho(\pi r^2)C_d\left(\frac{l_0 d\theta}{dt}\right)^2\right\} l_0 + \frac{m l_0^2 d^2\theta}{dt^2} = \left\{\frac{(bT_0 - a l_a d\theta/dt)}{(l_a d\theta/dt + b)}\right\} l_a. \tag{27}$$

We can solve this equation for $d\theta/dt$ numerically. Using reasonable estimates for the various constants in Hill's equation (see Daniel et al. 1989) and using dimensions indicated in the legend of Fig. 12, we find that the power stroke lasts 0.6s. While this value is near ones reported for rowing insects of similar size (Nachtigall 1960), we cannot take it too seriously. The time history of $d\theta/dt$ is sensitive to the geometry of propulsive muscles which, for our hypothetical creature, do not reflect reality. We simply use this example to point out how new analyses of locomotion can be used to predict more than just thrust and can, as well, examine the consequences to performance of variation in *both* morphological and physiological parameters.

7 Conclusions and New Directions in Studies of Aquatic Locomotion

Using only a few examples from the rich array of aquatic creatures, we explored the rules and phenomena which govern propulsion over a range of sizes. Momentum transfer, mediated by shear flows, accelerational flows, lift-generating velocity gradients, as well as combinations of these phenomena, was used to examine how thrust is produced and how the shape of aquatic organisms or the motions of their propulsors affect translation through their environment. Most of the theories we used in examining these various mechanisms arose from simplifications of the Navier–Stokes equations. In turn, recent modifications of these theories have been used to examine increasingly complex and yet biologically important problems, ranging from the mechanics of turning and acceleration to the peculiar hydrodynamic interactions that may exist between a propulsor and some nearby surface. It appears that, with the advent of computational methods for exploring challenging flow problems, we will soon be able to examine locomotor mechanisms for which the classic theories will not yield explicit solutions. Thus for animal motions of very large amplitude, for explicit integration of the mechanics of muscle and surrounding tissues, and for predictions of the thrust generated by creatures with complicated geometries at arbitrary Reynolds numbers, such new methods provide hope.

References

Alexander RMcN (1965) The lift produced by the heterocercal tail of selachii. J Exp Biol 43: 131–138

Alexander RMcN (1969) Orientation of muscle fibres in the myomeres of fishes. J Mar Biol Assoc UK 49: 263–290

Alexander RMcN (1982) Locomotion of animals. Blackie, London

Baba S (1972) Flexural rigidity and elastic constant of cilia. J Exp Biol 56: 459–467

Batchelor GK (1967) An introduction to fluid dynamics. Cambridge Univ Press, London

Berg HC (1983) Random walks in biology. Princeton Univ Press, Princeton, NJ

Blake JR (1972) A model for the micro-structure in ciliated organisms. J Fluid Mech 55: 1–23

Blake JR, Sleigh MA (1975) Hydromechanical aspects of ciliary propulsion. In: Wu TY, Brokaw CJ, Brennen C (eds) Swimming and Flying in Nature, vol 1. Plenum Press, New York: 185–210

Blake RW (1981a) Influence of pectoral fin shape on thrust and drag in labriform locomotion. J Zool Lond 194: 53–66

Blake RW (1981b) Mechanics of drag-based mechanisms of propulsion in aquatic vertebrates. In: Day MH (ed) Vertebrate locomotion. Symp Zool Soc Lond 48: 29–52

Blake RW (1983) Fish locomotion. Cambridge Univ Press, Cambridge

Bone QX, Trueman ER (1982) Jet propulsion of the calycophoran siphonophores *Chelophyes* and *Alypopsis*. J Mar Biol Ass UK 62: 263–276

Bone Q, Trueman ER (1983) Jet propulsion in salps (Tunicata: Thaliacea). J Zool Lond 201: 481–506

Breder CM (1926) The locomotion of fishes. Zoologica 4: 159–256

Brennen C (1975) Hydrodynamics of propulsion for ciliated microorganisms. In: Wu TY, Brokaw CJ, Brennen C (eds) Swimming and Flying in nature, vol 1. Plenum Press, New York: 235–51

Brennen C, Winet H (1977) Fluid mechanics of propulsion by cilia and flagella. Ann Rev Fluid Mech 9: 339–98

Cheung ATW, Winet H (1975) Flow velocity profile over a ciliated surface. In: Wu TY, Brokaw CJ, Brennen C (eds) Swimming and flying in nature. Plenum Press, New York: 223–34

Cheer AYL, Koehl MAR (1988) Paddles and rakes: fluid flow through bristled appendages of small organisms. J Theor Biol 129: 17–39

Childress S (1981) Mechanics of swimming and flying. Cambridge Univ Press, Cambridge

Chwang AT, Wu TY (1971) A note on the helical movement of micro-organisms. Proc R Soc Lond B 178: 327–346

Cox RG (1970) The motion of long slender bodies in a viscous fluid. Part 1.General theory. J Fluid Mech 44: 791–810

Daniel TL (1984) Unsteady aspects of aquatic locomotin. Am Zool 24: 121–134

Daniel TL (1985) Cost of locomotion: unsteady medusan swimming. J Exp Biol 119: 149–164

Daniel TL (1988) Forward flapping flight from flexing fins. Can J Zool 66: 630–638

Daniel TL, Meyhöfer E (1989) Size limits in escape locomotion of carridean shrimp. J Exp Biol 143: 245–265

Daniel TL, Webb PW (1987) Physical determinants of locomotion. In: Dejours P, Bolis L, Taylor CR, Weibel ER (eds) Comparative physiology: life in water and on land. Liviana, Padova pp 343–369

Daniel TL, Kingsolver JG, Meyhöfer E (1989) Mechanical determinants of nectar-feeding energetics in butterflies: muscle mechanics, feeding geometry, and functional equivalence. Oecologia 79: 66–75
DeMont ME, Gosline JM (1988) Mechanics of jet propulsion in the hydromedusan jellyfish, *Polyorchis penicillatus* II. energetics of the jet cycle. J Exp Biol 134: 333–345
de Pamphills ML, Adler J (1971) Fine structure and isolation of the hook-basal body complex of flagella from *Escherichia coli* and *Bacillus subtilis*. J Bacteriol 105: 384–395
Ellington CP (1984) The aerodynamics of hovering insect flight. VA vortex theory. Phil Trans R Soc B 305: 115–144
Fauci LJ (1990) Interaction of oscillating filaments: a computational study. J Comput Phys 86: 294–313
Fauci LJ, Peskin CS (1988) A computational model of aquatic animal locomotion. J Computational Physics 77(1): 85–108
Gal JM, Blake RW (1988a) Biomechanics of frog swimming I. Estimation of the propulsive force generated by *Hymenochirus boettgeri*. J Exp Biol 138: 399–411
Gal JM, Blake RW (1988b) Biomechanics of frog swimming II. Mechanics of the limb-beat cycle in *Hymenochirus boettgeri*. J Exp Biol 138: 413–429
Gray J, Hancock GJ (1955) The propulsion of sea-urchin spermatozoa. J Exp Biol 32: 802–814
Hancock GJ (1953) The self-propulsion of microscopic organisms through liquids. Proc R Soc Lond A 217: 96–121
Harper DG, Blake RW (1989) A critical analysis of the use of high-speed film to determine maximum acceleration in fish. J Exp Biol 142: 465–471
Hill AV (1938) The heat of shortening and the dynamic constants of muscle. Proc R Soc B 126: 136–195
Holwill MEJ (1966) Physical aspects of flagellar movement. Physiol Rev. 46: 696–785
Holwill MEJ (1977) Low Reynolds number undulatory propulsion in organisms of different sizes. In: Pedley TJ (ed) Scale effects in animal locomotion. Academic Press, New York
Hui CA (1988) Penguin swimming I: hydrodynamics. Physiol Zool 61: 333–343
Johnson RE (1980) An improved slender-body theory for Stokes flow. J Fluid Mech 99(2): 411–431
Johnson RE, Brokaw CJ (1979) Flagellar hydrodynamics. Biophys J 25: 113–127
Johnson IA (1983) Dynamic properties of fish muscle. In: Webb PW, Weihs D (eds) Fish biomechanics. Preager, New York, pp 36–67
Jordan CJ (1991) A model of rapid starting intermediate Reynolds number swimming: undulatory locomotion in the chaetognath *sagitta elegans*. J Exp Biol (submitted)
Keller Sr, Wu TY, Brennen C (1975) A traction-layer model for ciliary propulsion. In: Wu TY, Brokaw CJ, Brennen C (eds) Swimming and flying in nature, vol 1. Plenum Press, New York, pp 253–271
Kerfoot WC (1980) Evolution and ecology of zooplankton communities. Univ Press, New England
Koehl MAR (1981) Feeding at low Reynolds number by copopods. Lect Math Life Sci 14: 89–117
Koehl MAR, Strickler JR (1981) Copepod feeding currents; food capture at low Reynolds number. Limnol Ocenogr 26: 1062
Lighthill MJ (1971) Large-amplitude elongated-body theory of fish locomotion. Proc R Soc Lond 179: 125–138
Lighthill MJ (1975) Mathematical biofluiddynamics. Soc Ind Appl Math
Lighthill MJ (1976) Flagellar hydrodynamics. SIAM Rev 18(2): 161–230
Lighthill MJ, Blake RW (1990) Biofluiddynamics of balistiform and gymnotiform locomotion. Part 1. Biological background, and analysis by elongated body theory. J Fluid Mech 212: 183–207
Magnuson JJ, Prescott JH (1966) Courtship feeding and miscellaneous behavior of Pacific bonito (*Sarda chiliensis*). Animal Behav 14: 54–67
McMahon TA (1984) Muscles reflexes and locomotion. Princeton Univ Press, Princeton
Nachtigall W (1960) Über Kinematik, Dynamik und Energetik des Schwimmens einheimischer Dytisciden. Z vergl Physiol 43: 48–118
Nachtigall W, Bilo D (1965) Hydrodynamics of the body of Dytiscus marginalis (Dytiscidae. Coleoptera). In: Wu TY, Brokaw CJ, Brennen C (eds), Swimming and flying in nature. Plenum Press, New York, pp 585–595
O'Dor RK (1988) The forces acting on a swimming squid. J Exp Biol 137: 421–442
Oehme H (1977) On the aerodynamics of separated primaries in the avian wing. In: Pedley TJ (ed) Scale effects in animal locomotion. Academic Press, New York, pp 479–495

Packard A (1969) Jet propulsion and the giant fiber response of Loligo. Nature (Lond) 221: 875–877
Rayner JMV (1979a) A vortex theory of animal flight. Part 1. The vortex wake of a hovering animal. J Fluid Mech 91: 697–730
Rayner JMV (1979b) A vortex theory of animal flight. Part 2. The forward flight of birds. J Fluid Mech 91: 731–763
Sleigh MA, Blake JR (1977) Methods of ciliary propulsion and their size limitations. In: Pedley TJ (ed) Scale effects in animal locomotion. Academic Press, New York, pp 243–256
Taylor GI (1951) Analysis of swimming microscopic organisms. Proc R Soc Lond A 209: 447–461
Thomson KS (1976) On the heterocercal tail in sharks. Paleobiology 2: 19–38
Tucker VA (1977) Scaling in avian flight. In: Pedley TJ (ed) Scale effects in animal locomotion. Academic Press, New York, pp 497–509
Wainwright SA (1983) To bend a fish. In: Webb PW, Weihs D (eds) Fish biomechanics. Preager Press, New York, pp 68–91
Warner FD, Satir P (1974) The structural basis of ciliary bend formation. J Cell Biol 63: 35–63
Webb PW (1978) Fast-start performance and body form in seven species of teleost fish. J Exp Biol 74: 211–216
Webb PW (1984) Body form, locomotion and foraging in aquatic vertebrates. Am Zool 24: 107–120
Webb PW, Simms D, Schultz WW (1990) The effects of an air/water surface on the fast-start performance of rainbow trout (*Oncorhyncus mykiss*). J Exp Biol (in press)
Weihs D (1973a) The mechanism of rapid starting of slender fish. Biorheology 10: 343–350
Weihs D (1973b) Hydromechanics of fish schooling. Nature (Lond) 245: 48–50
Weis-Fogh T (1973) Quick estimates of flight fitness in hovering animals, including novel mechanisms for lift production. J Exp Biol 52: 17–25
Wu TY (1971a) Hydromechanics of swimming propulsion. Part 1. Swimming of a two-dimensional flexible plate at variable forward speeds in an inviscid fluid. J Fluid Mech 46: 337–355
Wu TY (1971b) Hydromechanics of swimming propulsion. Part 2. Some optimum shape problems. J Fluid Mech 46: 521–544
Wu TY (1971c) Hydromechanics of swimming propulsion. Part 3. Swimming of slender fish with side fins. J Fluid Mech 46: 545–568
Wu TY (1977) Hydrodynamics of swimming at low Reynolds numbers. Fortschr Zool 24(2/3): 149–169
Yates GT (1983) Hydromechanics of body and caudal fin propulsion. In: Webb PW, D Weihs (eds) Fish biomechanics. Preager Press, New York, pp 177–213
Yates GT (1986) How microorganisms move through the water. Am Sci 74: 358–365
Yates GT (1988) Optimum pitching axes in flapping wing propulsion. J Theor Biol 120: 255–276

Chapter 3

The Aerodynamics of Flight

G.R. Spedding

Contents

Departments of Aerospace and Mechanical Engineering, University of Southern California, Los Angeles, CA 90089-1191, USA

Symbols

a	vortex core radius
A	wake area
A_0	area swept by wings
$AR \equiv 2b/c$	aspect ratio
b	wing semi-span
c	wing chord
C	closed curve
C_L, C_D	lift, drag coefficient
c_l, c_d	force coeff./unit length
C_T	thrust coefficient
D	drag force
D'	drag/unit length
e	aerofoil efficiency
f	feathering parameter
$\mathbf{F}$	force vector
g	gravitational constant
$H \equiv h/b$	dimensionless height
$\mathbf{I}$	fluid impulse
J	advance ratio
k_0	arbitrary constant
k	reduced frequency, based on half chord
$K \equiv J^{-1}$	flapping frequency parameter
l	linear length scale
L	lift force
L'	lift/unit length
m	body mass
m_2	2nd moment of mass
n	frequency (Hz)
P	power
$q \equiv 1/2\rho U^2$	dynamic pressure
$Q \equiv W/S$	wing loading
r	spanwise location turning radius
R	total wing length vortex ring radius
R'	dimensionless ring radius
Re	Reynolds number
s	wake element spacing surface coordinate
S	surface area

S_d	wing disc area
T	stroke period thrust
$\mathbf{u}$	local velocity vector
u, v, w	velocity components in x, y, z
$\mathbf{u}_w$	local wing section velocity
$\mathbf{u}_t$	wingtip velocity
U	mean, steady velocity
v_2	2nd moment of virtual mass
w_i	induced velocity
W	body weight (= mg)
α	angle of attack
α_i	induced angle of attack
α_0	body angle
β	stroke plane angle
γ	wing positional angle
δ	small change
ζ	ratio of projected wake areas
η	propulsive efficiency
θ	glide angle
λ	wavelength
μ	viscosity
ν	kinematic viscosity
ξ	vorticity
ρ	density
σ	wing shape factor spatial wake correction
σ_0	correction to momentum jet
τ	downstroke ratio temporal wake correction shear stress
ϕ	stroke amplitude
ϕ_{hf}	fore/hind-wing phase angle
ω	radian frequency
Γ	circulation
Γ_0	total circulation
Θ	proportional feathering
Λ	dimensionless amplitude
Φ	roll angle
Ω	reduced frequency, based on semi-span

1 Introduction

1.1 Flight in Animals

It is usually estimated that there are around 10^6 extant animal species in the world, of which approximately three-quarters are insects. The Pterygota, or winged insects, comprise the vast majority of these species, and save for the orders of lice and fleas, nearly all members of this subclass have functional wings at some stage during their life cycle. Flight has evolved quite independently in the birds (class: Aves), numbering around 8000 species, and in bats, whose 850 or so species, although less conspicuous, are the most numerous, and second most widely distributed order of the class mammalia. In terms of number of species, nonflying animals might therefore be considered unusual, and the capacity for flight seems to confer certain advantages, if measures of success include the degree of dispersal and speciation.

One of the chief benefits appears to be a reduction in the cost of purposefully moving mass from A → B, and a dimensionless cost of transport, defined as the energetic cost of moving a unit weight a unit distance, is indeed generally lower for a flying animal of any given mass, *for those animals which must support their body weight*. The favourably low cost of transport is accompanied by a high rate of energy expenditure per unit time, and there are considerable incentives to improve the efficiency of flight, when it occurs. The adaptations are physiological, structural, behavioural and developmental, as well as aerodynamic, and at once place severe constraints on the operating range of the animal, while at the same time driving the evolution of a remarkable diversity of forms. It is one of the implicit assumptions behind much of the animal flight literature that energetic, and ultimately, aerodynamic considerations have a strong effect on the structure and ecology of the individual. It will be the *only* consideration in this chapter, others in this book supplying the remaining components.

In contrast to most instances of aquatic propulsion, fluid dynamic forces in flight must also support the body weight, and the (generally) nonlinear interaction between these two requirements of lift and thrust generation leads to certain interesting complications in the analysis. In common with the previous chapter, the absence of continuous rotating mechanical devices has resulted in the development of sophisticated reciprocal oscillating systems. The consequences of the periodic accelerations of the beating wings are profound; large inertial forces, significant unsteady effects and gross departures from standard linearised models may result. The degree of this departure, as measured by the relative frequency and amplitude of the beating motion, forms the basis of organisation of the chapter.

1.2 Omissions

The discussion will be focussed on animals that are capable of sustained flapping flight and weight support. The aerodynamic requirements place considerable demands on the mechanics, physiology and behaviour, and the responses of actively flying animals to these pressures allows some kind of unified treatment of the adaptations. Flying squirrels, lizards, frogs, snakes and fish could be considered as low aspect ratio gliders, but will not be specifically mentioned. The evolution of flight will not be covered, and papers by Caple et al. (1983), Rayner (1985a, 1986), Norberg (1985) and Pennycuick (1986, 1988) should be consulted for introductions to the vigorous bird and bat flight evolution debate. The usual, but somewhat arbitrary cutoff is also applied so that human-powered and human-designed flight systems are ignored. Since this book specifically concerns animal mechanics, the fluid dynamics of plants and inanimate objects in the environment are excluded. Fortunately, there are several fascinating books which deal with such topics, for example, those by Hertel (1966), Vogel (1981), Lugt (1983), and Ward-Smith (1984). A number of excellent reviews of various aspects of animal flight have been published in recent years. The papers in Ellington (1984, I–VI) are the current point of reference for research in insect flight aerodynamics. Pennycuick (1989a) provides clear explanations and advice on many practical issues in bird flight analysis. Aspects and applications of mechanical models of bird and bat flight have been reviewed by Rayner (1986, 1988) and Norberg and Rayner (1987), and a thorough account ranging from theoretical to ecological studies of vertebrate flight has been published by Norberg (1990).

1.3 The Modelling Enterprise

It is one of the theses of this chapter that a nontrivial understanding of animal flight must be based on some kind of aerodynamical model. Ultimately, such a model must itself be based on the known physics of the flow, which, for our purposes, reduces to the application of Newton's laws to an incompressible, homogeneous fluid. Specification of the initial and boundary conditions is, in principle, sufficient for the determination of the velocity and the pressure everywhere in the flow, by direct solution of the Navier–Stokes (NS) equations. However, even if we disregard the enormous problems in achieving adequate spatial and temporal resolution, the results of such an effort may not be particularly useful because it would not necessarily be obvious how they would generalise to other conditions, and the physical processes most responsible for some observed phenomenon in the flow field would still not be evident. We would only know that the NS equations produced it, but we already knew that since that is the system we programmed in the first place, and no physical insight has been gained. An essential part of the scientific endeavour involves the deliberate abstraction of some subset of the full physical system and the construction of a model based upon the judicious selection of appropriate simplifications. A useful model makes testable predictions, the success *or* failure of which provides new information concerning the initial assumptions. Parametric tests allow the relative significance of the model components to be assessed, and refinements in the model are derived from better physical approximations.

2 Some Nondimensional Numbers

2.1 Lift and Drag Coefficients

Regardless of the complexity of the flow or of the surface over which it is moving, there only two basic mechanisms by which a force may be communicated to a body moving through a homogeneous fluid. They are: (1) the *pressure*, $p(s)$, and

(2) the *shear stress* distribution, $\tau(s)$, over the surface. By convention, the resultant force, **F**, integrated over the whole surface, is decomposed into two orthogonal components, $L \equiv$ Lift, the component of **F** normal to the freestream velocity, U, and $D \equiv$ Drag, the component of **F** parallel to U. The dimensionless coefficients of lift and drag are then

$$C_L \equiv \frac{L}{q_\infty S}, \quad \text{and,} \quad C_D \equiv \frac{D}{q_\infty S}, \tag{1}$$

where q_∞ is the dynamic pressure and S is the surface area. The equivalent sectional lift and drag coefficients are the normalised forces per unit span, at a given chord, c:

$$c_l \equiv \frac{L'}{q_\infty c}, \quad \text{and,} \quad c_d \equiv \frac{D'}{q_\infty c}. \tag{2}$$

The total drag force is the sum of the contributions from the normal pressure and tangential stresses,

$$D = D_n + D_\tau. \tag{3}$$

D_n is known also as the form drag. When there is a mean lift on the body, it is conventional to express the sum of D_n and D_τ, minus the lift-induced drag, as the profile drag, D_{pro}. This decomposition of **F** is quite simple in conventional aeronautical applications involving steady motion in a uniform flow, but is less straightforward, and not necessarily particularly helpful, when the direction and magnitude of the fluid velocities at the surface vary greatly in space or time.

2.2 Reynolds Number

The Reynolds number, $Re \equiv Ul/\nu$, is the *only* dimensionless parameter required for specification of the dynamics of incompressible flow fields with uniform density, *for a given set of boundary and initial conditions. Re* emerges directly from the NS equations governing the motion of a fluid, when the pressure, inertial and viscous terms are in equilibrium. The triply-infinite number of possible solutions where U, l, μ and ρ ($\nu = \mu/\rho$) are chosen such that Re is unchanged, are said to be dynamically similar, determined by the ratio of the inertial and viscous terms. A dimensionless drag coefficient, as defined above, will be equal for geometrically similar bodies at the same value of Re. The chord Reynolds number, Re_c ranges from $\approx 10^1$ for the smallest insects considered here, to $\approx 10^6$ for the larger birds, a difference of 5 orders of magnitude. Certain aspects of the fluid mechanics can be expected to vary considerably across this range.

2.3 Reduced Frequency

While dynamical similarity under similar boundary and initial conditions may be assured for different flows when Re is of the same order, other dimensionless parameters enter the problem when specification of the boundary and initial conditions involves a third parameter, in addition to U and l, such as a frequency, for example. It will prove convenient to define reduced frequencies based on the mean half-chord, $c/2$, and on the semi-span, b, as:

$$k \equiv \frac{\omega \bar{c}}{2U}, \quad \Omega \equiv \frac{\omega b}{U}. \tag{4}$$

k and Ω are related via the aspect ratio, $AR \equiv 2b/c$, by $\Omega = kAR$. The circular frequency ω is an independent kinematic parameter (it does not vary simply as a consequence of variations in the steady freestream velocity, U or the lengthscale, l) and dynamical similarity of any two flows requires that both Re and k (and when appropriate, Ω) be the same. The frequency parameter Ω is proportional to the relative magnitudes of the spanwise and streamwise components of vorticity, and is a measure of the departure from the quasi-steady limit, when $k, \Omega \to 0$.

The spanwise variation in amplitude of a wing beating at frequency n (in Hz), may be taken into account, defining the ratio K, with ϕ, the wing stroke amplitude, and r, the spanwise distance from the wing root to the point of interest, as

$$K \equiv \frac{2\phi nr}{U}. \tag{5}$$

Its inverse, $J = K^{-1}$, is the advance ratio discussed by Ellington (1984, III), when $r = R$, the wing length, and so J is the ratio of the forward flight velocity to the mean wingtip velocity. We see from Eq. (4) that $J = \pi/(\phi\Omega)$, and so for typical wingbeat amplitudes where $\phi \cong \pi/2$, $J \cong 2/\Omega$.

3 Classical Aerodynamics and the Performance of Animal Wings

3.1 *2-D* Aerofoils

The forces on an aerofoil can be discussed in terms of the vorticity and circulation of the fluid, either on the aerofoil itself, or in the wake (Fig. 1). The vorticity of a fluid element,

$$\boldsymbol{\xi} = \nabla \times \mathbf{u}, \tag{6}$$

is equal to twice the local angular velocity of the fluid. There may be components of this in all space coordinates, and so it is a vector quantity. When $\boldsymbol{\xi} \neq 0$, the flow is *rotational*. The circulation, Γ, is the line integral of the fluid velocity around any closed curve, C,

$$\Gamma = -\oint_C \mathbf{u} \cdot d\mathbf{x}. \tag{7}$$

An intuitive, physical interpretation of this quantity is less obvious than for ξ, but when combined with the definition of the vorticity, it may be thought of as a measure of the strength of ξ, integrated over the area enclosed by C. The *Kutta–Joukowski* theorem of lift states that the lift per unit span, L', of a body is directly proportional to the circulation:

$$L' = \rho U \Gamma. \tag{8}$$

It just remains to select a value of Γ. For a given U, there is a unique value of Γ which ensures that infinite fluid velocities are not required at the trailing edge. This condition, when the separation point occurs *exactly* at the trailing edge, is known as the *Kutta condition*, and it effectively determines the lift on the aerofoil, which, for the purposes of lift calculation, may be replaced by a point vortex having this value of Γ.

Kelvin's circulation theorem, which follows from a consideration of Eq. (7) in a homogeneous, incompressible, inviscid fluid, states that the circulation around a circuit, C, will have the same value when measured over the same fluid elements comprising C, at any time, as C is followed in the flow,

$$\frac{D\Gamma}{Dt} = 0. \tag{9}$$

Let us define a large box around an aerofoil at rest, around which Γ is initially zero (Fig. 1). As the aerofoil begins to move, accelerating to some steady finite velocity, extremely high velocity gradients, and hence ξ, are generated at the trailing edge as the flow adjusts and the separation point begins to move towards the trailing edge. While the aerofoil moves on, the vorticity will be left behind, close to where it was created. The circulation, Γ, on the aerofoil must be balanced by an equal and opposite circulation, $-\Gamma$, in the wake, according to Eq. (9). When the Kutta condition is satisfied and the aerofoil has attained a steady lift, determined by Γ_0, the flow leaves the trailing edge smoothly and no more vorticity is shed into the flow. Where it started, the shed vorticity will tend to roll up into one concentrated vortex, with circulation $\Gamma_w = -\Gamma_0$, and at all times, the total circulation, defined over the whole box, remains at zero. The lift on the aerofoil can be calculated from the vorticity in the wake according to Eqs. (6), (7) and (8). Further thought experiments along these lines show that any changes in lift on the aerofoil will necessarily be accompanied by the shedding of vorticity in the wake, and so knowledge of the strength and distribution of vortex elements in the

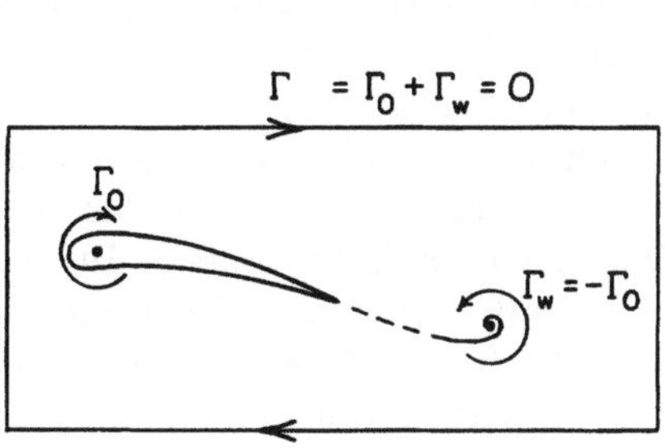

Fig. 1. The generation of bound (Γ_0) and wake (Γ_w) vorticity for an impulsively started aerofoil

wake may be used to deduce the circulation and forces on the wing. Batchelor (1967) can be consulted for an introduction to the basic fluid dynamical concepts.

Thin Aerofoil Theory. When the aerofoil thickness and camber (degree of curvature) are everywhere small, and the angle of attack, α, is small, the aerofoil can be modelled as a vortex sheet along the chord line (defined as the straight line connecting the leading and trailing edges), where the total circulation, Γ_0, is the value required to meet the Kutta condition at the trailing edge, and streamlines flow around the true cambered profile. Simple expressions may be derived for the lift coefficient and the centre of pressure showing that the centre of pressure is at the quarter chord point, and that the lift is a linear function of angle of attack. If the aerofoil is symmetric, the line of $c_l(\alpha)$ passes through the origin. Agreement with wind tunnel data for a real symmetric aerofoil is good up until a critical value of α, when the flow separates and the aerofoil stalls. When the aerofoil section shape is more complicated, the surface geometry can be defined with a number of vortex panels; the numerical solution for the circulation distribution required to satisfy the Kutta condition at the trailing edge proceeds along the same principles as before. Concise descriptions of these techniques appears in Anderson (1984), and a modern collection of aerofoil data may be found in Althaus (1980).

Reynolds Number, Flow Separation and Transition to Turbulence. Thus far, attached, high *Re* flows at small α have been implicitly assumed. In practice, for $Re < 10^6$, aerofoil performance is completely dictated by the comparatively poor resistance to separation of the laminar boundary layer. The physics of separated flows is a research topic unto itself; Lissaman's (1983) fine review contains discussion of the effects of flow separation on the performance of low *Re* aerofoils, Cheng and Smith (1982), and Cheng (1985) have reviewed and extended the theoretical analysis of laminar separation, and aspects germane to animal flight are considered by Lee and Cheng (1990). Depending on the presence or absence of flow separation and reattachment, on the location of these points on the aerofoil surface, and on the occurrence of transition to turbulence in the attached boundary layer, or in the separated free shear layer, a broad spectrum of behaviour can be observed over a critical range of $10^4 < Re < 10^6$. Two examples serve to illustrate the problem. Figure 2a shows the superior lift/drag polar of a cambered thin plate over both a flat plate, and a thin aerofoil at $Re_c = 4 \times 10^4$. Following a threefold increase in Re_c (Fig. 2b), the thin aerofoil now performs best, and with appreciably higher c_l than any of the cases at the lower Re_c. The Re_c of a bird with a mean chord of 0.1 m, gliding at $10\,\mathrm{m\,s^{-1}}$ would be about 7×10^4. It is somewhat exasperating to note how, from the theoretical point of view, animal locomotion often appears to occur in the least convenient of all possible parameter ranges, as also noted in Chap. 2. Although laminar, attached flows can be maintained over at least portions of carefully designed aerofoils (cf. Liebeck 1978) at the top of the *Re* range above, they may be considerably disrupted by small disturbances in the flow. It is common to deliberately introduce a small perturbation at the desired transition location, so that the resulting turbulent boundary layer, by virtue of the increase in momentum

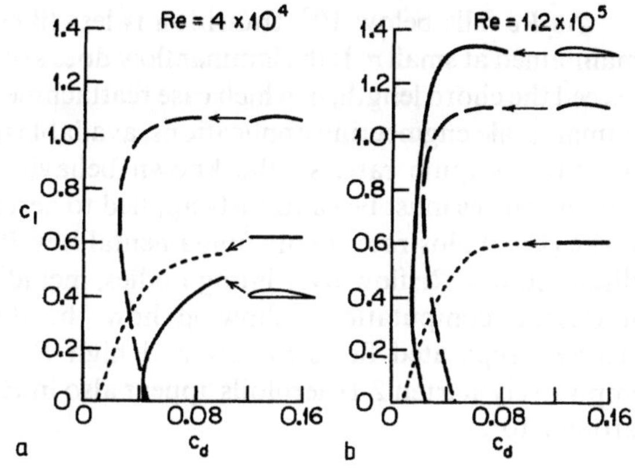

Fig. 2 a, b. Lift-drag polars for smooth aerofoils and flat plates at different Reynolds numbers. (After Jones 1990 and Schmitz 1960)

transfer from the freestream, is less unstable and more likely to withstand adverse pressure gradients along the aerofoil. Similarly, transition to turbulence in the free, separated shear layer increases the likelihood of downstream reattachment, when a *laminar separation bubble* results. For natural flight at moderate values of Re, turbulent flow might be the default condition, when small nonuniformities in the surface geometry (a rumpled feather) or equally small amounts of turbulence in the freestream (a gust of wind) are not unusual. Other factors influencing transition include the geometric shape of the aerofoil surface, the presence of auxiliary lifting elements, translational and rotational accelerations of the surface, and even the previous steady state of the boundary layer, as demonstrated in Fig. 3, where considerable hysteresis is observed for $c_l(\alpha)$ of a thick and thin aerofoil section at $Re_c = 10^5$. Birds and bats, operating in the $10^4 < Re < 10^6$ domain, will encounter and generate flows which are very sensitive to changes promoting or inhibiting separation and/or transition to turbulence, and the performance of their wings could fluctuate accordingly.

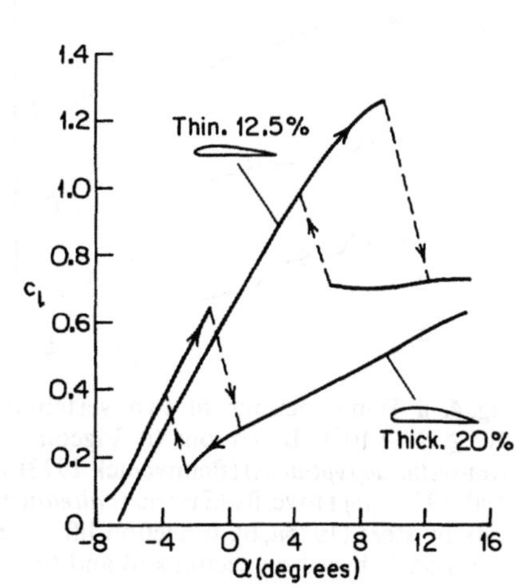

Fig. 3. Hysteresis in $c_l(\alpha)$ curves for thin and thick aerofoils at $Re = 10^5$. The formation and bursting of laminar separation bubbles is responsible for dramatic fluctuations in the lift. (After Jones 1990 and Schmitz 1960)

As Re falls below 10^3, transition is less likely, and fully laminar flow may be maintained at small α. If the laminar flow does separate, the bubble length is likely to exceed the chord length, in which case reattachment cannot occur. Due to the lack of human-scale engineering applications, available technical data on lifting surfaces at $Re < 10^3$ is quite rare, so the known behaviour of fluids at these Re in other circumstances must be cautiously applied to aerofoil and flat plate geometries. Van Dyke (1982) illustrates many fundamental low-Re fluid flows, and Lugt (1983) has discussed low Re flow over lifting bodies, including the results of some interesting numerical computations, showing how the distribution of vorticity differs in strongly separated flows at low and high Re. Intriguing flow visualisations of impulsively started *2-D* aerofoils appear also in Hertel (1966), where Re_c is probably around 10^3.

Animal Aerofoil Sections. Figure 4a shows the approximate wing sectional geometry for a bird, a bat, and two insect species, covering a range of Re_c from 10^5 to 10^2. All sections have appreciable camber, and the transition from smooth profile to flat plate with decreasing Re_c is consistent with the performance noted in Fig. 2. The pigeon section profile tapers into a very thin, extended trailing edge, bearing some resemblance to the trailing edge extensions added to low Re_c ($\approx 5 \times 10^5$), high lift aerofoils analysed by Ito (1989). The thin, tapered section is also similar in appearance to certain of the high-lift aerofoil sections reverse-engineered by Liebeck (1978), particularly when transition occurs close to the leading edge. By contrast, the bat wing membrane and supporting structures appear more like a cambered flat plate. The elastic wing membrane deforms in a passive aeroelastic response to the fluid pressures during flight, but the degree of camber may be altered by changing the tension in the membrane. In similar experiments on the gliding flight of both

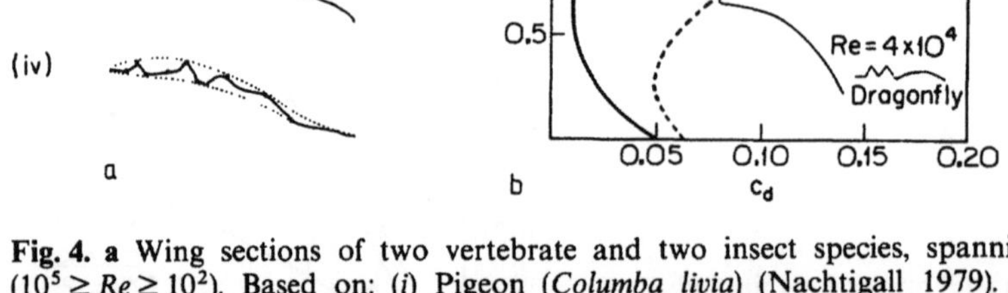

Fig. 4. **a** Wing sections of two vertebrate and two insect species, spanning a range of Re ($10^5 \geq Re \geq 10^2$). Based on: (*i*) Pigeon (*Columba livia*) (Nachtigall 1979). (*ii*) Dog-faced bat (*Rousettus aegyptiacus*) (Pennycuick 1973). (*iii*) Dragonfly (*Aeschna interrupta*) forewing (Newman et al. 1977). (*iv*) Hoverfly (*Syrphus balteatus*), showing smooth envelope profile used in comparison tests by Rees (1975a, b). **b** Section lift-drag polars for a modified Wortman aerofoil (Ito 1989), and models based on sections (*i*) and (*iii*)

species (Pennycuick 1968, 1971a), Re_c can be calculated as 2×10^5 for the pigeon, and 3×10^4 for the bat at the preferred gliding speeds, an interesting correspondence with the data in Fig. 2. Both insect wing sections have noticeable corrugations, which have been interpreted as promoting transition, separation and reattachment of the boundary layer for the dragonfly at $Re_c \cong 10^4$ (Hertel 1966; Newman et al. 1977). The fluid dynamics must differ somewhat at the lower $Re_c = 4.5–9 \times 10^2$ for the hoverfly profile, for which Rees (1975b) found that lift/drag polars were similar to those for an aerofoil shaped like a smooth envelope around the corrugated sheet. In this case, the primary function of the corrugations is seen as a structural adaptation for maintaining spanwise stiffness with little added wing mass inertia, and without any great aerodynamic performance reduction.

Figure 4b compares the section polars of a modified high-lift aerofoil with profiles based on sections (i) and (iii) for the pigeon and dragonfly. The performance of the pigeon section is quite respectable for the Re_c, and that of the dragonfly is rather extraordinary. Both $(L/D)_{max}$ and the slope of $c_l(c_d)$ are high, and the kink in the curve of the dragonfly section was attributed to the development of a large separation bubble entirely covering the corrugations at $\alpha \cong 8°$. The benefits of this separation bubble appear to include extended high lift regions at higher α, and the accessibility of a range of c_d at almost constant c_l. Manoeuvrable, hunting flight performance might thus be improved. Confirmation of these results would be reassuring, since Azuma and Watanabe (1988) did not find such $c_l(c_d)$ profiles in tests with miniature gliders constructed with real dragonfly hindwings.

3.2 Finite Wings

The pressure difference cannot be maintained at the tips of a finite wing, and a *rotational* motion is produced as the fluid flows around the tips. The result is thus a streamwise component of ξ, which rolls up into two trailing vortices. A vertical cut taken through the wake (Fig. 5a), illustrates the mean downward flow induced by the tip vortices. The effective angle of attack is correspondingly reduced (Fig. 5b) and the vector sum of the freestream velocity and the induced *downwash velocity* w_i, gives rise to a local velocity vector which is no longer aligned with U. Now, the local lift vector, defined as the component of the force normal to the local velocity, has a backwards component compared to the original; it adds to the total drag. This drag component is called the *induced* drag and is entirely due to the influence of the vortex-induced motion. Since these vortices were produced as a result of the pressure difference across the wing, the induced drag is an inevitable consequence of the lift. The total drag on a finite wing can now be expressed as the sum of the skin friction, pressure and induced drags [cf. Eq. (3)].

$$D = D_\tau + D_n + D_i. \quad (10)$$

D_τ is the drag due to viscous shear stress. The fact that the integrated sum of D_n over the wing surface is nonzero may also be traced to the effect of viscosity and flow separation, and the two terms are often lumped together as D_{pro}. Obviously, D_{pro} varies with Re, but it varies also with anything causing measurable changes in the

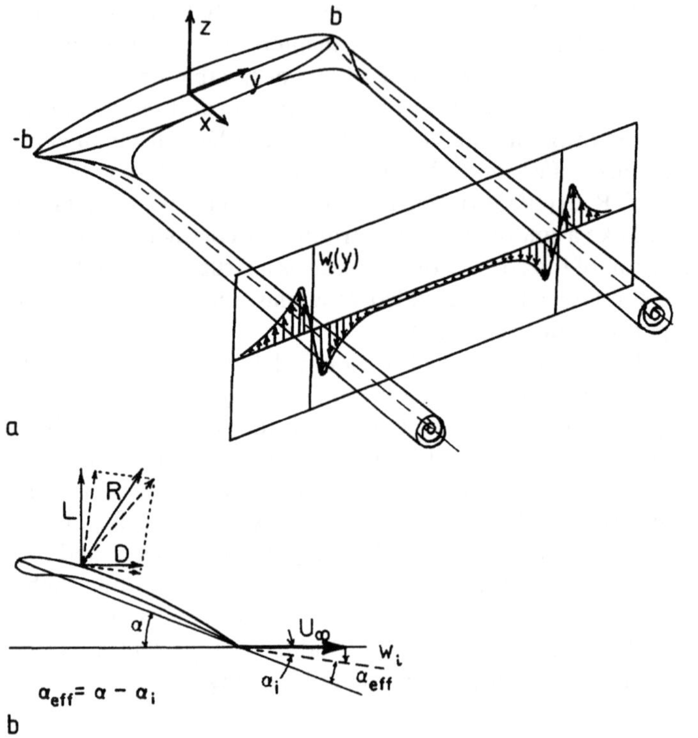

Fig. 5. **a** Wingtip trailing vortices behind a fixed wing in steady motion. y is the spanwise distance along the quarter chord line, and the total span is $2b$. The velocity profile in the vertical plane cut in the wake shows the spanwise distribution of velocity normal to the direction of motion $w_i(y)$, due to the vortex system. **b** The effect of the vertical velocity w_i is to reduce the effective angle of attack and reorient the lift vector

boundary layer flow. It is extremely hard to calculate accurately for all but the most simple flows. In practice, it is frequently estimated from empirical coefficients derived from wind tunnel tests. The remaining component, D_i, depends on the distribution of wake vorticity, which in turn depends on $\Gamma(y)$.

The Lifting Line. The action of a wing with an arbitrary, continuous $\Gamma(y)$ distribution can be analysed as a *lifting line* composed of an infinite number of coincident vortex filaments of different strengths. The wake is therefore also composed of an infinite number of line vortices and so initially has the form of a vortex sheet shed at the trailing edge. The induced drag may then be calculated, for any U, given only the form of the circulation distribution on the wing, $\Gamma(y)$. Apart from the requirement that $\Gamma(-b)=\Gamma(b)=0$ at the wingtips, we are free, in principle, to choose any convenient $\Gamma(y)$. One possible form is an *elliptic* distribution, when an expression for C_{D_i} can easily be derived,

$$C_{D_i}=\frac{C_L^2}{e\pi AR} \qquad (11)$$

where $e \leqq 1$ is the aerofoil efficiency factor. $e = 1$ when the loading is elliptic, which is therefore the optimum load distribution for minimum wake energy loss for a given lift on a planar wing in steady motion. Elliptic loading can be achieved by constructing a wing with an elliptical planform. Alternatively, the local wing section profile may vary along the span, as may the local wing twist. In practice, small deviations from elliptic loading have a rather small influence on the induced drag, and high AR, rectangular wings work tolerably well, for example. Intuitively, it seems perfectly reasonable that C_{D_i} should be proportional to the square of the lift, and inversely related to AR, since it is a physical consequence of the pressure difference at the wingtips. A more complete analysis would note that there are conflicting requirements between minimising induced drag and mean bending moments (long/short span), and between reducing both skin friction drag and maintaining sufficient structural strength (small/large chord). Jones (1990) has reviewed these requirements for fixed wings and reports that the induced drag may be reduced by about 10% over the elliptic wing, for the same integrated bending moment, by using a slightly longer wing with a more pointed tip. There is much room for further research combining both structural *and* aerodynamic considerations.

Applications of the Steady, Straight, Lifting Line. Lifting line theory provides the foundation for a large number of models in aeronautics, natural or otherwise. The numerical solution of a lifting line problem for arbitrary $\Gamma(y)$ distributions is quite straightforward, and instead of describing any particular case in detail, it is only necessary to understand the basic recipe. A wing is first divided into a number of strips at equally spaced spanwise stations, and an initial $\Gamma(y)$ is assumed. α_i is calculated at each station, y_0, and from the geometric angle of attack at that station, the effective angle of attack is given by $\alpha_{eff} = \alpha - \alpha_i$ (Fig. 5), and the section lift coefficient is determined from empirical curves for that profile; any nonlinear $c_l(\alpha)$ may thus be used. Given c_l, a new value of Γ is calculated. If it differs substantially from the old value, these steps are iterated until convergence. This algorithm is commonly used for high aspect ratio, straight wings, and derivatives of it have been widely applied to animal flight. However, *it is known to break down* whenever significant spanwise (three-dimensional) flows occur over the wing, in which case the validity of the use of *2-D* sectional lift coefficients becomes questionable.

A related technique, similar in spirit to the vortex panel analysis in *2-D*, is the lifting surface method. The wing is covered by regular arrays of vortex filaments, whose strengths vary with both x and y, which, together comprise a *lifting surface*. The induced downwash at any point on this surface is an integral of the vorticity over the lattice and the wake, and the system is solved so that the fluid velocity is everywhere tangential to the lifting surface. A variation on this theme is to cover the wing with a lattice of horseshoe vortices. The normal velocities are given by the velocity induced from all the filaments in the *vortex lattice*, and the solution proceeds as before. These are still essentially two- dimensional techniques, but the extension to *3-D* flows and geometries is relatively straightforward in principle. They are variously known as *vortex doublet* or *panel* techniques; the surface is covered with a network of vortex patches, panels or lifting lines, and at control points on the body surface a system of linear algebraic equations is solved such that normal velocities are zero. The problem of mapping a complicated surface with an appropriate lattice mesh is often nontrivial in itself, and extremely sophisticated panelling algorithms have been developed for mapping arbitrary and/or time-varying geometries.

Animal Wing Systems. The bird, bat and insect wing planforms of Fig. 6 illustrate half of the challenge of animal flight aerodynamic analysis: how can the classical analysis be adapted and modified for application to these diverse forms? The second

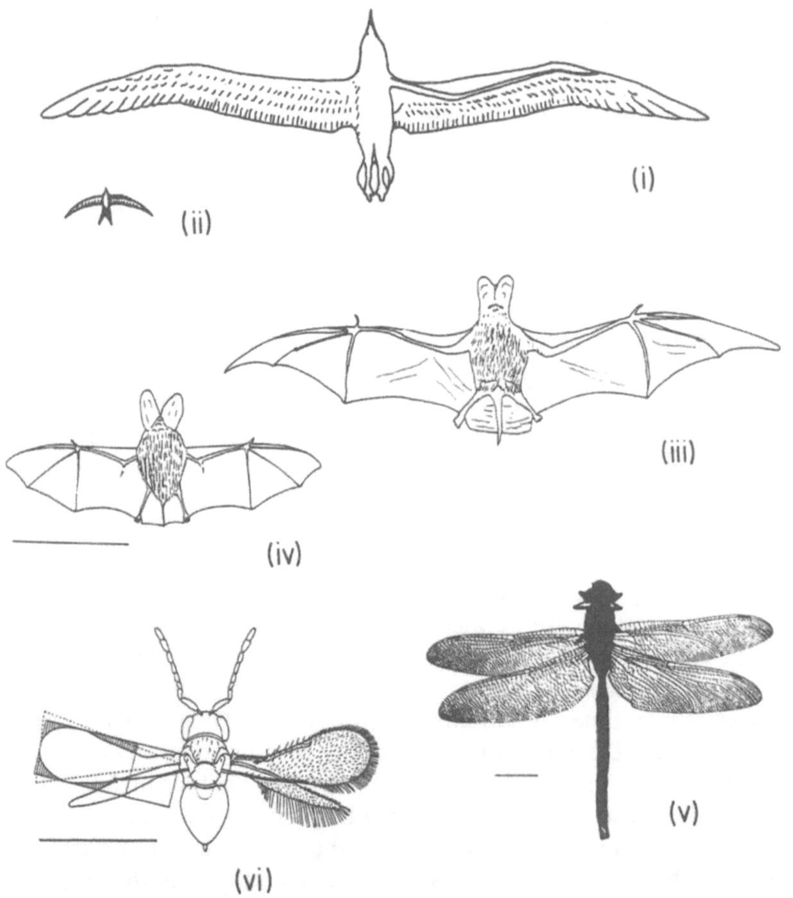

Fig. 6 (i)–(vi). The wing planforms of various birds, bats and insects. **(i)** *Diomedea* sp. and **(ii)** *Apus apus*, redrawn from Herzog (1968) to the same scale. The wing planform of *Apus* is strikingly similar to the lunate tail shape discussed in Sect. 5.5, and in Chap. 2. **(iii)** *Plecotus auritus* and **(iv)** *Otomops martiensseni*, traced from photographs in flight by U. Norberg, in Norberg and Rayner (1987) *Scale bar* 0.1 m. **(v)** *Aeschna palmata*, slightly modified from shadowgraph appearing on front cover of Somps and Luttges (1985). *Scale bar* ≈ 1 cm. **(vi)** *Encarsia formosa*, from Weis-Fogh (1973). *Scale bar* ≈0.5 mm. Re_c during flapping flight ≈ 1.5×10^3 and 15 for *Aeschna* and *Encarsia*, respectively. In the entire figure, the wingspan ranges from 3.4 m in the *Diomedea*, to 1.4 mm in *Encarsia*

part is to satisfactorily account for the fact that most of these wing systems experience significant oscillations for most of the time.

There have been a number of attempts to measure the planforms of animal wings in order to find general expressions for the wing shape, either to compare them with theoretical predictions, or for later use in strip-wise aerodynamic analysis. Oehme and Kitzler (1975), for example, measured the spanwise chord distribution of a number of species of birds and found that most could be approximated by a single general function, which was not an ellipse, but could be approximated by one for analytical convenience. Weis-Fogh (1973) found that many insect planforms could be approximated by a small number of simple functions, and Ellington (1984, II) has since measured the moments of mass and area for 18 insect species, with the remarkable result that the normalised wing shapes were found to lie within 5% of Beta distributions determined by one single parameter. The implication is that some strong, but as yet poorly understood, set of constraints is acting to produce such uniformity. Tantalisingly, reasonable agreement was also noted for the hummingbird *Amazilia*, and for the bat, *Plecotus*. It is likely that more is to be learned in this area.

One can also determine the wing properties empirically by measuring the $C_L(C_D)$ polar curves in wind tunnels for the complete wings, or three-dimensional models. The drag measurement therefore includes the induced drag, D_i. Figure 7 for insect wings shows that, as Re_c decreases, the maximum possible lift coefficient decreases and the sharpness of the stall decreases also. In fact, the *Drosophila* wings could hardly be said to stall at all, the drag simply continues to increase, up until

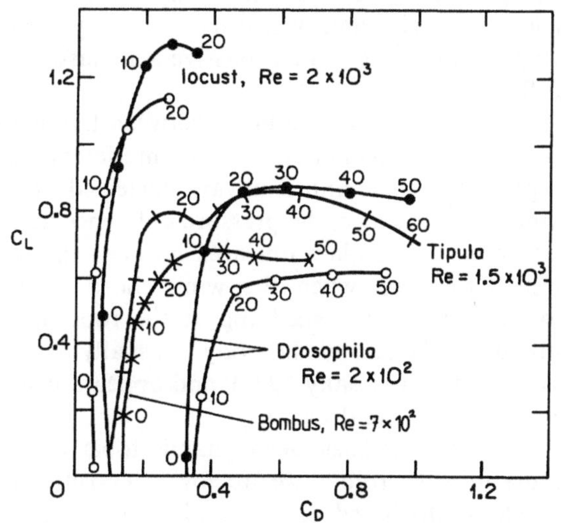

Fig. 7. Lift-drag polars for whole insect wings: Locust, *Schistocerca gregaria*, (forewing) from Jensen (1956); crane-fly, *Tipula oleracea*, from Nachtigall (1977); fruitfly, *Drosophila virilis* from Vogel (1967); bumblebee, *Bombus terrestris*, from Dudley and Ellington (1990b). *Open* and *closed circles* are for flat and cambered wings respectively

$\alpha = 50°$. At low Re_c, wings are less efficient at generating lift, but are also much less sensitive to changes in α, and will continue to function at very large α. Obtaining wind tunnel measurements of bird wings that are at all representative of the free flight performance is hampered by the complicated shape and response of the wings and feathers, and Tucker (1987) has shown that polar curves of fixed bird wings may look very different from those calculated from wind-tunnel tests on live gliding birds, due to active alteration of span and local section profiles. These difficulties would be even more acute for bat wings.

In general, animal wings are not planar, and a wing with fore or aft sweep will also be nonplanar (with respect to the freestream) when placed at some nonzero angle of attack. Cone (1962a) showed that certain nonplanar lifting systems had lower induced drag than the classical elliptically loaded planar form, and the efficiency factor e in Eq. (11) can similarly be increased, up to 1.07 behind swept wings (Burkett 1989). The large local upwash velocities associated with the sweep necessitate high values of wing twist near the tips to avoid stall there, but this effect can also be mitigated by changes in the local lifting line curvature (Cheng 1976), a geometric adaptation which does not depend on α, and which appears to have been adopted by the swift in Fig. 6(ii). Further discussion of the lunate planform in oscillating wings can be found in Chap. 2 and briefly in Sect. 5. Induced drag reduction by slotted wingtips or wingtip sails has been reviewed by Spillman (1987) and Jones (1990). The mechanism is to extract useful energy from the wingtip vortex by immersing smaller aerofoil sections in the rotating flow at the appropriate angles of attack. The cost-benefit analysis here is complicated, since the observed drag reductions could also be achieved just by increasing the aspect ratio, but it appears that induced drag reduction is effective for a wing limited in span, and for a given structural weight, when the individual winglets have variable incidence. The separated primary feathers of many bird species automatically have this property by virtue of their passive aeroelastic response to the instantaneous local airflow.

The presence of the body between the wings may be considered as another instance of nonplanarity, but the interference drag between the wings and body is probably small, and is always neglected. If there is no vortex shedding at the wing/body interface, the pressure difference across the wings is maintained across the body, whose planform area between the wings is thus included in the wing area. Otherwise, when the wing chord is small at the base, there must be considerable vortex shedding at the root, which will be reflected in the wake structure. The rest of the body is usually assumed to do no useful aerodynamic work (but cf. Csicsaky 1977), and appears instead as an extra component in the drag, the parasite drag, D_{par}.

Given these basic aerodynamic tools, and an understanding of the wings as conventional aerofoils, we are now in a position to discuss their practical operation, which we do based on three kinematic regimes of gliding, forward flight, and hovering. These correspond roughly to domains of $k = 0$, finite, nonzero k, and $k = \infty$, respectively, though it will become clear that the analytical domains do not exactly respect these boundaries.

4 Gliding

4.1 Basic Principles

Figure 8 shows the force balance on a body gliding at constant speed, U, at an angle θ with the horizontal. The lift and drag forces are

$$L = mg\cos\theta, \quad D = mg\sin\theta, \tag{12}$$

and the glide angle is determined by the ratio of lift to drag

$$\theta = \tan^{-1}\left(\frac{D}{L}\right). \tag{13}$$

The total drag on the body is the sum of the pressure and skin friction drags with the vortex-induced drag Eq. (10). From Eq. (11), the induced drag is

$$D_i = \frac{L^2}{\frac{1}{2}\rho U^2 b^2 e\pi}, \tag{14}$$

and, if the skin friction and normal pressure drags on the wings and body are combined and expressed as a profile drag coefficient (Sect. 2.1), the total drag is

$$D = \tfrac{1}{2}\rho U^2 S C_{D_{pro}} + \frac{L^2}{\frac{1}{2}\rho U^2 b^2 e\pi}. \tag{15}$$

The first term increases with U^2, while the second term, representing the energy lost in the vortex wake, decreases with U^2. At some point therefore, there will be a value of U, U_{md}, at which the total drag is minimised; if $L \approx W$ is fixed, and ignoring variations in b, S and $C_{D_{pro}}$, this point is reached when

$$U_{md} = \left(\frac{2L}{\rho b}\right)^{1/2}\left(\frac{1}{S C_{D_{pro}} e\pi}\right)^{1/4} \tag{16}$$

Lighthill (1975) noted that this defines a speed above which gliding is stable when any further increases in U will increase the magnitude of the first term in Eq. (15) which will consequently tend to reduce U. On the other hand, when $U < U_{md}$, the second term (induced drag) dominates, and increases in U will reduce the total drag, so stable gliding is not possible. As $L \sim m$, U_{md} will decrease as the mass decreases, and a simple dimensional analysis of Eq. (16) shows that $U_{md} \sim l^{1/2}$, where l is a linear dimension. Geometrically similar gliders with linear dimensions different by a factor of 10, should have characteristic gliding speeds different by

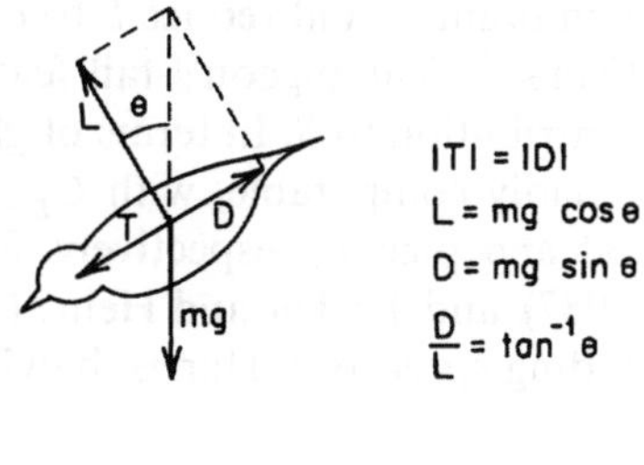

Fig. 8. The force balance on a bird gliding at constant airspeed, U, and glide angle, θ

a ratio of $10^{1/2}$, or just over 3. An alternative perspective of the minimum gliding speed may be obtained in terms of the maximum lift coefficient, since when θ is small, $L \approx mg$ [Eq. (12)], we may write

$$mg = \tfrac{1}{2}\rho U^2 S C_L, \quad \text{or} \quad \frac{mg}{S} = \tfrac{1}{2}\rho U^2 C_L. \tag{17}$$

Clearly, as U drops, C_L must increase in order to still support the weight, but C_L cannot be increased indefinitely, as flow separation occurs above a certain critical angle of attack (Sect. 3). Equation (16) shows that U_{md} may be increased either by reducing the wing area, S, or by reducing the span, b (which could also reduce S), and so animals gliding at high speed ought to reduce their wingspan, if they are able to do so. The ratio of weight to wing area, $mg/S \equiv Q$, is the *wing loading*, and is of fundamental importance in flight mechanics. According to Eq. (17), any characteristic flight speed, U^*, will be proportional to the square root of the wing loading. A low wing loading implies low gliding speeds, or, alternatively, lower freestream velocities are sufficient for continuous soaring.

4.2 Animal Gliders

The lifting line analysis is simple to apply to animal gliders, and permits testable predictions to be made concerning possible flight speeds, the geometry of the wing, and scaling laws. There is also some evidence that the lifting line assumptions behind Eq. (14) are appropriate. Quantitative flow visualisation experiments on the wake of a gliding kestrel (Spedding 1987a) showed that the wake was indeed composed of a trailing vortex pair, and calculations of Γ from velocity profiles such as that in Fig. 5 were very close to theoretical predictions. The transverse wake element spacing and the vortex core diameter were also in accordance with predicted values, and the inferred values of e and C_L were 0.96 and 1.16, respectively. This value of C_L is quite reasonable for the $Re_c \cong 4 \times 10^4$, and with $e \cong 1$ a circulation distribution close to elliptic is implied. This is consistent also with the absence of any additional concentrations of ξ in the wake caused by significant $\delta\Gamma$ on the wingspan.

The capacity to substantially vary the wing planform geometry distinguishes animal wings from many of their human-engineered counterparts, and has been observed to vary with flight speed, just as predicted in Sect. 4.1, by Pennycuick (1968, 1971a) in experiments on pigeons and bats gliding in a tilted wind-tunnel. As U increased from 8 m s^{-1} to 22 m s^{-1}, the pigeon was able to reduce b to 37% of the maximum value, whilst the bat, obliged to avoid collapse of the elastic wing membrane, could reduce b to only 83% of the maximum as U ranged from 5.5 to 10 m s^{-1}. The pigeon's tail feathers were unfurled at low U, providing an extra contribution to S. In terms of gliding performance, the two species were otherwise broadly comparable, with $C_{L\,max} = 1.3$ and 1.5, and $(L/D)_{max} = 6$ and 6.8, for the bat and pigeon, respectively. Similar measurements have been made by Tucker (1987), and Tucker and Heine (1990) who reported a $C_{L\,max}$ of 1.6 at the minimum gliding speed of a Harris' hawk.

Although both locusts and butterflies have been observed to glide frequently, their range of gliding speeds might be restricted by comparison because of their inability to substantially alter b or S during flight.

The larger gliding animals generally operate in an Re domain (10^3–10^5) where the estimation of the skin friction and normal pressure drag coefficients is highly uncertain. In the absence of flow separation, integration of the boundary layer equations gives a result for the dependence of total friction drag on Re_l as

$$D_\tau = k_0 \rho U_\infty^2 S Re_l^{-1/2}, \tag{18}$$

where Re_l is based on an appropriate streamwise lengthscale, and k_0 depends on the shape. A turbulent boundary layer grows faster than a laminar one, when Re_l enters Eq. (18) as $Re_l^{-1/5}$. These relationships have been well confirmed by experiment. The magnitude of the normal stresses on a body with a growing boundary layer depend on the boundary layer thickness, and so, all other things being equal, the normal pressure drag coefficient will also vary with $Re_l^{-1/2}$ or $Re_l^{-1/5}$. The value of D_n is in practice determined by the presence or absence of separation in the flow, and it is usually measured empirically.

The current status of measurement of D_τ and D_n on the wings and bodies of birds has been discussed by Pennycuick et al. (1988) and Tucker (1987, 1990). The observed drag coefficients on the wings and body, $C_{D_{pro}}$ and $C_{D_{par}}$, appear to decrease across the critical Re range, implying that early transition to turbulence in the boundary layer either prevents separation altogether, or moves the separation point further downstream, or assists in its subsequent reattachment. The measurements of D_{par} were extremely sensitive to the smoothness of the body surface, and drag coefficients were reduced by approximately 15% by Pennycuick et al. when the feathers were flattened with hairspray, and by 40% by Tucker when a smooth model with the same contour replaced the dead body. D_{pro} depends also on the local wing section profile and changes in b and S with flight speed, U. The relationship between any two of these parameters is not simple; in general, both C_L and $C_{D_{pro}}$ tend to *decrease* with increasing U for an intact gliding bird, but changes in wing geometry as measured by b, for example, introduce considerable variation.

It is possible that aerofoil operation around Re_{crit} plays a critical role in the gliding performance capabilities of soaring birds, but much remains to be accurately measured and understood. As a practical matter, given the tremendous uncertainties in the drag coefficient measurements, an appropriate procedure might be to first compute a value of Re_l, and then use it in a form of Eq. (18), with a suitable value of k_0 to account for experimental results when available.

4.3 Gliding Performance

Horizontal Distance Travelled. Maximising the ratio of horizontal distance travelled to vertical distance climbed is equivalent to minimising the glide angle, θ, and inspection of Eq. (13) shows that the L/D ratio is inversely related to θ. Given that $L \approx mg = W$ is a constant, then the minimum glide angle θ_{min} corresponds to gliding with the minimum possible total drag, D_{min}. The expression (16) for U_{md} thus also defines a best glide speed, U_{bg}. Rearranging Eq. (16) slightly,

$$U_{bg} = U_{md} = \left(\frac{2W}{\rho S}\right)^{1/2} \left(\frac{1}{ARC_{D_{pro}} e\pi}\right)^{1/4}. \tag{19}$$

W has been substituted for L and U_{bg} is expressed in terms of the wing area, S, and the aspect ratio AR. The best glide speed increases with W, and decreases with both S and AR. If the objective is to extract energy from the environment and glide without loss of height, then it is useful to be capable of this for as low a value of U as possible, and there are two possibilities for minimising U_{bg}; have wings with large area, S, or have long narrow wings and increase AR. These contrasting strategies are illustrated in Fig. 9, from Pennycuick (1983), who compared the gliding performance and wing planform of the frigatebird, pelican and vulture, all of which were observed to make extensive use of thermal soaring. The frigatebird has a very low wing loading ($Q = 37\,\text{Nm}^{-2}$), as well as a higher aspect ratio ($AR = 13$), compared with both the pelican and the vulture ($Q = 58$ and $55\,\text{N m}^{-2}$, $AR = 10$ and 6, respectively), and so manages to minimise both of the terms in Eq. (19). It is distinguished by almost constant gliding and soaring in trade wind thermals for days on end, rarely coming in to land. The shorter, slotted wings of the pelican and vulture probably represent a balance between adaptations for landing and take-off and induced drag reduction for wings with limited span and low moment of inertia.

Air Time. The cross-country performance of a glider may alternatively be measured by energy consumption per unit time. In steady flight, the kinetic energy of the glider is constant, and so the rate of energy loss required to overcome drag (and generate a wake) is balanced by a reduction in potential energy, occurring at a rate mgU_s, where U_s is the vertical downward velocity, so

$$mgU_s = DU. \tag{20}$$

mg is a constant, and DU can only be minimised through U_s. The total drag was given in Eq. (15), and substituting an approximate expression of the form of Eq. (18),

Fig. 9. Silhouettes of three gliding birds. From top: *Fregata magnificens, Pelecanus occidentalis, Coragyps atratus.* (Pennycuick 1983)

assuming a turbulent boundary layer, for the skin friction and normal pressure drag components of $D_{pro} + D_{par}$, U_s may be written in the form

$$U_s = k_0 \rho U^3 S^2 Re^{-1/5} + \frac{2W}{\rho U S A R e \pi}. \tag{21}$$

The first term in Eq. (21), due to profile drag on the wings and body, increases as S^2, while the second term from the lift-induced drag is proportional to S^{-1}, so U_s increases with S. The second term is inversely proportional to AR so a cross-country glider should have as high an aspect ratio as is structurally possible, compatible with the other flight requirements. Albatrosses have aspect ratios of around 15, while modern composites and specialised function allow $AR > 20$ in almost all man-made gliders. Equation (21) depends only on the wing geometry through S and AR, for a glider of a given weight moving with a given airspeed, U. The graph of $U_s(U)$ is called a glide polar (Fig. 10a). Because birds, and to a lesser extent bats also, have the ability to modify b and S at different U, they will occupy an envelope on this plot determined by the family of glide polars attainable from varying the wing planform. Similarly, changes in profile camber, or the recruitment of additional high lift devices, such as those discussed in Sect. 3.2, will extend the domain of available $U_s(U)$ ratios, as discussed by Tucker (1987).

Turning Radius. Glider pilots and birds (cf. Cone 1962b; Pennycuick 1975), but presumably not nocturnal bats, make extensive use of atmospheric thermals, which are most often columns of circulating and rising air, but may also be buoyant, rising vortex rings. Both are created by differential heating on the ground by solar radiation, and both are limited in horizontal extent, supplying an incentive to be able to maintain a small turning radius. In a steady, nonaccelerating turn, the centrifugal force, mU^2/r, is balanced by the component of the lift vector projected onto r, so, if Φ is the angle between the lift and the vertical,

$$\frac{mU^2}{r} = L \sin \Phi. \tag{22}$$

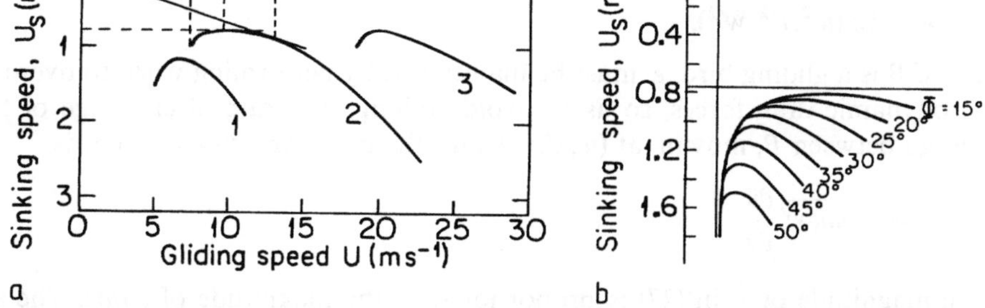

Fig. 10. a Glide polars for the fruit bat (*1*), *Rousettus aegyptiacus*, the white-backed vulture (*2*), *Gyps africanus*, and the ASK-14 motor glider (*3*). (Data from Pennycuick 1971a, b). **b** Family of glide polars generated for *Gyps africanus* turning at different bank angles Φ. (After Pennycuick 1975)

r is constrained by possible values of Φ and L. For convenience, one can set an upper limit for $\sin\Phi = 1$ (practically speaking, it will be much lower), and the highest value of L is determined by $C_{L_{max}}$, and so the limit on r is,

$$r_{min} = \frac{2m}{\rho S C_{L_{max}}}. \tag{23}$$

If $C_{L_{max}}$ is constant, r_{min} depends only on m/S, equivalent to the wing loading (for constant g). The component of lift opposing mg is now $L\cos\Phi$, and U_s in a turn will be related to the straight-line U_s by

$$U_{s_t} = U_s \cos\Phi. \tag{24}$$

A family of glide polars may therefore be constructed for a range of Φ (Fig. 10b). Pennycuick (1971b) examined the gliding performance of the white-backed vulture by following individuals around in a small glider, noting that these thermal soarers have comparatively low AR wings, but large wing areas. According to Eq. (23), this would appear to make them ideally suited for small radius turns, and comparisons with a hypothetical albatross-shaped vulture showed them better able to remain inside small thermals than if they had the higher AR, but lower S wing planform. However, this would be offset by the albatross-vulture's improved cross-country performance between thermals, and it seems that the difference cannot be accounted for purely in terms of aerodynamic soaring performance.

Dynamic Soaring. Lighthill (1975) outlined a surprisingly simple analysis of dynamic soaring, which we use here. Consider a body, B, in a fluid where there is a mean shear of the u component in the vertical direction, z, such as at a wall boundary layer, or over an air/sea interface. B might equally be a fluid element in a turbulent shear flow, or an albatross in a wind gradient. B has velocity

$$\mathbf{u} = \bar{u}(z) + u', v', w', \tag{25}$$

where the primed components are the fluctuating velocities in the turbulent flow, or the components of velocity of the bird, in (x, y, z). Overbars are mean quantities, and the mean shear $\bar{u}(z)$ varies only with z. The kinetic energy of B *relative to the velocity* gradient is

$$e' = \tfrac{1}{2}m(u'^2, v'^2, w'^2), \tag{26}$$

and if B is a gliding bird, e' must be maintained by performing work to overcome aerodynamic drag forces, so as to avoid stalling. The rate of exchange of total energy between B, moving at (u', v', w'), and the mean velocity gradient is

$$\dot{P} = -m\overline{u'w'}\frac{\partial \bar{u}}{\partial z}. \tag{27}$$

The magnitude of $\dot{P}$ in (27) is proportional to the magnitude of $\delta\bar{u}/\delta z$. The sign of the product $\overline{u'w'}$ may be positive, implying a transfer of energy from the mean flow to the fluctuating part (the bird). This will occur when u' and w' have opposite signs; when flying into the shear, u' is negative so w' should be positive, and, conversely, when u' is positive, w' should be negative. The required behaviour of flying upwards into wind shear, and downwards away from it has often been noted, and has been subject to quite elaborate analyses (e.g. Cone 1964; Wood 1973), but more accurate measurements of albatrosses gliding in wind shear by Pennycuick (1982) revealed that the magnitude of $\delta\bar{u}/\delta z$ would not be sufficient to maintain the kinetic energy, except very close to the water surface. Instead, there must be some conversion of kinetic into potential energy, so that the bird decelerates as it gains height, deriving most energy from slope lift in waves at the surface, much as proposed by Wilson (1975). Note that the mechanism of dynamic soaring probably still pertains, and the flight path is chosen with the sign of Eq. (27) still in mind, but that the mean shear cannot by itself supply sufficient energy. Also, the variable wing geometry that allows birds to glide at low speeds and high lift coefficients also relaxes somewhat the condition (26), more readily allowing e' to fall.

5 Forward Flapping Flight

5.1 The Principles of Lift and Thrust Generation by a Beating Wing

In the absence of any energy input from the environment, the beating wings of an animal in steady flight must provide both lift and thrust required to maintain height and forward speed. We therefore begin by outlining the kinematic basis for generating both force components from an oscillating wing section. As Lighthill (1969) noted, in order to provide weight support, some asymmetry must be introduced into the combined pitching and heaving motion of a lifting surface,

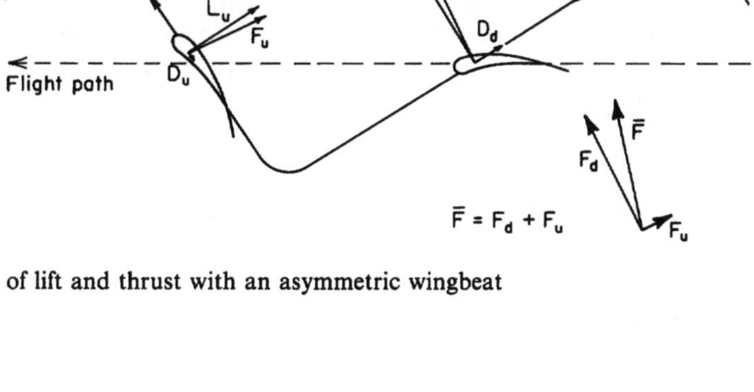

Fig. 11. The production of lift and thrust with an asymmetric wingbeat

and it is sufficient just to add a constant angle of attack to the otherwise symmetric motion of a thrust-generating wing. The easiest way to imagine doing this for a flying animal is to tilt the whole body, and/or the wings relative to the body, so that they beat in a plane inclined at an angle, β, to the horizontal. The result is analysed in Fig. 11. Owing to $\beta \neq 0$, the wings sweep forward, with respect to the wing root, on the downstroke, and backwards on the upstroke. Both the incident velocity, $\mathbf{u}$, and the geometric angle of attack, α, are larger on the downstroke than on the upstroke, and so the magnitude of the resultant force, $\mathbf{F}_d$, is greater than that produced during the upstroke, $\mathbf{F}_u$. If the time spent on up- and downstrokes is equal (the downstroke ratio, $\tau = 0.5$), and the L/D ratio is constant, Fig. 11 is a graphically accurate demonstration that the mean resultant force, $\bar{\mathbf{F}} = \mathbf{F}_d + \mathbf{F}_u$, averaged over the wingbeat, can have positive components of lift and thrust. Actually, this figure includes a second strategy for introducing the required asymmetry, as α is even further reduced during the upstroke, following a strong pitch-up rotation (or supination) at the end of the previous half-stroke. This contributes to the reduction in magnitude of $\mathbf{F}_u$. There is considerable scope for exerting fine control over the magnitude and direction of $\mathbf{F}$, via changes in α (it may even be negative), aerofoil section camber, and τ. The proportional feathering parameter, Θ, (Lighthill 1969) was introduced in Chap. 2, and it controls the local incidence angle compared to the local relative wind, effectively determining the balance between thrust and efficiency. The complicated kinematics, together with the highly variable wing geometries encountered in nature (recall Figs. 4 and 6), conspire to defeat attempts to include all possible kinematic and geometric parameters in a theoretical analysis, and the focus turns to formulating a simple model problem which can be solved, but which still captures some of the essential physics. Generally, one may proceed either by modelling the beating wing, or by constructing a simple model of the wake (recall Fig. 1).

5.2 Blade Element Analysis

The classic reference for this approach is the work of Jensen and Weis-Fogh 1956; Weis-Fogh and Jensen 1956; Weis-Fogh 1956; Jensen 1956) on the desert locust, *Schistocerca gregaria*. One of the advantages of dealing with locust wings is that their geometry and kinematics are comparatively simple, a prerequisite for this type of analysis. Figure 12a introduces the coordinate system for describing the flapping wing of an animal moving at forward speed U. The wings beat in a plane, at an angle β to the horizontal, through a total angle ϕ. The kinematic analysis starts by examining the wing position as a function of time, $\gamma(t)$. Assuming the wingbeat is sinusoidal, then

$$\gamma(t) = \bar{\gamma} + \tfrac{1}{2}\phi \sin \omega t, \tag{28}$$

where $\omega = 2\pi n$ is the radian frequency of the wingbeat. The angular velocity is

$$\dot{\gamma}(t) = \frac{\omega}{2}\phi \cos \omega t,$$

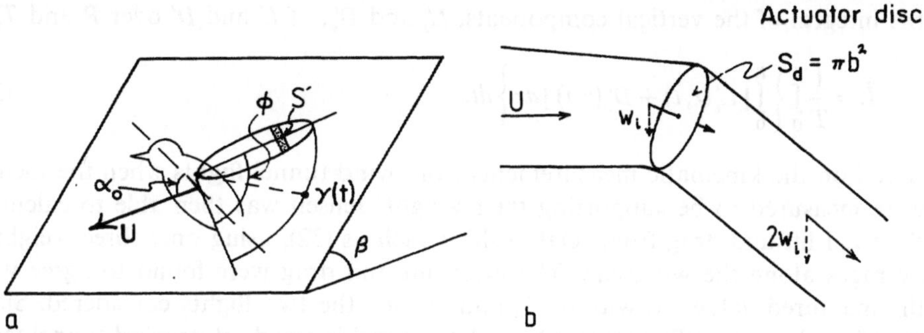

Fig. 12 a, b. Two alternative perpectives for the aerodynamic analysis. **a** The kinematics of a flapping wing. **b** Wake momentum flux

so the velocity of the wing at a span section, r, and at time t is

$$\mathbf{u}_w(r,t) = r\dot{\gamma}(t) = r\frac{\omega}{2}\phi\cos\omega t. \tag{29}$$

The local incident velocity, $\mathbf{u}(r,t)$, at a wing section will depend on the resultant, $\mathbf{u}_n(r,t)$, of $\mathbf{u}_w(r,t)$, and U, the forward flight velocity, and also on $w_i(r,t)$, the downwash velocity, which alters the local angle of attack (Sect. 3.2, Fig. 5),

$$\alpha_{\text{eff}} = \alpha - \alpha_i, \tag{30}$$

where

$$\alpha_i = \tan^{-1}\left[\frac{w_i(r,t)}{\mathbf{u}_n(r,t)}\right]. \tag{31}$$

Some means of estimating $w_i(r,t)$ in (31) must be found. Commonly, it is assumed to be uniform across the span, as in the elliptically-loaded lifting line (Sect. 4.3), with a value estimated from actuator disc models (Sect. 5.3). Jensen and Weis-Fogh actually did not do this. Instead, locust wings were placed in the wall boundary layer of a wind tunnel, and the lift and drag on the entire wing in a shear flow similar to that produced by flapping about the wing root were measured. The drag thus included the induced drag, which did not have to be estimated separately.

Now, given α_{eff}, provided that all *3-D* effects can be ignored, and if steady-state aerofoil properties can be used, then c_l may be estimated from empirical $c_l(\alpha)$ data. Given $\mathbf{u}$ and c_l, the local section lift and drag forces are:

$$\left.\begin{aligned} L'(r,t) &= \tfrac{1}{2}\rho\mathbf{u}^2 c_l S' \\ D'(r,t) &= \tfrac{1}{2}\rho\mathbf{u}^2 c_d S' \end{aligned}\right\} \tag{32}$$

S' is just the wing chord, $c(r)$ at the span r, and the primes indicate that L' and D' are the components of the aerodynamic force per unit length. The mean lift is the integral of the vertical components, L'_v and D'_v, of L' and D' over R and T,

$$\bar{L} = \frac{1}{T}\int_0^T\left\{\int_0^R [L'_v(r,t) + D'_v(r,t)]\,dr\right\}dt. \tag{33}$$

Based on the kinematic measurements from wind tunnel flights when the locusts were measured to be supporting their weight, Jensen was then able to calculate the total lift and drag from relationships such as (32), using only three weighted averages along the wingspan. The mean lift and drag were found to agree with the measured values to within 3% and 7% for the two flights considered. Since the lift and drag coefficients had been determined in steady state wind-tunnel tests, these results were construed as convincing evidence that the aerodynamics were consistent with steady-state principles, contrary to many beliefs at the time. It was further argued that this would be *expected* to be the case, given the wingbeat kinematics and angles of attack measured during flight. The frequency parameters k, Ω and K may be calculated for a locust forewing with mean chord, $\bar{c} = 0.93$ cm, semi-span, $b = R = 5.21$ cm, forward speed, $U = 350\ \mathrm{cm\,s^{-1}}$, stroke angle, $\phi = 83°$ (1.45 rad), and wingbeat frequency, $n = 17.5$ Hz ($\omega = 110\ \mathrm{rad\,s^{-1}}$):

$$k = \frac{\omega\bar{c}}{2U} = 0.15, \quad \Omega = \frac{\omega b}{U} = 1.64, \quad K = \frac{2\phi n R}{U} = 0.755.$$

k and Ω are not exceptionally small, and the advance ratio, $J = K^{-1} = 1.32$ is relatively large, but certainly not huge. Although the authors at the time had little alternative but to proceed and use the aerodynamic forces calculated in this way to estimate the power required of the flight muscles to generate the necessary torques at the wing root, we shall later see that *3-D* and unsteady corrections can nevertheless be significant in this domain.

Further simplifications of Eq. (33) are possible. If the changes in local wing twist, sectional geometry and chord length are such as to maintain a constant value of c_l and c_d along the span, then the coefficients may simply be replaced with mean values over the whole span and wingstroke. If a constant w_i across the span is also assumed, then the problem reduces simply to finding some manageable formulation of the wingbeat kinematics and wing planform geometry. Examples of this type of approach may be found in Norberg (1976a) for bats, Pennycuick (1975) for birds, and Dudley and Ellington (1990a, b) for the bumblebee. In steady horizontal flight, the vertical components of the integrated aerodynamic forces during each wingbeat cycle must support the weight, and so one may write equations of the form:

$$\int_0^T W\,dt = \rho\bar{C}_L\int_0^T\left\{\int_0^R \mathbf{u}^2(r,t)S'(r)\cos\psi\,dr\right\}dt + \rho\bar{C}_D\int_0^T\left\{\int_0^R \mathbf{u}^2(r,t)S'(r)\sin\psi\,dr\right\}dt, \tag{34}$$

where the angle ψ is a much simplified representation of the required projection geometry calculations. Here it plays a role quite analogous to a projection onto the horizontal plane of the vertical impulse of the wake. This analogy should become clearer in Sect. 5.4. A similar equation for the horizontal force projections can be written, and if $\mathbf{D}_{par}$ and $\mathbf{D}_{pro}$ can be estimated, the system may be solved for $\bar{C}_L$ and $\bar{C}_D$. The argument now shifts slightly; if reasonable values of $\bar{C}_L$ and $\bar{C}_D$ are computed, consistent with the known or likely performance of the aerofoil wings, then the initial assumptions of steady-state performance are supported. If not, some other mechanism must be significant, and unsteady effects of various kinds may be proposed. Just by itself, this is a curiously circular sort of argument, of little interest unless the numerical results are then used in subsequent calculations. The controversy over this proof by contradiction, summarised by Ellington (1984, I, IV, V), seems endless, and we shall return to it in the hovering flight discussion (Sect. 7.2). Note, however, that regardless of whether *3-D* and/or unsteady corrections are required for an accurate *aerodynamic analysis*, their relative importance in the wing aerodynamics is a slightly different issue than whether or not they *need* to be incorporated in a general flight model, and it is easy to imagine a wing which accelerates and rotates at the extremes of the wingbeat, being dominated by unsteady effects there, but overall having a performance roughly equivalent to an unspectacular mean lift coefficient averaged over the whole wingbeat. Indeed, direct measurements of the *instantaneous* lift forces on the locust-in-a-wind-tunnel experiment by Cloupeau et al. (1979) showed that the amplitudes of the fluctuations in aerodynamic lift were about two times those calculated by Jensen, while the *time-averaged* values were approximately the same, implying that unsteady effects are significant. Moreover, Dudley and Ellington's (1990b) wind tunnel experiments on the bumblebee (in free flight) for $U = 0 - 4.5\,\mathrm{m\,s^{-1}}$ indicated that the minimum required C_L values *always* exceeded 1, while the wind-tunnel tests showed the wings to be incapable of $C_L > 0.8$ in steady flow. The same frequency parameters can be calculated as for the locust, and at the top speed of $4.5\,\mathrm{m\,s^{-1}}$ one finds:

$$k = 0.40, \quad \Omega = 2.65, \quad K = 1.52.$$

The reduced frequencies are substantially higher than for the locust, and it is no longer true that $k \ll 1$. It will later be demonstrated (Sect. 5.5) that at these values of k and Ω, one *expects* unsteady corrections to be significant, and that the conclusion of Dudley and Ellington, that at no flight speed can the aerodynamics of the bumblebee be satisfactorily accounted for on steady-state principles, is as predicted on theoretical grounds.

A detailed, steady, lifting surface analysis of the wing of a house sparrow in mid-downstroke has been performed by Hummel and Mollenstadt (1977). Quite modest values of $\bar{C}_L =$ from 0.3–0.7 were reported, although the authors suggested that values of 0.7 are close to the maximum for such a wing profile at $Re_c \cong 10^4$. In general, blade element/quasi-steady lifting line methods will be valid for birds, bats and insects in steady forward flapping flight at sufficiently low Ω. The chief disadvantage is the requirement of, at least, detailed kinematic and morphological data, and preferably additional steady-state aerofoil measurements for a number of wing section profiles. The analysis can become cumbersome and tedious, and it may be difficult to generalise the results. Moreover, just what constitutes a small enough Ω remains unclear, as there are indications that $\Omega \approx 2.0$ is too high, even from a rough, practical viewpoint. With further increases in Ω, the basic aerodynamic premises and assumptions are almost certainly invalid.

5.3 Momentum Jet

The momentum jet, or actuator disc model, is also a steady-state model, but it is a dramatic simplification of the situation (Fig. 12b). Nothing remains of the original animal apart from a circular disc, whose diameter is equal to the wingspan, 2*b*. The disc applies a uniform downward acceleration to the fluid, which reaches a final vertical velocity of $2w_i$. The mass flux through the disc is $\rho U S_d$, and the rate of generation of downward momentum is the product of the mass flux times the velocity, $2w_i$, to which it is accelerated. In steady flight, this force balances the weight:

$$W = \rho U S_d \cdot 2w_i, \quad \text{or,} \quad w_i = \frac{W}{2\rho U S_d}. \tag{35}$$

The second expression gives a value for the induced velocity across the disc. It may be used as a first estimate to correct for the induced downwash in the blade element theory [Eq. (31)]. w_i is constant and so the power required to generate an upward force equal to the weight, *W*, is

$$P_i = W w_i = \frac{W^2}{2\rho U S_d e}. \tag{36}$$

e is the aerofoil efficiency introduced before, and it is used to account for the fact that air is not accelerated uniformly and steadily across S_d. Equation (36) is exactly equivalent to Eq. (11) derived from the lifting line equations, after substituting for the drag and lift coefficients, and noting that $S_d = \pi b^2$. There are no parameters which reflect the fact that the wings are actually beating, except indirectly via the increased thrust required to achieve the airspeed *U*.

Note how the perspective has changed, from considering the kinematics of the wingbeat and lifting properties of the aerofoil surfaces to calculating the momentum flux in the wake model, albeit a highly simplified one. The principal advantage of such a model is its simplicity and ease of manipulation. The power requirement can be calculated given only the body mass and the wingspan, and no other inputs are necessary. The disadvantage is that we may have thrown the baby out with the bath water; the model cannot reflect any changes in wing area, aspect ratio, wingbeat frequency, section geometry or kinematics, except by ad hoc adjustments of the factor *e*. Furthermore the wake model is not realistic.

5.4 Vortex Wake Models

Vortex Loops. In a sense, the momentum jet model is a vortex wake model; an infinitely thin, cylindrical sheet of vorticity surrounds the uniform jet, separating it from the outer undisturbed fluid. As w_i is spanwise constant there is no vorticity

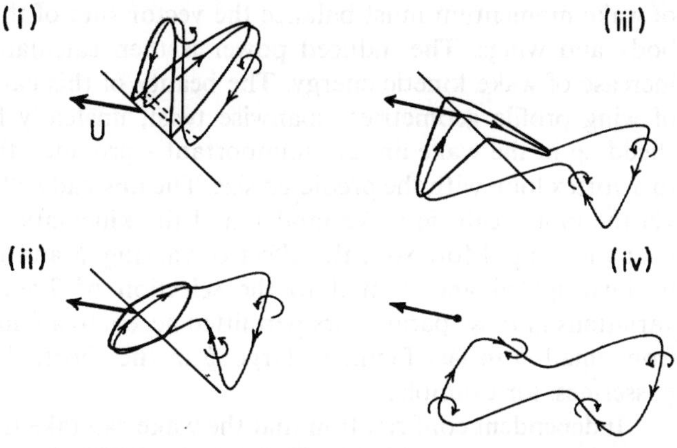

Fig. 13 (i)–(iv). A conjectural mechanism for the origin of a closed vortex loop behind a flapping wing pair. In each of the four timesteps, labelled frames (i)–(iv), the wing pair and wake are seen in oblique view, moving into the page with velocity U. The stick-like body takes no part in the aerodynamics. Vorticity shed during the downstroke is assumed to roll up immediately, and can therefore be represented as a single, curved line-vortex. The true behaviour of such a convoluted vortex is not known; here, local fluctuations in curvature decrease in magnitude with time. At the end of the downstroke, the wings vanish, since they contribute no further to force generation

inside the cylinder, which contracts in diameter as w_i increases. Figure 13 represents an attempt to construct a more realistic wake model from first principles, considering the motion of a rigid, beating wing pair on an infinitely thin body in steady forward motion. In the first frame, the wings are just beginning to move on the downstroke. According to the principles outlined in Sects. 3.1–3.2, the circulation at each span location, r, depends on the local chord, $c(r)$, and on the relative velocity, $\mathbf{u}(r)$, which, ignoring wake induced velocities for the moment, is the vector sum of the forward speed, U, and the local flapping velocity, $\mathbf{u}_w = 2\phi nr$. The local circulation may therefore fall towards the wing root, as indicated by the dotted lines in the wake, but if this $\delta\Gamma$ is assumed to be small, the starting vortex shed at the wing trailing edge could have an initial configuration something like frame (i). As the wings continue during the downstroke [frame (ii)], the tip vortices link the bound vorticity on the wing with the convoluted starting vortex which convects and deforms according to its own induced velocity field. At the end of the downstroke, the circulation on the wings drops to zero and a stopping vortex shed at the trailing edge completes a closed vortex loop which drifts away in the wake. Now, suppose the upstroke does no aerodynamic work whatsoever (the wings have disappeared in the last frame), and suppose that the loops of frame (iv) may reasonably be approximated as elliptical, planar, small-cored vortex rings. In this case, the wake is composed of a series of vortex loops, one created on every downstroke. This is the basis of the flight model for forward flight introduced by Rayner (1979a, b). The energy of each ring is determined by its size and circulation. The streamwise and spanwise ring dimensions are determined by the wingbeat kinematics and geometry, and by the form of $\Gamma(y)$ on the wings. The ring circulation is computed by enforcing the force balance condition that the rate of generation of wake momentum must balance the vector sum of the weight and drags on the body and wings. The induced power is then calculated from the mean rate of increase of wake kinetic energy. The beauty of this calculation is that the details of wing profile geometries, spanwise twist, unsteady lift generation and vortex shedding in the wake are all unimportant – provided that the result is equivalent to a vortex loop with the predicted size. The unsteady effects are taken into account via the more realistic wake model, and the kinematic data requirements are not overwhelming. Moreover, the effect of varying ϕ and downstroke ratio, τ, could be investigated and related to the selection of flight style. It was found that variations in these parameters permitted access to a family of $P(U)$ curves, passing from small ϕ in fast flight, to large ϕ in the short, slow flights typical of small passerines, for example.

Independent confirmation that the wake can take the form of a series of closed vortex loops was provided by Kokshaysky (1979), from wake visualisation experiments on small passerines in a flight cage. Further quantitative work, using multiple-flash stereo photogrammetry and *3-D* wake reconstruction techniques (Spedding et al. 1984; Spedding 1986) showed that the only detectable concentrations of wake vorticity left in the wakes of the pigeon and jackdaw in slow flight were indeed almost-circular vortex rings (Fig. 14), but direct measurements of the wake momentum found only approximately half of that required for weight support. It was also demonstrated (Spedding 1986) that the wake circulation measurements were not internally self-consistent, implying that its structure, or time history, might be more complicated than suspected. This complexity is particularly evident in the initial and final roll-up of the convoluted vortex sheet [frames (i) & (iii)]. Mutual annihilation of viscous, finite-core vortices of opposite sign, followed by vortex reconnection, has been documented in interacting vortex rings, and it is reasonable to enquire whether it could occur so as to reduce the size and/or strength of the final, measured ring structure here. Saffman (1990) has

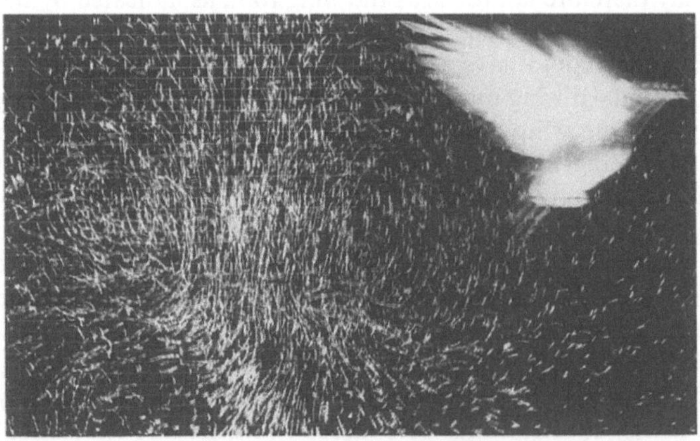

Fig. 14. Vortex ring in the wake of a slow flying pigeon. The flow is traced by multiflash images of neutrally buoyant, helium-filled soap bubbles

proposed a model which allows the time scales of these events to be estimated, governed by two parameters, ε and μ:

$$\varepsilon = \frac{a}{R}, \quad \mu = \frac{\pi v'}{\Gamma_0 \varepsilon^{1/2}}.$$

ε is the ratio of core radius, a, to ring radius, R, and μ is like an inverse Reynolds number, where $v' = 2\pi v$, and Γ_0 is the circulation. The characteristic times in seconds for cancellation and reconnection are T_c^* and T_r^*, respectively, estimated as:

$$\left.\begin{aligned} T_c^* &= \frac{5\pi R^2}{\Gamma_0}\varepsilon^{5/4}\left(\ln\frac{1}{\mu}\right)^{1/2} \\ T_r^* &= \frac{2\pi R^2}{\Gamma_0}\varepsilon^{1/2} \end{aligned}\right\}$$

Substituting the appropriate values for the pigeon wake, one arrives at time scales, $T_c^* \cong 0.4$s, and $T_r^* \cong 0.2$s, with the same order of magnitude as the stroke period ($T = 0.15$s). Initially, local regions of opposing, deformed loops will have smaller values of R, and consequently, shorter local time scales, T_c^* and T_r^*. The analysis is limited to cases where the separation distance, $z \ll R$. Firm conclusions cannot be drawn from this approximate analysis, but it does suggest that such viscous physical processes could significantly affect the dynamics.

A qualitatively similar wake structure has been reported in noctule and long-eared bats (Rayner et al. 1986), and together with Kokshaysky's result, it appears that this type of wake is characteristic of slow flight at high Ω ($\Omega = 5.8$ and 4.1 for the pigeon and jackdaw respectively) in birds and bats. Such a model predicts that less energy will be consumed by increasing the amplitude of the wingbeat, rather than the frequency, as U decreases, and this seems to be the strategy adopted in nature.

The closed-loop wake can only occur if the upstroke performs no aerodynamic work, and consequently leaves no trailing vortices. This is most simple to envisage in the slow flight of birds and bats when the wings are folded close to the body on the upstroke, whose function may be interpreted either as doing as little as possible (including minimising inertial and profile power requirements) for as short a time as possible, or as a means for effectively removing the bound vorticity built up on the wing during the downstroke. The upstroke often consumes a substantial fraction of the wingbeat period, and the latter interpretation may shed some light on the complex twisting and rotational motions. Insect wings cannot fold to the same extent during flight, and it is well established that they operate over both downstroke and upstroke (e.g. Jensen 1956; Vogel 1967; Nachtigall 1974), generating useful forces (with lift and/or thrust components) at either positive or negative angles of attack. The wake structure in this event will be different, the exact form depending on whether the starting and stopping vortices of successive wing strokes are merged or not (e.g. Brodskii and Ivanov 1984). Wake photographs behind a dragonfly in a wind tunnel (Azuma and Watanabe 1988) clearly show continuous tip vortices, but the cross-stream elements are hard to identify. Experiments by Luttges and co-workers (summarised in Luttges 1989) have demonstrated complicated vortex wakes composed of a number of discrete coherent vortex structures behind tethered dragonflies and model wings in a wind-tunnel for values of Ω around 0.9–2.7, but one would be hard-pressed to formulate a general model based on such intricate smoke trail patterns. Presently, a vortex theory based on a simple wake model for forward flight in insects is lacking, although the LCM model, considered later, does incorporate a simple, fixed wake model into a blade element type of calculation.

There is no shortage of evidence from the wingbeat kinematics of both birds (e.g. Brown 1963) and bats (e.g. Norberg 1976a) in medium-speed flight, that at least part of the wing is aerodynamically active on the upstroke, and a conceptual model different from Fig. 13 is required.

The Roller-Coaster. As we have already remarked, the crucial requirement, in order to support the weight and provide thrust for steady forward flight, is that the wingstroke be, in some sense, asymmetric. The resultant aerodynamic force,

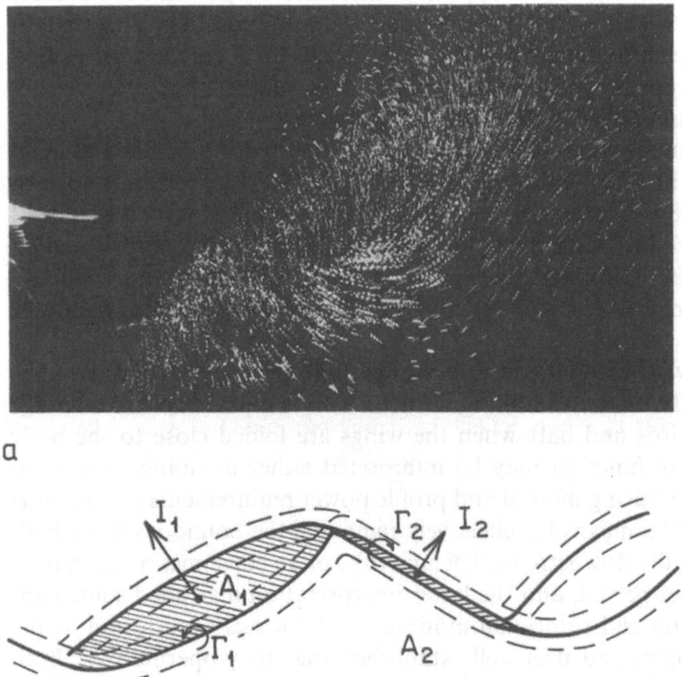

Fig. 15. a The wake of a kestrel in cruising flight. The flight was from right to left. The trailing vortex behind the starboard wing is marked by bubbles trapped inside the core. In the section created during mid-downstroke, the scarcity of bubbles has created a cutout, where the axial core flow, and behind this, the outer, induced, recirculating flow can be clearly seen. **b** Wake model based on **a**. Two line-vortices with constant circulation enclose planar areas A_1 and A_2 traced out by the extended wingtips, and folded primaries during the downstroke and upstroke, respectively

F, on a single beating wing with circulation distribution $\Gamma(y)$ over the semi-span, b, can be expressed as

$$\mathbf{F} = \rho \mathbf{u} \int_0^b \Gamma(y) dy, \tag{37}$$

and the orientation of **F** may be altered by changes in the wing stroke geometry, effectively changing the direction of **u**, while its magnitude may be modified via changes in $\Gamma(y)$, or in the semi-span, b. For any $\delta\Gamma$ on the wings, there must be an increment of $-\delta\Gamma_w$ in the wake, corresponding to the shedding of vorticity. If, on the other hand, b is gradually reduced, to b', say, while maintaining the same Γ_0, vortex shedding will be confined mostly to regions near the wingtips, where the largest adjustments in $\Gamma(y)$ occur, and the generation of concentrated spanwise vortices can be avoided. A nonzero net force may be generated according to the difference in impulse of the two segments of the wake. The wake of a kestrel in cruising flight has been shown (Spedding 1987b) to consist of two continuous undulating vortices, without detectable cross-connections between them, as confirmed by measurements of invariant circulation along the wake vortices. At the beginning and end of the flight, transverse starting and stopping vortices will be generated, thus completing the roller-coaster track. In cruising flight, the kestrel flexes the primary feathers on the upstroke, while the secondary feathers remain loaded, and the tip vortices appear to be shed at the corner between the two. Figure 15a shows an example wake photograph. Because the wake circulation, $\Gamma_w = \Gamma_0$, is constant, a very simple vortex wake model may be constructed, as in Fig. 15b, and the induced power requirement may be calculated from it. The impulses, $\mathbf{I}_1$ and $\mathbf{I}_2$, of the wake segments with areas A_1 and A_2, created during the downstroke and upstroke respectively are,

$$\mathbf{I}_1 = \rho\Gamma_0 A_1 \quad \text{and} \quad \mathbf{I}_2 = \rho\Gamma_0 A_2. \tag{38}$$

The reaction force is equal to the time rate of generation of the wake impulse, $\mathbf{F} = d\mathbf{I}/dt$, and so, if the planar wake segments make positive angles ψ_1 and ψ_2 with the horizontal, the average lift and thrust ($\equiv \bar{D}$ in steady, equilibrium flight) forces, integrated over one wingbeat period, T, are:

$$\left.\begin{aligned} \bar{L} &= \frac{1}{T}\rho\Gamma_0(A_1\cos\psi_1 + A_2\cos\psi_2) \\ \bar{D} &= \frac{1}{T}\rho\Gamma_0(A_1\sin\psi_1 - A_2\sin\psi_2) \end{aligned}\right\}. \tag{39}$$

The power requirement may be calculated by writing out the usual relationships for the force balance, when L must balance the weight minus any vertical components of the extra-lift drags, and D balances the remaining $\mathbf{D}_{par}$ and $\mathbf{D}_{pro}$ on the body and wings. If $\mathbf{D}_{pro}$ and $\mathbf{D}_{par}$ can be estimated, Eq. (39) can be solved for Γ_0 and one more unknown, which may be ψ, determined by the wingbeat kinematics, or $A_{1,2}$, determined also by the wing *shape*. The ratio of $A_1 : A_2$ in Fig. 15 depends on the ratio of the flexed wingspan to the outstretched wing span, and since the span is reduced mostly by flexion of the primary feathers, simple estimates of the flight performance can be calculated in terms of this ratio of wing planforms. For example, if $\psi_1 = \psi_2$, and the ratio of the upstroke to downstroke wake areas, $A_2/A_1 = \zeta$, ($\zeta \leqq 1$), then the lift: drag ratio may be expressed as

$$\frac{L}{D} = \left(\frac{1+\zeta}{1-\zeta}\right) \cot \psi. \tag{40}$$

L/D depends only on ζ and ψ, and it will increase with increasing ζ, and with decreasing ψ. When the L/D ratio is of concern, such as in long distance migration, the relationship above would predict that a sensible strategy would be to reduce the wing beat amplitude and flexion of the wing as far as is possible while still generating the required thrust. Alternatively, if ζ and ψ can be measured, then the values of L/D, or of D_{pro} and D_{par} implied or assumed in a flight model may be checked. Pennycuick (1989b) has done this to find significantly higher L/D ratios than predicted by simple actuator disc theory in the foraging flight of cormorants.

The roller-coaster model has a convenient balance between complexity and realism on the one hand, and simplicity and tractability on the other. The wingbeat amplitude and frequency and wing shape are represented as parameters affecting the wake shape, but since the circulation of the wake vortices is constant, the calculations are simple to perform. A comparable wake has been reported in the medium speed flight of bats (Rayner et al. 1986), and together with the observation that the cruising flight of a kestrel seems unremarkable and similar to that of many other birds in medium speed flight, it is likely that this wake structure (and hence valid application of the roller-coaster model) may be quite widespread. If the kinetic energy of the vortex wake essentially represents lost energy, then the roller-coaster wake without transverse vortex elements reflects a more efficient transport of material per unit distance than if the cross-stream elements were present. On these grounds, one might expect this wake to be present in *most* instances of long distance commuting or migrating flight.

Insect wings cannot flex in this fashion to reduce the wingspan, and so the required asymmetry must be achieved either from variations in the effective angle of attack, or the relative time of the down- and upstrokes. In either case, cross-stream vortices will be shed, and an accurate model should reflect this.

Swifts have curved, high aspect ratio, relatively rigid wings and may be unable to generate a roller-coaster wake. If so, their wings should be interpreted as extreme specialisations for soaring and high manoeuvrability, but perhaps accompanied by a different strategy for reduction of wake energy losses, via the parabolic lunate planform, which we shall discuss later. Hummingbirds have developed rigid wings with a low moment of inertia and extreme mobility of the shoulder joint, and their hovering flight resembles more that of insects than other birds. This specialisation for hovering apparently excludes them from the roller-coaster wake mode of flight, and is correlated with a rather limited record of nonstop migration. Altough the noctule of Rayner et al. (1986) was reported to be generating this wake, the limited ability of bats to reduce the wingspan beyond a minimum required for maintaining tension in the wing membrane, may also be predicted to limit their use of this mode flight, and correlates with a similar lack of notable migratory performance.

These issues have been discussed by Pennycuick (1988), who also speculates as to the possible wake patterns in extinct flying animals.

Rayner (1986) has incorporated this model wake structure into a quasi-steady lifting line analysis, where the form of $\Gamma(y)$ was taken to be that described by Jones (1980), and Γ_0 and ϕ were solved for a given frequency, n. $P(U)$ curves could be constructed from those combinations of n and ϕ that minimised the mechanical power requirement at each U. The inertial forces and wing root bending moments were also computed, and inertial components were found to be large, comparable to the aerodynamic forces in magnitude, as reported also by Bilo et al. (1984). Inertial power requirements have customarily been assumed to be negligible in forward flapping flight, but in the light of these results, this might not be correct for birds and bats executing large amplitude wing motions at low flight speeds.

5.5 Unsteady Lifting Line and Relatives

LCM. The Local Circulation Method has been developed and applied to the forward flight of dragonflies by Azuma et al. (1985) (LCM I), and refined by Azuma and Watanabe (1988) (LCM II). It combines the approach of the blade element analysis of wing sections with a more realistic and complete analysis of the modifying effects of the unsteady (time-varying) wake. The wing planform and initial circulation distribution is approximated by the superposition of a series of elliptical load distributions of diminishing size, each operating in the appropriate local shear flow. The wake model consists of the trailing and transverse (LCM II) vortices fixed by the path of the wing tip and trailing edge (Fig. 16a), and an iterative procedure is applied for the wake corrections to the circulation distribution, along the lines of Sect. 3.2. Nonlinear, empirical $c_l(\alpha)$ curves were used to compute the forces and moments on the wings at each blade element. Since the computations are quite lengthy, the wake attenuation coefficients were actually calculated only at $0.75R$.

The LCM I method was applied to the slow climbing flight of a dragonfly, for which k and Ω, as defined here, were approximately 1.6 and 16. Both wings operated well away from the linear $c_l(\alpha)$ range for a substantial portion of the wingstroke. The peak interference of the forewing wake on the hindwing was in mid-upstroke of the hindwing, and the wings operated with a phase difference so that the pitching moment was close to zero and a small net horizontal force was generated. First-order unsteady effects were accounted for and shown to be significant improvements over actuator disc-based estimates of constant downwash over the span, but no other special high-lift mechanisms requiring large scale separated flows were required to balance the forces in the analysis. The conclusions were similar for the LCM II study of the free flight of a dragonfly in a wind tunnel ($0.2 \leqq k \leqq 1.2$, $3.2 \leqq \Omega \leqq 12$). It is hard to gauge the significance of the additional corrections in LCM II since the two flight behaviours, and the reported curves of the total aerodynamic and inertial torques are quite different. The comparison of LCM II with a standard blade element calculation (constant w_i) at the highest flight speed, $U = 3.2\,\mathrm{m\,s^{-1}}$, and consequently lowest k and Ω ($=0.2$ and 3.2)

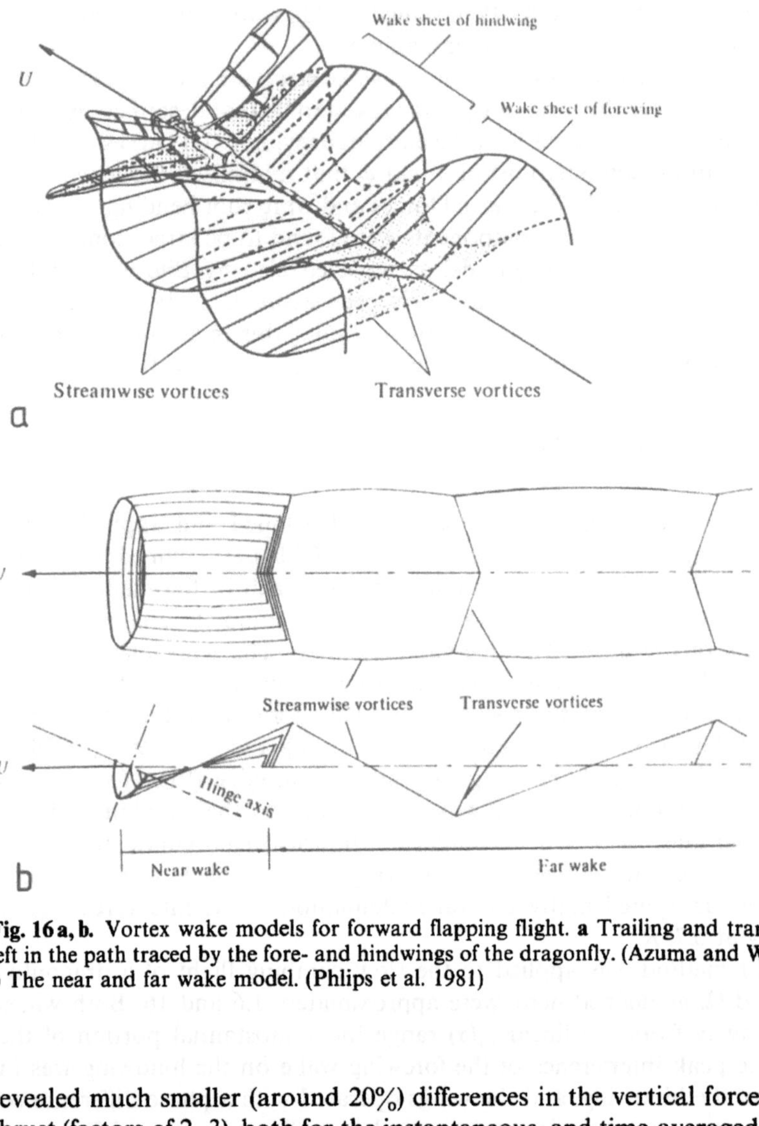

Fig. 16 a, b. Vortex wake models for forward flapping flight. **a** Trailing and transverse vortices left in the path traced by the fore- and hindwings of the dragonfly. (Azuma and Watanabe 1988). **b** The near and far wake model. (Phlips et al. 1981)

revealed much smaller (around 20%) differences in the vertical forces than in the thrust (factors of 2–3), both for the instantaneous, and time-averaged values. Note that neither correction can be considered negligible. The estimated power requirements were well within the range for insect flight muscle.

Vortex Lattice and Lifting Line Methods. This group of techniques attempts to extend the standard lifting line of Sect. 3.3 to oscillating or flapping wings with geometry and kinematics closer to that found in nature. In contrast to the modified blade element approach, detailed kinematic data are not taken from experiment, but analytical forms of the planform geometry and wing motion are prescribed. Usually these take the form of linearised, small departures from the straight-edged, planar wing in steady motion, but the capacity to account for larger and more realistic perturbations is growing with increasing sophistication of the analysis. The considerable advantage is that detailed parametric studies may be made of the effect of changing lifting surface geometries and kinematics. Also, the magnitude of the unsteady and *3-D* corrections to the quasi-steady analysis can be explicitly calculated, lending a more solid basis for the sometimes rather speculative discussions of the likely contributions of unsteady effects. Even when the strong nonlinearities of the true complex geometry and kinematics of natural flight exceed the domain of the model calculations, they can still provide a baseline estimate of the lower order effects, upon which further corrections can be superimposed. It is common to express the results in terms of the propulsive efficiency, $\eta \equiv U\langle T\rangle/\langle P\rangle$, where $\langle T\rangle$ and $\langle P\rangle$ are the time-averaged thrust, and the power required, respectively.

1. *Quasi-vortex lattice.* Lan (1979) formulated a version of a doublet lattice method for an oscillating flat plate wing, where, instead of representing a wing in steady uniform flow by a continuously-loaded lattice (Sect. 3.2), the oscillating surface was covered by a time-varying doublet lattice, and the system was solved for zero normal velocity at the lifting surface, as usual. When the pitch axis was fixed at the trailing edge, the sensitivity of a swept wing to phase differences in pitch and thrust was much reduced compared to an unswept rectangular wing. The swept wing always produced positive thrust while the rectangular wing occasionally produced a negative thrust i.e. a drag. This effect was more pronounced as k increased from 0.15 to 0.75. The interaction of pairs of rectangular wings oscillating in tandem was investigated, and the efficiency, η, and thrust coefficient, $\bar{C}_T$, were plotted as a function of phase difference, ϕ_{hf}, between the fore and hind wings, for different reduced frequencies and different gaps between the wing chords. ϕ_{hf} for optimum η and $\bar{C}_T$ decreased, both as k increased and as the separation distance decreased. On average, the predicted optimum ϕ_{hf} for energy extraction by the hindwings from the wake of the forewings was 45° and 90° for maximum $\bar{C}_T$ and η respectively. Azuma and Watanabe (1988) reported values of ϕ_{hf} between 51° and 93° for their dragonfly in free flight for $U = 0.7\,\mathrm{m\,s^{-1}}$ and $3.2\,\mathrm{m\,s^{-1}}$ respectively. The smaller ϕ_{hf} to maximise $\bar{C}_T$ rather than η is consistent with the changes in ϕ_{hf} of dragonflies in various flight maneouvres noted by Alexander (1984), and also with the qualitative trends reported by Ruppell (1989).
2. *2-D unsteady motion.* The basic principles of a high AR wing should be accessible to a *2-D* analysis, and detailed performance analysis and comparisons of *2-D* rigid and flexible aerofoils in unsteady motion have been made, following the lead of Lighthill (1970). These have been applied to problems of aero- and hydrodynamics, and the previous chapter may be consulted for further details. The more simple analysis allows a quite rigorous and thorough approach towards optimising the kinematics of the motion. Yates (1986) has considered in detail the optimum location for the pitch axis in a *2-D* analysis of a heaving and pitching fin or wing. The analysis included the physiologically inspired quantities of mean square pitching moment and mean square power requirement for maintenance of lift, which are of obvious relevance to animal flight. When

a mean lift force requirement was considered, the optimum pitch axis location always moved towards the quarter chord location. This corresponded closely to the presumed torsion axis at the location of the larger bones in vertebrate wings. Detailed static tests of torsional rigidity (Ennos 1988a, b) and estimation of the forces acting in Dipteran flight (Ennos 1989) indicate that the torsion axis typically lies at $0.15c$, while the centre of mass is located at about $0.3c$. It was noted that if the aerodynamic centre of presure is at $0.25c$, then local torques about the torsion axis will be created by both aerodynamic and inertial forces set up along the span during the wingbeat. The changes in local pitch could also explain the tip to base torsion waves observed in insect wings.

3. *Quasi-steady lifting line* ($k \ll 1$). Betteridge and Archer (1974) and Archer et al. (1979) applied the classical lifting line analysis to rigid and flexible wings in small amplitude flapping flight, and showed increased η with increasing k, and predicted high values of local wing twist, to maintain an elliptic loading profile. Based on similar model assumptions, Jones (1980) re-examined the optimum load distribution required to minimise the induced drag for a given wing root bending moment, and found that a 10% increase in thrust could be obtained, for the same η, for a $\Gamma(y)$ slightly reduced close to the base, and increased at the tip, compared to the elliptic distribution. Recall from Sect. 3.2 the preferred reduced tip loading for induced drag reduction in the steady lifting case. On these grounds, one would predict that specialised steady gliders/soarers will have more pointed wingtips than those that resort to flapping flight more often. This would appear to be the case, as we shall see in Fig. 18.
4. *3-D unsteady lifting line.* For the cases of most relevance to natural flight, a subset of three of the domains identified by Cheng (1976) in the asymptotic analysis can be identified for similar wings of high *AR*, characterised by the relative magnitudes of the chord length, c, the semi-span, b, and a wavelength in the wake, $\lambda = 2\pi U/\omega = \pi c/k$:

$$
\begin{array}{lll}
\text{I} & c \ll b \ll \lambda & \Omega \ll 1 \\
\text{II} & c \ll b = O(\lambda) & \Omega = O(1), \quad k \ll 1 \\
\text{III} & c = O(\lambda) \ll b & k = O(1).
\end{array}
$$

Domain I includes the limit as $\Omega \to 0$, and the flapping frequencies may be regarded as sufficiently small for the *3-D* problem to be treated as quasi-steady. In domain II, the flow next to the wing is nearly quasi-steady, while the velocity induced by the unsteady wake vorticity is as large as that induced by the trailing vortices. The roller-coaster kestrel wake was observed for $k = 0.26$, $\Omega = 2.32$, and we have already argued that this constitutes a range of some importance. Cheng and Murillo (1984) and Karpouzian et al. (1990) developed and examined the performance of an unsteady lifting line with centreline curvature for $\Omega = O(1)$, showing that the balance of *3-D* corrections with the leading order thrust terms was responsible for an optimum degree of sweep (K_{opt}) for maximum η, when a reasonable, nonzero mean thrust was required. The difference in η between the *2-D* quasi-steady calculations, and those with *3-D* and unsteady corrections amounted to about 10% at K_{opt}, for $\Omega = 1.5$. Around these values of K, the parabolic lunate planform was superior to a V-shape when no motion was allowed at the root. The identified optimum K would appear to be similar to the planforms of the swift and swallow families [Fig. 6(ii)], but this has not been rigorously checked. In domain III, Cheng (1976) showed that the global effects of the unsteady far wake were self-averaging, and that a *2-D* analysis is therefore superior for a straight wing (adding justification to the *2-D* investigations outlined above), but that local *3-D* corrections remained important for swept wings, as is true also of the quasi-steady domain I.

A *3-D* unsteady lifting line theory for planar, rigid, unswept wings of high AR has been developed by Phlips et al. (1981). It is based on a simplified model of the vortex wake, which is divided into near and far wake regions (Fig. 16b). Transverse vortices are shed at the extremes of the wingstroke, where they roll up, and are represented by a single line-vortex. Streamwise vortices are shed at the trailing edge and are assumed to remain in the path traced by the wings. In the far wake, they collect into trailing vortex lines. This model simplifies the calculation of wake induced velocities on the flapping wing, and the normal force on the wing can be calculated assuming the standard linear $c_l(\alpha)$ relation (Sect. 3.1). These assumptions effectively restrict the analysis to moderate values of k. When the wingbeat amplitude, ϕ, exceeded approximately 60°, significant departures from the steady state calculations were reported, particularly for $k > 1$. The effect of the induced velocity field in the unsteady calculation was to increase the value of the mean lift coefficient ($\approx 20\%$ for $\phi = \pi/2$, $k = 2$), but the mean thrust coefficients were reduced slightly, so that the efficiency, η fell with increasing k. This is a first order unsteady theory only, as spanwise flow along the wing and flow separation are ignored, but the clear implication is that unsteady terms cannot be neglected for $k \geqq 1$.

5.6 Unsteady Effects in Forward Flapping Flight

Some interim conclusions, summarising and generalising the results of the experimental and theoretical work above, can be stated. Note first that there are two contributors to unsteadiness in the aerodynamics. The first is that due to the kinematics of the lifting surfaces, and is measured by the frequency parameters k, Ω and K. The mean errors in time-averaged lift force, pitching moment, and aerodynamic torque calculations from quasi-steady analysis might be expected to be below 10% for $K \leqq 0.75$ ($J \geqq 1.5$), but *detailed* analysis and understanding of the effects of the wing kinematics and geometry require unsteady and *3-D* corrections in this domain, corresponding approximately to domain II above, where $k \ll 1$, $\Omega = O(1)$. In the cases of insect flight, ϕ varies little, and while the time-averaged corrections for $k \approx 0.1, \Omega \approx 1$ might be small, instantaneous forces and moments are unlikely to be correctly represented in the quasi-steady, *2-D* analysis. This covers most of the range of insect forward flight.

The second contributor to unsteadiness on the aerofoil comes from time varying shedding of vorticity at separation points on the lifting surface other than the trailing edge, about which the lifting-line models and current generation of experimental techniques can say very little. Various high-lift mechanisms may be postulated, as in the hovering flight section, Sect. 6.4. Strongly separated flows could greatly affect the time-varying forces and moments on the wing. However, for the moment, the first order corrections due to the vortex wake can be used as lower bounds estimates for the unsteady effects. Azuma et al. (1985), using familiar logic, have argued that strongly separated flows need not be postulated for the dragonfly, as the forces balance correctly with only wake vortex corrections. Be that as it may, the actual details of the flow and the effects of unsteady separation on the instantaneous forces on a flapping animal wing for $k \geqq 0.1$ are not currently known with any great precision or certainty.

5.7 Flapping Flight Performance

The Aerodynamic Power Requirement. Just about any flight model generates a characteristic U-shaped curve of the total aerodynamic power requirement, P, with flight speed, U. The reasons for this may be found in Eq. (15), which in turn is based on the fundamental lifting-line analysis of Sect. 3.2. The combined profile drags on the wings and body increase by some constant times U^2 and so the power required, $P \doteq DU$, rises with the third power of the velocity. By contrast, the induced drag is proportional to $1/U^2$. Any model – blade element, momentum jet, vortex ring, roller-coaster, LCM, unsteady lifting line – will consequently generate a generic, U-shaped power curve similar to Fig. 17, given some reasonable set of data, and the cost of transport, $C = P(mg \cdot U)$, will usually also possess a well-defined minimum. The references for each flight model may be consulted for detailed comparisons. Two general points may be made.

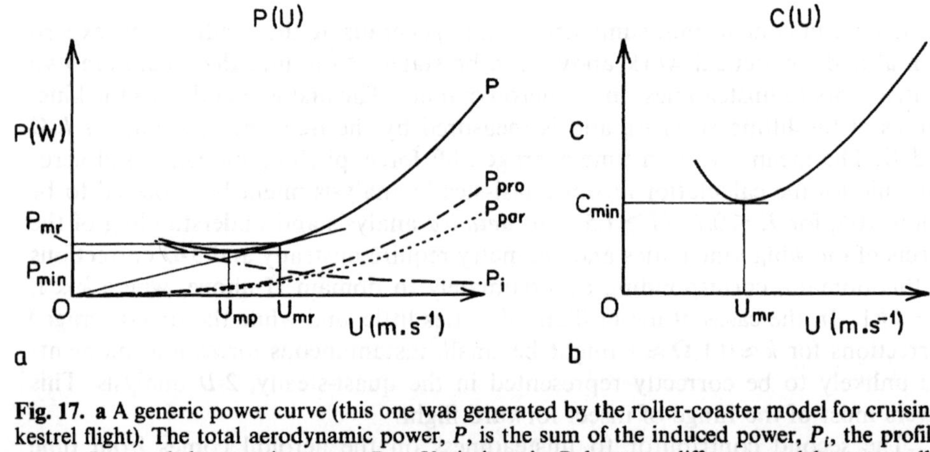

Fig. 17. **a** A generic power curve (this one was generated by the roller-coaster model for cruising kestrel flight). The total aerodynamic power, P, is the sum of the induced power, P_i, the profile power, P_{pro}, and the parasite power P_{par}. Variations in P_i between different models are usually small compared to differences in the calculation of P_{pro} and P_{par}. **b** The dimensionless cost of transport, C, has a well-defined minimum at U_{mr}

1. *Scale effects.* Equation (17) in Sect. 4.1 may be rearranged to show that a characteristic flight speed, such as that which maximises C_L, for example, is related to the square root of the wing loading, i.e.

$$U_{C_{L,max}} \propto \left(\frac{mg}{S}\right)^{1/2}.$$

In geometrically similar animals, since $m \sim l^3$ and $S \sim l^2$, then $U_{C_{L,max}} \sim l^{1/2}$. The dimensional power is $F.U$ and as $F \sim m$, then the mass-specific power requirement is also proportional to $l^{1/2}$. The maximum mass-specific power output of vertebrate and insect muscle actually decreases somewhat at contraction frequencies, $n < 10$ Hz (Weis-Fogh and Alexander 1977), and thus will decrease slightly with m, so at some size, the power required must exceed the power available from the flight muscles. The implications of these types of scaling relations have been explained by Penycuick (1975), who derived them from his momentum jet model. The power available, relative to any characteristic power requirement, is larger for small flying animals than larger ones. Small birds can hover or fly very slowly for small periods, while larger ones cannot. The larger birds have a very limited range of flight speeds available to them for continuous flapping flight, and show extreme specialisations for taking maximum advantage of gliding and soaring techniques. Deviation from geometric scaling, as larger gliding birds have higher aspect ratio wings, is clear in Fig. 18. Moreover, all the larger species in the left column of Fig. 18, albatrosses and the giant petrel, were found to possess a mechanical lock capable of resisting wing elevation above the horizontal. On the other end of the scale, a hummingbird with $m \cong 5$ g is capable of continuous aerobic hovering flight, although the largest of these, *Patagona gigas* ($m \cong 20$ g), appears barely capable of this. One should be wary of generalising across phyla, but these simple scaling laws are sufficient to account for the fact that almost all insects can hover continuously.

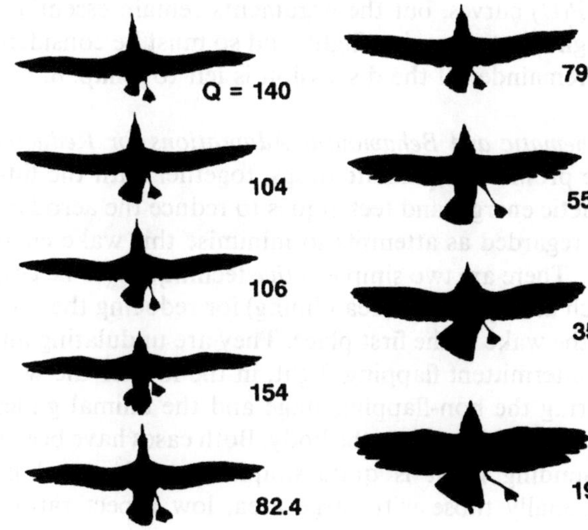

Fig. 18. Allometry in the wings of Procellariiformes. The outlines are rescaled by wingspan, $2b$. AR and $2b$ vary from 15 and 3 m in the wandering albatross at *top left*, to 8.04 and 0.39 m in Wilson's petrel. *bottom right*. The wing loading, Q, is shown for each species. (After Pennycuick 1982)

2. *The shape of the power curve.* The second point concerns the agreement between aerodynamic power requirements and metabolic measurements. In general, metabolic measurements result in U-shaped power curves (consult Norberg (1990) for a sample), but the shapes are often flatter than predicted. From the aerodynamics point of view, two potential contributors to such a trend might be noted: (1) *Re*-associated changes in the flow over the wings and/or body cause C_{Dpro}, C_{Dpar} to decrease significantly with U. (2) The flight style changes with U. (1) has been discussed in Sect. 3.1. Transition to turbulence around Re_{crit} can result in a strong inverse dependence of C_{Dpro} on U, and soaring birds appear to operate very close to that value. The most detailed measurements to date also indicate that, at high *Re*, C_{Dpar} is almost constant with increasing U, or falls slightly with *Re*. Even at the lower *Re* of insect flight, Dudley and Ellington (1990b) measured reductions in C_{Dpar} of approximately 40 and 25% as *Re* increased from 1×10^3 to 1×10^4, for body inclination angles (α_0) of 60° and 0° respectively, in their bumblebee experiment. The magnitude of the reduction and its dependence on α_0 is reasonably consistent with the expected behaviour of separated flows over bluff bodies at this *Re*. The result also suggests that changes in α_0 with U should be properly accounted for. (2) remains a possibility. D_{pro} on the wings could fall with increasing U if the wing span, $2b$, and so the wetted surface area, S_w, are reduced by the live animal, although the concomitant increase in D_i, together with associated changes in wing twist and local angle of attack, complicate the picture. While certain wingbeat parameters, such as the stroke amplitude ϕ, and the downstroke ratio, τ, are allowed, if not required, to vary with U in various flapping flight models (Rayner 1979b), others, such as the wingspan, b, are typically held fixed. The availability of families of glide polars with variable geometry wings is well established (Tucker 1987), but similar effects of systematic variations in wing planform geometry in flapping flight at different speeds have yet to be investigated. This could reduce the penalty for flight away from U_{mp} and would also contribute to scatter in the metabolic measurements. Both postulated mechanisms in (1) and (2) could flatten out the $P(U)$ curves, but the arguments remain essentially extrapolations from steady gliding to flapping flight, and so must be considered somewhat speculative. The remainder of the discussion is left to Chap. 8.

Kinematic and Behavioural Adaptations for Reducing Energy Consumption. Both the profile and parasite drags, together with the lift-induced drag, generate wake kinetic energy, and techniques to reduce the aerodynamic energy expenditure may be regarded as attempts to minimise this wake energy loss.

There are two simple *active* techniques (to be contrasted with *passive* measures such as improving streamlining) for reducing the total amount of energy deposited in the wake in the first place. They are undulating and bounding flight, both forms of intermittent flapping flight; in the former the wings are extended into the flow during the non-flapping stage and the animal glides, while in the latter they are folded tightly against the body. Both cases have been considered by Rayner (1985b). Bounding flight is quite simple to understand in principle. For small birds, especially those with large area, low aspect ratio wings, the wing profile drag comprises a large fraction of the total aerodynamic drag, and at high flight speeds the increase in induced drag required to generate sufficient force for steady flight in only a portion of the wingbeat is more than offset by the significant reduction in wing drag during that time. For this to be a viable option, the flight speed must be high so the relative cost of induced drag is low. Careful considerations reveal several subtleties in the cost-benefit analysis, however, including the idea that muscle efficiency might be optimised at a fixed flapping frequency. Rayner's paper should be consulted for details. Bounding flight is common in passerines and also in hummingbirds, among others, but has not been recorded in bats. An intriguing variation on this theme has been reported by Betts and Wootton (1988) where the fast flying, low aspect ratio skipper butterfly (*Idmon* sp.) was reported to have its wings closed above the abdomen for $2T$ in intermittent flight.

A realistic analysis of undulating flight is quite complicated. It appears that when the wing loading is low and/or aspect ratio is high, it is more economical to alternate powered climbs, and gains in potential energy, with slower glides close to the minimum drag speed, than to fly at a constant flapping rate. This can be true when the maximum range speeds for flapping and gliding are different, and switching between the two is accompanied by potential and kinetic energy exchange. The allometric relationships manifest in Fig. 18 result in undulating flight being commonly observed in the larger seabirds, and it ought to be noted that this type of behaviour pattern also allows for the potential exploitation of updrafts, either in a systematic fashion such as when pelicans flap in the intervals between wave soaring, or on a more opportunistic basis. Undulating flight has been observed in bats (Thomas et al. 1990) and in butterflies (Betts and Wootton 1988), both of which have quite low wing loading.

There are further possibilities for energy extraction from the wakes of other flying colleagues by selecting a position such that the wing experiences the updraft created in their wakes, but not the downdraft (Lissaman and Schollenberger 1970; Hingdon and Corrsin 1978). Formation flight in birds has often been interpreted in this way, but the positioning accuracy required is quite considerable, and the likely benefits of attempting to track a complex undulating wake profile in flapping flight are not immediately obvious. One could perhaps consider the incentives to remain *away* from downdrafts, commensurate with the behavioural advantages of remaining in a flock, as restricting the domain of sensible locations. Formation flight has not been reported for insects.

Finally, the interaction of a trailing vortex pair with a wall boundary, such as the ground or sea surface, causes a reduction in the induced downwash on the wings, and consequently reduces the drag. The cancellation of normal velocities at the impermeable surface can be modelled by the addition of an imaginary mirror image vortex system an equal distance beneath the surface, where the induced velocities will therefore be exactly zero. For a simple horseshoe vortex, the ratio, R, of induced drag in ground effect to induced drag out of ground effect may be estimated by

$$R = \frac{C_{D_{i,g}}}{C_{D_i}} = \frac{(16H)^2}{(1 + 16H)^2}, \tag{41}$$

where $H = h/2b$ is the ratio of the flying height to wingspan. For entirely plausible values of H of around 0.2 (if $AR \approx 10$, this is 2 mean chord widths), $R \cong 0.6$, a substantial reduction. Aircraft must routinely take this into account on take-off and landing, and various sea birds, such as brown pelicans, clearly profit by flying at or below these values of H. Hainsworth (1988) has measured typical gliding heights of pelicans in ground effect, and Blake (1983) has discussed the topic with reference to bird flight in some detail.

6 Hovering

6.1 The Flow Regime

Hovering flight can be defined as the ability to remain airborne in a fluid with zero or negligible mean relative velocity between the body and the fluid. Since the forward speed of the body contributes nothing to the flow over the wings, hovering is the most energetically demanding of flight styles. From the scaling arguments presented in the forward flight analysis, this mode of flight will clearly be restricted to smaller flying animals, mostly insects and hummingbirds, and the latter may be considered as honorary insects for our purposes. Momentary hovering is typically unavoidable in landing and take-off and must be accomplished by most flying animals that cannot rely on special mechanisms involving either the storage of potential energy (dropping from a height) or the extraction of reliable sources of kinetic energy in the environment (soaring). Hovering and slow forward flight have been instrumental in enabling insects to occupy microniches not available to larger flying animals. From an analytical perspective, some simplifications arise since the form drag on the body may be neglected, but the importance of accurate estimates of the lift-induced power, and by implication, realistic models of the wake vorticity distribution, is greatly increased.

Hovering flight may be defined as all cases where $\Omega \geq 20$, following the pragmatic example of Ellington (1984, III). Since the reduced frequency parameter Ω is infinite when $U = 0$, it is no longer particularly helpful in distinguishing amongst the different hovering regimes. A dimensionless amplitude may be defined by the number of mean chord lengths travelled by the mid-point on the wing, $R/2$:

$$\bar{\Lambda} \equiv \frac{\phi R}{2\bar{c}}. \tag{42}$$

This is also a measure of the mean distance of a wing chord from the unsteady influence of its own starting vortex. Similarly, the instantaneous position from the starting point at the end of a half-cycle can be expressed as the ratio, $\lambda = \gamma r/c$. In comparisons with *2-D* experiments, we let $\Lambda = h/c$, where h is the oscillation amplitude of a chord motion given by $x = h \sin \omega t$.

6.2 Quasi-Steady Analysis

Normal Hovering. To provide weight support to the animal, there has to be a horizontal component of the wing motion, and in the absence of any net horizontal body motion, the wings themselves must therefore beat in a horizontal plane. Acting within the mechanical constraints of the musculo-skeletal system, the simplest way of achieving this is for the body to tilt toward the vertical, and this indeed is the mode of hovering for hummingbirds and for several orders of insects analysed in classic papers by Weis-Fogh (1972, 1973) and Ellington (1984, I–VI). Figure 19 shows the course of one wingbeat cycle in a hovering sphingid moth. The time is spent approximately equally on the morphological up- and downstrokes, at the ends of which the wings twist and rotate so that the stiffened anterior part of the wing functions as the aerodynamic leading edge throughout. The wing motion relative to the body is often assumed to follow a figure-of-eight path, as reported for the hummingbird (Stople and Zimmer 1939), but the measurements of Ellington (1984, III) on the kinematics of insects *in free flight* show that this is not always so. Weis-Fogh (1973) introduced some very elegant simplifications to the basic blade element analysis leading to Eq. (34). The instantaneous lift δL on a wing section (blade element) δS can be expressed as:

$$\delta L(r,t) = \tfrac{1}{2}\rho \times C_L(r,t) \times \delta S(r) \times r^2 \left(\frac{d\gamma}{dt}\right)^2 . \tag{43}$$

First, based on observations of the kinematics, C_L was assumed to be constant over both the span, R, and over the stroke period, T. δS is determined by the

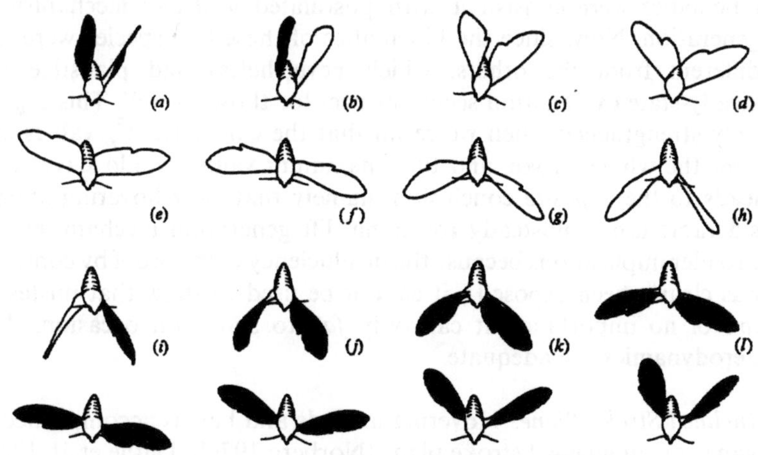

Fig. 19a–p. Normal hovering in *Manduca sexta*, as traced from every 10th frame of high speed cinefilm at ≅ 3450 frames s^{-1}. The view is from above, and the undersides of the wings are in *black*. (Weis-Fogh 1973)

second moment of area of the wing, for which analytical expressions were derived for simple wing shapes. The mean lift over one wing stroke is thus

$$\bar{L} = \tfrac{1}{8}\rho\omega^2\phi^2\bar{C}_L \times \int_0^R c(r)r^2\,dr + \frac{1}{T}\int_0^T \cos^2 \omega t\,dt, \tag{44}$$

and the first integral is equal to $\sigma c R^3$, where σ is the wing shape factor, and the second evaluates exactly to 1/2. Since $\bar{L}$ must balance half the weight if there are two wings, the mean lift coefficient is

$$\bar{C}_L = \frac{8W}{\rho\omega^2\phi^2\sigma c R^3}. \tag{45}$$

This is an extremely simple expression for estimating the mean lift coefficient required to support the weight of a hovering animal, given just the kinematic parameters in the denominator. In the majority of cases, normal hovering could be satisfactorily explained on steady-state principles, and the required lift coefficients appeared to be quite modest (from 0.8 at Re_c 1600, to 1.2 at Re_c 6700). However, all of the cases where the wingbeat was inclined significantly from the horizontal required abnormally large values of $\bar{C}_L$ and the steady state hypothesis had to be rejected. This included the bat *Plecotus auritus*, the butterfly *Pieris napi*, the hoverflies (Syrphinae) and the now famous chalcid wasp, *Encarsia formosa*.

These arguments have been re-evaluated by Ellington (1984, I–VI), based on meticulous kinematic and morphological measurements. In addition to pointing out certain methodological inconsistencies, he noted that the ladybird (*Coccinella 7-punctata*) and the crane-fly (*Tipula obsoleta*) would require mean lift coefficients of 1.7 and 1.2, respectively, too high at $Re_c \cong 10^3$. However, the measured wingbeat kinematics *were* consistent with postulated *unsteady* mechanisms for extra lift generation. Now, since the kinematics of these two species were not appreciably different from the others, which nevertheless had plausible $\bar{C}_L$ values, the steady-state explanation seems suspect for those as well. This argument, which is only strengthened when we recall that the calculated $\bar{C}_L$ values are *mean* values over the whole stroke, and that maximum values would have to be higher still, leads to the *opposite* conclusion, namely that *most* hovering animals make considerable use of unsteady rotational lift generation mechanisms. There are even broader implications because the insufficiency of the proof by contradiction method has clearly been exposed – it cannot be used to show that unsteady mechanisms are of no importance, it can only *fail* to show, on occasion, that steady-state aerodynamics are adequate.

Inclined Stroke Plane. Hovering in birds and bats is accomplished by beating the wings on an inclined stroke plane (Norberg 1976b; Dathe et al. 1984). The wingtip traces out a path with an average angle of 20–30° to the horizontal, and the wing is flexed and folded during the upstroke so that, although the upstroke may generate some lift (Dathe and Oehme 1978) it is much less than on the downstroke, and a crucial asymmetry is introduced into the contribution to weight support.

Table 1. Mean downstroke lift coefficients and dimensionless amplitudes for horizontal and inclined stroke plane hovering

Species	Reference	β(degrees)	$\bar{C}_{Ld}$	$\bar{\Lambda}$
Bombus hortorum	EL84	−15	1.1	3.4
Episyrphus balteatus (HF07)	EL84	−2	1.17	3.4
(HF08)		21	3.93	2.3
(HF08)		32	5.08	2.4
Aeschna juncea	NA75	60	3–4*	2.5
Amazilia f.f	WF72	11	2.3*	3.6
Ficedula hypoleuca	NU75	30	6.0*	1.98
Columba livia	DA82	38	2.6	4.6
Larus ridibundus	DA82	41	2.0	4.4
Plecotus auritus	NU76	30	4.3*	3.2

References: WF82, Weis-Fogh (1972); EL84, Ellington (1984, VI); NA75, R. Norberg (1975); NU75, U. Norberg (1975); DA82 Dathe (1982); NU76, U. Norberg (1976b). The asterisked quantities were recalculated in Ellington (1984).

In insects, inclined stroke plane hovering is characteristic of hoverflies and dragonflies, where α_{eff} can remain high on the upstroke, but the vertical component of **F** will still be reduced. Conventional aerodynamic analysis shows that this places unreasonable demands on the lifting performance of the wing on the downstroke, so that all animals using inclined stroke plane hovering must rely to some extent on unsteady lift generation mechanisms. Table 1 summarises some data from the literature and shows that estimated lift coefficients are uncomfortably or impossibly high wherever ß increases much beyond 15°. The value of 2.3 recalculated by Ellington for the hummingbird in the horizontal stroke plane suggests that here too the quasi-steady analysis fails.

6.3 Vortex Models

Theory. Specifically to remedy the deficiencies in the quasi-steady analysis, Ellington (1978, 1980) introduced and refined (1984, V) a vortex theory of hovering flight. The principle is to construct a reasonable and solvable mathematical model of the wake vorticity distribution, based on the wing beat kinematics. The wake must be allowed to convect with its own self-induced velocity, and the mechanical power requirement is then computed from the mean rate of increase of wake kinetic energy. The procedure is actually quite similar to that of the original momentum jet model, only now a more accurate wake model is required. There are two obvious corrections to be made to the steady, uniform momentum jet: (1) spatial corrections to account for the fact that the wings do not sweep over the whole disc area, and,

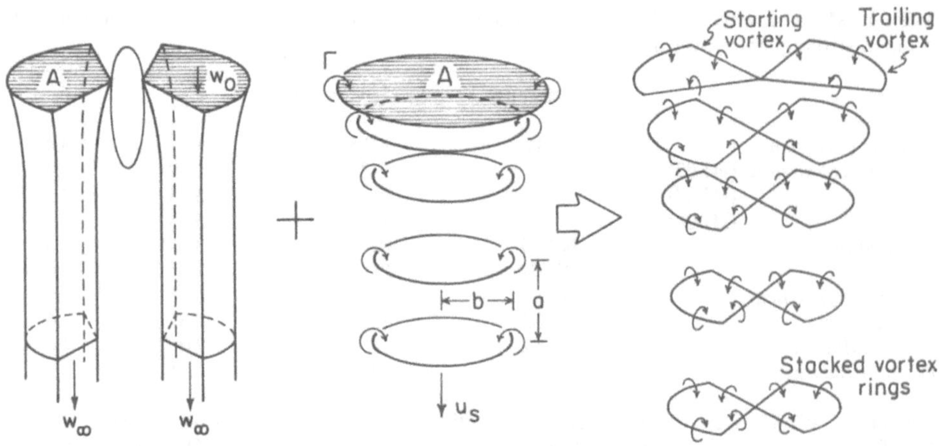

Fig. 20. The nonuniformities in space and time of the vortex wake of a hovering animal are modelled by the superposition of two independent corrections to the uniform momentum jet. (Assembled from Ellington 1978, 1980)

(2) temporal corrections to include the time-varying circulation distribution. In the limit of a wing pair beating through an angle of 2π, at an infinite frequency, we should recover the ideal momentum jet. A simplified illustration of these corrections is given in Fig. 20. Physically, since vorticity can only be created at the wing surface, and since the wings are oscillating in time with angular velocities falling to zero twice per wingbeat (when vorticity must be shed into the wake), the wake should be composed of a series of stacked, closed vortex loops, one completed beneath each wing on every half-cycle. Mathematically, the first order effect of these two small perturbations can be expressed as the addition of two linearly independent corrections to the steady momentum jet model for the power:

$$\bar{P}_i = \bar{P}_{i,mom}(1 + \sigma + \tau). \qquad (46)$$

σ is a *spatial* correction which depends on the ratio of circulation required to support the weight over the area actually swept by the wings ($A_0 = \phi R^2 \cos \beta$), to the circulation required if the whole disc ($A = \pi R^2$) were used. It is a positive constant, vanishing to zero when $A_0 = A$. τ is a *temporal* correction which may be expressed as $0.079 s^2$, where s is a parameter defining the wake element spacing, related to the ratio of the distances a/b in Fig. 20. When the spacing is zero, the wake is continuous and τ is zero also. Ellington (1984, V) should be consulted for details. There are two immediate benefits from this analysis. First, one has a method for calculating the induced power which takes reasonable account of unsteady aerodynamic effects because it is based on a reasonable model of the vortex wake. This analysis does not require assumptions concerning quasi-steady section lift coefficients. Second, a more accurate estimate of the true induced velocities and explicit corrections to the steady-state models enable *quantitative* estimates of the effect of these steady-state assumptions to be made. Ellington estimated that the corrections were around 20% for normal hovering, and up to 60% for the inclined stroke plane cases.

Rayner (1979c) formulated a model for hovering animal flight based on a wake of stacked, coaxial, small-cored vortex rings. Provided that the core radius is small compared to the ring radius, the system is effectively determined by two parameters; the feathering parameter, $f = w_i^2/|\mathbf{u}_t|^2$ is the square of the ratio of the steady induced velocity in the momentum jet to the mean wingtip velocity, and, typically, is almost an order of magnitude smaller in insects than in birds. R' is a measure of the ring size with respect to the span, b; it depends on $\Gamma(y)$ on the wing, and also on the area actually swept out by the wings, i.e. on ϕ. Reasonable values seem to be around 0.75. The correction σ_0 to the actuator disc induced power requirement ($P_i = \sigma_0 P_{i,mom}$) could be written

$$\sigma_0 = \frac{0.95}{R'} + \frac{1.2}{R'^5} f. \tag{47}$$

For insects, σ_0 is close to 1, and corrections to the steady momentum jet model were estimated to be typically 10–15%. Calculations for Norberg's *Ficedula* data, on the other hand, showed $\sigma_0 = 1.8$, a considerable difference from the momentum jet model, implying significantly increased wake energy losses. This is related to the increased wake element spacing, which in turn derives from the inactive morphological upstroke. The implication is that "avian" hovering places unattainable demands on the aerobic power output of the flight muscles of most birds, which, therefore, do not hover, except momentarily. In general, where the vortex models of Ellington and Rayner overlap, as in the case of hovering insect flight, the numerical results and conclusions are quite similar. A detailed comparison appears in Ellington (1984, V).

It should be noted that the vortex wake models are not free of assumptions and simplifications. Although it is true that no steady-state wing section lift coefficients need be invoked, the generation of the wake elements with a certain size, R', say, with respect to the wing shape and kinematics *does* implicitly assume a certain circulation distribution on the wing in order to create this pattern of wake vorticity. This capacity for ignoring the complex details of the circulation and lift on the wing itself is precisely one of the strengths of vortex wake models, but the validity of the assumed wing circulation or core vorticity distributions is difficult to confirm. In view of the problems in interpreting the existing quantitative wake velocity measurements, the presumed roll-up of shed vorticity into simple wake elements with predicted R' cannot yet be regarded as verified. Despite these reservations though, the closed vortex loop wake models are easily the most realistic, solvable, formulations available for slow and hovering flight, and overall corrections due to complex wake geometries and deformations may yet prove to be small.

Experiment. Maxworthy (1979) constructed a simple *3-D* mechanical model of a pair of rigid wings which rotated through an angle of 2π in a horizontal plane, like an idealised hovering wing pair (the geometry is similar to the insert of Fig. 21). When the wings start and finish pressed together, the final structure in the wake is a large vortex ring, with a smaller one of opposite-signed circulation inside it, shed from the inboard trailing edge of each wing. This, together with observations of vortex rings in the wakes of birds in slow flight, indicates that complete vortex rings may result when the stroke amplitude, ϕ, is high. On the other hand, when the wing tapers to very small chord lengths at the root, or is separated from the body by a small gap, there is likely to be strong vortex shedding in the neighbourhood of the root, where the normal velocities are also small. Dathe et al. (1984) have observed single closed vortex loops beneath models of isolated rectangular wings beating in a horizontal plane, and this is to be expected when there is no possibility of mutual wing interference effects. In this respect, the flow is just as postulated in Ellington's vortex model, and will probably be the form of the wake when ϕ is small so that the wings operate in effective isolation, such as in the inclined stroke plane hovering of the hoverflies, for example.

It is extremely hard to visualise the wake of a real hovering animal. Ellington (1978, 1980) presented streak photographs of a tethered crane-fly (*Tipula paludosa*) and a butterfly (*Pieris brassicae*). Large vortex-dominated motions can be distinguished over the wings of *Pieris*, but the *Tipula* wake structure was indistinct. Coherent vortex structures could be seen in phase-averaged photographs of neurogenically forced wing motions in tethered dragonflies (*Libelluala luctuosa*) by Somps and Luttges (1985), but the overall picture was incomplete.

If the vortex theories can confirm that unsteady mechanisms must be operating in nearly all cases of hovering flight, they do not indicate what those mechanisms might be. In the absence of definitive experimental evidence from the real animal, physical mechanisms must be postulated based on the known behaviour of fluids, or on mechanical models.

6.4 High-Lift Mechanisms

Dynamic Stall. An impulsively started aerofoil must move through more than four chord lengths to reach 80% of the asymptotic circulation. This is known as the Wagner effect, which reduces the lift initially available to an accelerating wing (cf. Chow and Huang (1982) for solutions for flat-plate and finite thickness aerofoils, and Auerbach (1987) on the difficulties in satisfactory modelling of even this geometrically-simple, separated flow). If, however, the aerofoil is started at a geometric angle of attack higher than the steady-state stalling angle, α_{crit}, large transient circulations can be built up during the first few chord lengths of travel, when the lift exceeds the steady-state values. Both rotational and translational accelerations can produce this effect up until $\lambda \approx 4$. Dynamic stall (see McCroskey (1982) for a review) may be accompanied by leading edge separation and flow reattachment before the trailing edge, forming a separation bubble whose increase in circulation is added to the normal wing circulation. Eventually, the bubble moves away and convects downstream over the wing surface, with a very sharp drop in lift, but if the wing is oscillating continuously, the unsteady forces may remain favourable through much of the cycle. Favier et al. (1982) have shown that coupled, in-phase variations in U and α ($\Delta U \cong 0.68U$; $\Delta\alpha = 6°$) of an accelerating and pitching aerofoil produced *mean* lift coefficients over the oscillation cycle of 2.66 times the steady-state value, for $k = 0.59$, $\Lambda = 0.57$, and $Re_c = 8 \times 10^4$.

At lower Re_c, stall is less abrupt, and useful lift can be generated even at high α, so dynamic stall mechanisms might be less effective. However, the phenomenon of flow separation at a sharp edge is almost *Re*-independent. Large local decelerations in the flow and high concentrations of vorticity at the surface can cause separation even at $Re < 0.1$. The initial shear layer instability is inviscid, and while viscous effects dominate in the separated region itself, affecting stability and reattachment of the flow there, roll-up of the shed vorticity will still occur. Maxworthy's (1979) experiments on model wings showed almost identical separated flows at $Re_\omega \cong 30$ and $Re_\omega \cong 10^4$ ($Re_\omega = \dot{\gamma}c^2/\nu$) although the vortex cores were noncircular at the lower Re_ω. This spans almost the entire range of *Re* of interest in insect flight. The flow visualisations of impulsively started aerofoils at $16 < Re_a < 5.2 \times 10^4$ ($Re_a = \dot{v}^{1/2}c^{3/2}/\nu$) by Freymuth (1985) clearly demonstrate that the onset of flow separation is not particularly dependent on *Re*, even from a smooth profile, although the shape and structure of the separation vortices themselves can vary dramatically.

Wing rotation and acceleration at the ends of the wingbeat are likely to be of great importance in hovering flight over a broad range of length scales. Even when mean steady state lift coefficients appear adequate, they could not be achieved without some means of avoiding the Wagner effect, most likely through some dynamic stall process.

Wing Interference. Apart from the alteration of the effective incidence angle due to the unsteady induced flow from previous wingbeats, there are two possible mechanisms for wing interference in a solitary hovering animal. The first involves interactions between opposing wings as they approach one another at the ends of the wing strokes. An extreme example of this was discussed by Weis-Fogh (1973) and Lighthill (1973) based on wing motions observed in *Encarsia formosa*, which apparently operated with $\bar{C}_L \cong 2.3$, an impossible feat for steady aerofoils at $Re_c = 20$. The wing kinematics were distingushed by an unusually prolonged "clap" phase at the end of the upstroke, when the dorsal wing surfaces were pressed together. When they opened on the subsequent downstroke, they did so by rotation about their trailing edges which remained connected until the including angle approached 120°, when they parted with opposing sections translating away from each other. Air rushing into the opening gap could produce very large circulations around each wing, without any vortex shedding, since the circulation around each wing was equal in magnitude and opposite in sign, and the total around both wings (which were still connected) was still zero! This surprising result, a new *inviscid* mechanism of lift generation, showed that high instantaneous lift forces would be generated by such flows, and was consistent with the observed vertical accelerations of the body. The experiments of Maxworthy (1979), numerical simulations by Haussling (1979), and theoretical analysis by Edwards and Cheng (1982) revealed that very large separation vortices would be generated at the sharp leading edge, significantly enhancing the circulation, which continued to increase, even beyond including angles of 120°. Calculations from instantaneous force measurements and simultaneous quantitative flow visualisations over an opening wing pair at $Re_\omega = 3 \times 10^3$ by Spedding and Maxworthy (1986), showed *mean* lift coefficients of 6.9–8.5 over the wing opening time. It was not necessary for the wings to be completely closed to achieve this result. Significant performance enhancements are clearly achievable, though the wing opening time history can strongly affect the growth of circulation in the separation vortex, which is turbulent at this *Re*.

Ellington (1980, 1984, IV) has commented on the likely modifications of the basic clap and fling mechanics due to wing flexibility and departures from the idealised rigid wing kinematics, and Fig. 21a shows two wing sections approaching, but not touching, and then peeling apart as the wings flex when the leading edge initially begins to rotate. In the second variation (Fig. 21b), the wings meet at the trailing edge, but never at the leading edge, and separation is again preceded by a flexible peel motion. The available but incomplete experimental evidence suggests that these mechanisms ought to be successful in producing high unsteady lift forces, and that this type of motion is quite widespread in nature. The near or partial clap and fling motions of many birds and bats in take-off and landing manoeuvres and of many insects in hovering and rapidly accelerating flight may ultimately be crucial for the success of these species, which perform many wingbeats of essentially steady flight, but also absolutely depend on the ability to occasionally generate high transient lift forces.

The second case of wing interference is specific to insects and concerns the interaction between fore- and hindwing pairs. Wing interference was calculated to

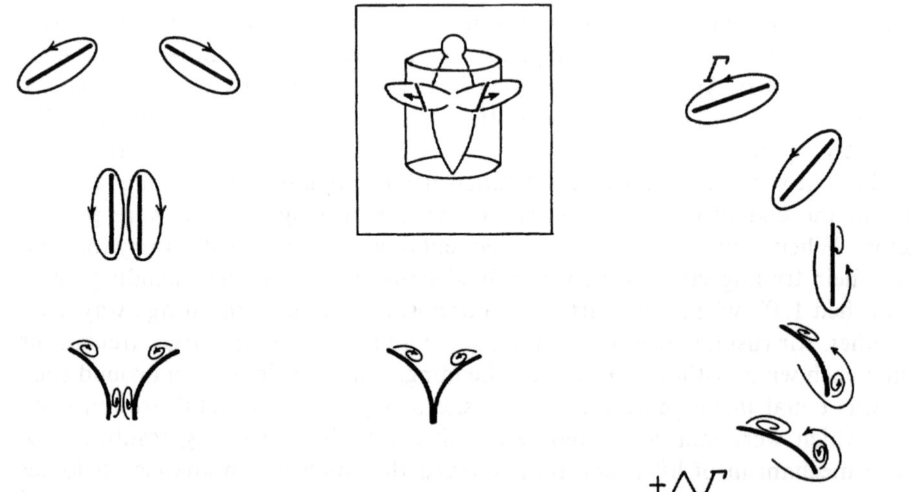

Fig. 21 a–c. Unsteady lift generation mechanisms. The plane of section is normal to the long axis of the wings, which thus appear as chord sections. If the wings beat in a horizontal plane, then the sections lie on the surface of an imaginary vertical cylinder which intersects the wings at mid-span (*inset*). **a** and **b** Modified versions of the clap and fling motion. **c** The flex mechanism for an isolated wing in rotation. (Ellington 1984, IV)

be responsible for only 2% corrections in Jensen's (1956) analysis of locust flight, but the phase relations between wing pairs in dragonflies nevertheless appear to conform to theoretical predictions based on optimised wake energy extraction (Sect. 5.4). Reavis and Luttges (1988) measured the lift and drag forces on a tethered dragonfly and correlated them with simultaneous flow visualisations. Leading edge separation vortices were produced over both wings and those produced by the forewing convected downstream to interact with the hindwing. Interpretation of the force records is complicated somewhat by the fact that the inertial contribution, which Azuma and Watanabe (1988) have demonstrated may be equal in magnitude to the aerodynamic force, cannot be isolated. Extremely complex flows were described by Saharon and Luttges (1989) for mechanical models of dragonfly wing pairs, and lift and thrust control were interpreted from the varying wing-vortex interactions produced with different phase angle differences. According to this view, both the magnitude and direction of the aerodynamic forces are dominated by the interaction of the lifting surfaces with coherent vortices shed during different phases of the wingbeat.

Isolated Rotation. From the discussion of dynamic stall, it is clear that single wings in isolation may generate high transient lift values by rotation alone. The most plausible mechanism has been suggested by Ellington (1984, IV), coined the "flex" (Fig. 21c). Consider a wing section towards the end of a stroke (top of the figure). Various arguments have been advanced that it is likely to have a separation bubble generated by rotational acceleration at the beginning of the stroke. Since the circulation must be reversed in sign before the next stroke, this vorticity must be shed with the same sign as the starting vortex for the next stroke. If the wing rotates as it decelerates towards the end of the stroke until the section is almost vertical, there will be vortex shedding, but now of opposite sign at the leading edge, while the circulation from the separation bubble is left around the trailing edge. If the wing rotates about this trailing edge, a new separation bubble may be formed at the leading edge without the generation of extra circulation at the stationary trailing edge. The previous separation vortex appears as a combined starting and stopping vortex. The wing flexion appears to make such kinematics possible for the hoverflies observed by Ellington, although all that is really required is that rotation should occur about the stationary trailing edge. Since the starting and stopping vortices are combined, a hoverer using the flex mechanism should have half as many wake vortices as one which does not, and they should be spaced twice as far apart in the vertical direction. The correction σ to the induced power requirement in Eq. (46) should be similarly affected, as the wake energy losses are reduced. Freymuth (1990) has published photographs of the wake of a *2-D* flat plate in combined translation and pitching oscillations, and one of his modes of oscillation is similar to that of Fig. 21c for a rigid plate rotating about the trailing edge. Figure 22 is redrawn based on his photographs. It is unfortunate that the flow over the surface cannot be seen, but the starting and stopping vortices of successive strokes are obviously combined during this motion. For $\Lambda = 3$ and $Re_\Lambda = \omega\Lambda c^2/\nu = 1.7 \times 10^3$, lift coefficients of 2.5–3 were estimated from the time averaged jet impulse. This is quite strong, though not conclusive, evidence for the

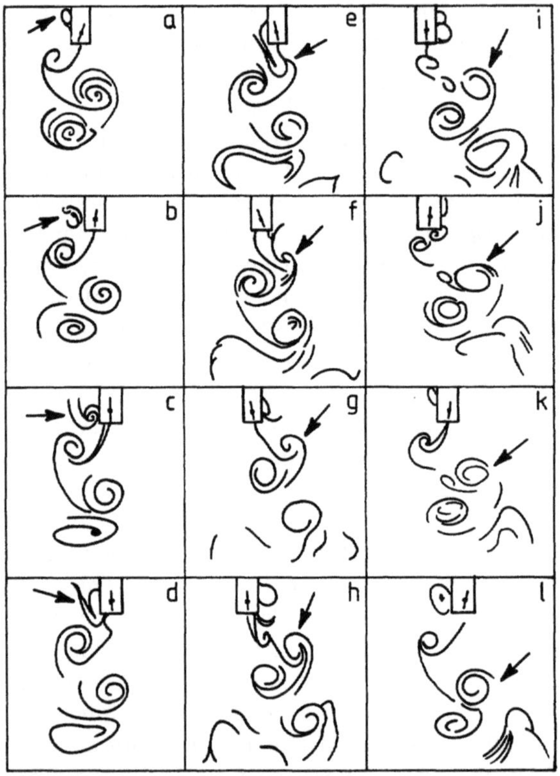

Fig. 22a–l. The wake of a rigid *2-D* plate in oscillating pitch and translation. The first frame is in mid-stroke, moving from left to right, and the time sequence proceeds by column for one complete oscillation cycle. The *arrow* follows the life history of a single vortex which begins as a leading edge vortex (**a–c**), and ends as a combined starting and stopping vortex (**f, g**), convecting away in the far wake thereafter. The plate itself could not be seen as it is obscured by the mounting strut, whose rectangular shilhouette is shown in outline here. The line inside this is a rough estimate of the location and orientation of the plate, based on information elsewhere in the paper. The pitching and heaving motions are combined with a 90° difference in phase so that rotation occurs approximately about the trailing edge at the ends of each stroke. The lines in the flow represent concentrations of $TiCl_4$ which originates on the surface of the plate, as does the vorticity. $\Lambda = 0.5$, $Re_\Lambda = 340$. (Redrawn from photographs in Freymuth 1990)

practicability of Ellington's flex mechanism, and it would be interesting to see clear quantitative demonstrations of this.

6.5 Power Requirements

The total mechanical power required for hovering flight is the sum of the power required to produce the aerodynamic forces,

$$\bar{P}_a = \bar{P}_i + \bar{P}_{pro}, \tag{48}$$

together with the power required to accelerate the wings to and fro during the wingbeat. This will have two components, one to accelerate the second moment of the wing mass, m_2, and the other to accelerate the second moment of the virtual mass, v_2, which is related to the mass of the air influenced by the accelerating wing. The virtual wing mass is usually calculated from the mass of an imaginary cylinder of air with diameter equal to the wing chord and length equal to R. The mean, mass-specific power required to accelerate the wing at the beginning of a wing stroke thus has the form,

$$\bar{P}^*_{acc} = \frac{\omega}{\pi}(m_2 + v_2)\left(\frac{d\phi}{dt}\right)^2_{max} \Big/ mg. \tag{49}$$

In the analysis of forward flight, this term was ignored since $d\phi/dt$ is usually small, but this is no longer true for hovering flight. Moreover, no useful net forces can be derived from this equation as $d\phi/dt$ is zero when averaged over the whole wingbeat cycle. Unfortunately, decelerating wings still require positive expenditure of metabolic energy, albeit only 1/10th of that required to accelerate them in the first place. If we make a charitable assumption that the deceleration occurs almost free of charge, then, in the absence of any elastic energy storage mechanism, the total specific power required is

$$\bar{P}^* = \bar{P}^*_a + 2\bar{P}^*_{acc}. \tag{50}$$

from the components in Eqs. (48) and (49). If there were perfect elastic energy storage, the kinetic energy of the wings would be converted at the end of every wing stroke to elastic strain energy, which would be returned to reaccelerate the wings on the following stroke, and $\bar{P}^*$ would equal $\bar{P}^*_a$ alone. Ellington calculated that $\bar{P}^*_a$ for his hovering insects was 17–19 W N^{-1}. If the maximum power output of insect fibrillar muscle is roughly that of vertebrate striated muscle, then the maximum specific power available is about 26 W N^{-1} (Weis-Fogh and Alexander 1977). This leaves almost no margin for accelerating the wings. Since this power requirement is large, estimated to be $\bar{P}^*_{acc} \approx 1.4$–$5.9\,\bar{P}^*_a$, there must be substantial elastic energy storage in hovering insect flight. Given the small power margin, the elastic efficiency must also be high. Deformations of the cuticular thoracic box, and of the protein resilin, and passive stretching of the fibrillar flight muscles all provide possible elastic storage mechanisms in insects, but equivalents do not appear to have been developed by vertebrates. Calculations similar to these for Weis-Fogh's hummingbird, *Amazilia*, indicated that without elastic storage [i.e. using the full form of Eq. (50)], the mechanical power requirement was about 21 W N^{-1}, which would require a conversion efficiency of around 24%. This is an entirely reasonable result, and it also indicates that the flight of hummingbirds is dominated by inertial power requirements. At high contraction rates, elastin has high internal damping and is not suitable as an elastic resonator (Weis-Fogh 1972), and the apparent failure of vertebrates to evolve a substitute has probably considerably limited their access to motions involving high frequency oscillation. Hummingbirds may well represent the practical limit. On the other hand, the same aerodynamic analysis leads to the conclusion that the considerable success of insects in slow or hovering flight, and the exploitation of microenvironments is due in significant measure to the combined elastic properties of the flight musculature, exoskeleton and specialised protein rubbers.

7 Concluding Remarks

This chapter has reviewed the current state of knowledge concerning aerodynamic mechanisms in animal flight, by examining the various analytical models and their predictions and consequences. These provide a framework within which the evolution and adaptation of natural flight systems can be understood. There is considerable scope for refinement and improvement of many of the model techniques, but even the simplest, the venerable actuator disc, has been shown to be successful, as judged by the ability to make useful, testable predictions of behaviour and morphology.

It is hoped that the steady/unsteady debate has now run its course. The broad spectrum of aerodynamic mechanisms and analytical techniques is poorly represented by insisting on a binary classification. Progress in theory and experiment in the last 10 years has allowed the problem to be recast in a more physical and mathematically-precise fashion, and the aerodynamic effects of the unsteady *kinematic* and *3-D* corrections have been explicitly calculated over a large fraction of the parameter space relevant to animal flight. In many cases, it ought to be possible to attempt an estimate of the magnitude (and sign) of at least the first order corrections. It does not seem unreasonable that these numerical estimates could be used to generate families of possible power curves, or curves with error bars or envelopes. Such calculations are conspicuously rare in the literature.

On the other hand, establishing the role of the dynamics of viscous-inviscid interactions in the boundary-layer and strongly separated regions is of critical importance. Fluctuating or periodic growth and bursting of separation regions is another source of unsteadiness in the flow (as opposed to the kinematics, measured by k, Ω). Related to this point, there is a clear need for further quantitative investigations of the fluid dynamics of simple bodies in the Reynolds number range 10^1–10^5, with rotational and translational accelerations appropriate for animal flight. In keeping with the introductory comments on scientific models, results for the most basic shapes are the more generalisable, starting with *2-D* flat plates. The accurate determination of the normal pressure and skin friction drags, even on simple fixed bodies and wing shapes, also remains a major stumbling block and source of uncertainity in the aerodynamic calculations.

Much remains to be discovered concerning time varying separated flows, stability and control, the effects of flexibility and porosity of membranous and feathered surfaces, and the role of passive and active changes in geometry of real animal wings. Aspects of manoeuvrability and agility, structural robustness and damage repair have also been neglected here. Some of these phenomena may be of interest only to a small number of fluid dynamicists, while others will have profound consequences for the balance of energetics of the organism. Ultimately the correct sorting category can only be assigned when the details are known sufficiently for accurate model simulations.

Note. An electronic accompaniment to this chapter is available in the form of *Mathematica* notebooks where analytical and numerical examples of the model equations may be manipulated and plotted. Readers interested in receiving a copy are invited to send a formatted Macintosh disc to the author.

Acknowledgments. I would like to thank my friends and colleagues, Professors FK Browand and HK Cheng for their diligent reviews of an earlier draft of this article.

References

Alexander DE (1984) Unusual phase relationships between the forewings and hindwings in flying dragonflies. J Exp Biol 109: 379–383

Althaus D (1980) Profilpolaren fur den Modelflug. NeckarVerlag, Karlsruhe

Anderson JD (1984) Fundamentals of Aerodynamics. McGraw-Hill, NY

Archer RD, Sapuppo J, Betteridge DS (1979) Propulsion characteristics of flapping wings. Aero J 83: 355–371

Auerbach D (1987) Experiments on the trajectory and circulation of the starting vortex. J Fluid Mech 183: 185–198

Azuma A, Azuma S, Watanabe I, Furuta T (1985) Flight mechanics of a dragonfly. J Exp Biol 116: 79–107

Azuma A, Watanabe T (1988) Flight performance of a dragonfly. J Exp Biol 137: 221–252

Batchelor GK (1967) An Introduction to Fluid Dynamics. Cambridge Univ Press, Cambridge

Betteridge DS, Archer RD (1974) A study of the mechanics of flapping wings. Aero Q 25: 129–141

Betts CR, Wootton RJ (1988) Wing shape and flight behaviour in butterflies (Lepidoptera: Papilonoidea and Hesperioidea): a preliminary analysis. J Exp Biol 136: 271–288

Bilo D, Lauck A, Nachtigall W (1985) Measurement of linear body accelerations and calculation of the instantaneous aerodynamic lift and thrust in a pigeon flying in a wind tunnel. Biona Rep 4: 87–108

Blake RW (1983) Mechanics of gliding in birds with special reference to the influence of the ground effect. J Biomech 16: 649–654

Brodskii AK, Ivanov VD (1984) The role of vortices in insect flight. Zool Theor 63: 197–208

Brown RHJ (1963) The forward flight of birds. Biol Rev 38: 460–489

Burkett CW (1989) Reductions in induced drag by the use of aft swept wing tips. Aero J 93: 400–405

Caple G, Balda RP, Willis WR (1983) The physics of leaping animals and the evolution of preflight. Am Nat 121: 455–467

Cheng HK (1976) On lifting-line theory in unsteady aerodynamics. Univ So Cal Aero Eng Rep USCAE 133

Cheng HK (1985) Laminar separation from airfoils beyond trailing-edge stall. Univ So Cal Aero Eng Rep USCAE 141

Cheng HK, Murillo LE (1984) Lunate-tail swimming propulsion as a problem of curved lifting line in unsteady flow. Part 1. Asymptotic theory. J Fluid Mech 143: 327–350

Cheng HK, Smith FT (1982) The influence of airfoil thickness and Reynolds number on separation. J Appl Math Phys (ZAMP) 33: 151–180

Chow C-Y, Huang M-K (1982) The initial lift and drag of an impulsively started airfoil of finite thickness. J Fluid Mech 118: 393–409

Cloupeau M, Devillers JF, Devezeau D (1979) Direct measurements of instantaneous lift in desert locust; comparison with Jensen's experiments on detached wings. J Exp Biol 80: 1–15

Cone CD (1962a) The theory of induced lift and minimum induced drag of nonplanar lifting systems. NASA TR R-139
Cone CD (1962b) Thermal soaring of birds. Am Sci 50: 180–209
Cone CD (1964) A mathematical analysis of the dynamic soaring flight of the albatross. V Inst Mar Sci SSR 50: 1–104
Csicsaky M (1977) Body gliding in the zebra finch. Fortschr Zool 24: 275
Dathe HH (1982) Efficiency calculations on the hovering flight of the pigeon (*Columba livia*) and the black-headed gull (*Larus ridibundus*). Zool Jb Physiol 86: 209–242
Dathe HH, Oehme H (1978) Typen des Ruttelfluges der Vogel. Biol Zbl 97: 299–306
Dathe HH, Oehme H, Kitzing H (1984) On the configuration of the wake of hovering birds. Zool Jb Physiol 88: 387–403
Dudley R, Ellington CP (1990a) Mechanics of forward flight in bumblebees. I. Kinematics and morphology. J Exp Biol 148: 19–52
Dudley R, Ellington CP (1990b) Mechanics of forward flight in bumblebees. II. Quasi-steady lift and power requirements. J Exp Biol 148: 53–88
Edwards RH, Cheng HK (1982) The separation vortex in the Weis-Fogh circulation-generation mechanism. J Fluid Mech 120: 463–473
Ellington CP (1978) The aerodynamics of normal hovering flight: three approaches. In Schmidt-Nielsen K, Bolis L, Maddrell SHP (eds) Comparative Physiology – water, ions and fluid mechanics. Cambridge Univ Press, Cambridge, pp 327–345
Ellington CP (1980) Vortices and hovering flight. In: Nachtigall W (ed) Instationare Effekte an schwingenden Tierflugeln. Steiner, Wiesbaden, pp 64–101
Ellington CP (1984) The aerodynamics of hovering insect flight. I. The quasisteady analysis. II. Morphological parameters. III. Kinematics. IV. Aerodynamic mechanisms. V. A vortex theory. VI. Lift and power requirements. Phil Trans R Soc Lond B 305: 1–181
Ennos AR (1988a) The importance of torsion in the design of insect wings. J Exp Biol 140: 137–160
Ennos AR (1988b) The inertial cause of wing rotation in Diptera. J Exp Biol 140: 161–169
Ennos AR (1989) Inertial and aerodynamic torques on the wings of Diptera in flight. J Exp Biol 142: 87–95
Ennos AR, Wootton RJ (1989) Functional wing morphology and aerodynamics of *Panorpa germanica* (Insecta:Mecoptera). J Exp Biol 143: 267–284
Favier D, Maresca C, Rebont J (1982) Dynamic stall due to fluctuations of velocity and incidence. AIAA J 20: 865–871
Freymuth P (1985) The vortex patterns of dynamic separation: a parametric and comparative study. Prog Aerospace Sci 22: 161–208
Freymuth P (1990) Thrust generation by an airfoil in hover modes. Exp Fluid 9: 17–24
Hainsworth FR (1988) Induced drag savings from ground effect and formation flight in brown pelicans. J Exp Biol 135: 431–444
Haussling HJ (1979) Boundary-fitted coordinates for accurate numerical solution of multibody flow problems. J Comp Phys 30: 107–124
Hertel H (1966) Structure, Form and Movement. Reinhold, NY
Herzog K (1968) Anatomie und Flugbiologie der Vogel. Fischer, Stuttgart
Higdon JJL, Corrsin S (1978) Induced drag of a bird flock. Am Nat 112: 727–744
Hummel D, Mollenstadt W (1977) On the calculation of the aerodynamic forces acting on a house sparrow (*Passer domesticus*) during downstroke by means of aerodynamic theory. Fortschr Zool 24: 235–256
Ito A (1989) The effect of trailing edge extensions on the performance of the Gottingen 797 and the Wortmann FX 63-137 aerofoil sections at Reynolds numbers between 3×10^5 and 1×10^6. Aero J 93: 283–289
Jensen M (1956) Biology and physics of locust flight. III. The aerodynamics of locust flight. Phil Trans R Soc Lond B 239: 511–552
Jones RT (1980) Wing flapping with minimum energy. Aero J 84: 214–217
Jones RT (1990) Wing Theory. Princeton Univ Press, Princeton
Karpouzian G, Spedding GR, Cheng HK (1990) Lunate-tail swimming propulsion. Part 2. Performance analysis. J Fluid Mech 210: 329–351
Kokshaysky NV (1979) Tracing the wake of a flying bird. Nature (Lond) 279: 146–148
Lan CE (1979) The unsteady quasi-vortex-lattice method with applications to animal propulsion. J Fluid Mech 93: 747–765

Lee CJ, Cheng HK (1990) An airfoil theory of bifurcating laminar separation from thin obstacles. J Fluid Mech 216: 255–284
Liebeck RH (1978) Design of subsonic airfoils for high lift. J Aircraft 15: 547–561
Lighthill MJ (1969) Hydromechanics of aquatic animal propulsion – a survey. Ann Rev Fluid Mech 1: 413–446
Lighthill MJ (1970) Aquatic animal propulsion of high hydromechanical efficiency. J Fluid Mech 44: 265–301
Lighthill MJ (1973) On the Weis-Fogh mechanism of lift generation. J Fluid Mech 60: 1–17
Lighthill MJ (1975) Mathematical Biofluiddynamics. Chapter 8: Animal flight. SIAM, PA, pp 151–178
Lighthill MJ (1977) Introduction to the scaling of aerial locomotion. In: Pedley TJ (ed) Scale effects in animal locomotion.Academic Press, London, pp 365–404
Lissaman PBS (1983) Low-Reynolds-numbers airfoils. Ann Rev Fluid Mech 15: 223–239
Lissaman PBS, Schollenberger CA (1970) Formation flight of birds. Science 168: 1003–1005
Lugt HJ (1983) Vortex Flow in Nature and Technology. Wiley, NY
Luttges MW (1989) Accomplished insect fliers. I: Gad-el-Hak M (ed) Frontiers in experimental fluid mechanics. Lecture Notes in Engineering 46. Springer, Berlin Heidelberg New York Tokyo, pp 429–456
Maxworthy T (1979) Experiments on the Weis-Fogh mechanism of lift generation by insects in hovering flight. Part 1. Dynamics of the 'fling'. J Fluid Mech 93: 47–63
McCroskey WJ (1982) Unsteady airfoils. Ann Rev Fluid Mech 14: 285–311
Nachtigall W (1974) Insects in Flight. McGraw-Hill, NY
Nachtigall W (1977) Die aerodynamische Polare des Tipula-Flugels und eine Einrichtung zur halbautomatischen Polarenaufnahme. In: Nachtigall W (ed) The Physiology of Movement; Biomechanics. Fisher, Stuttgart, pp 347–352
Nachtigall W (1979) Der Taubenflugel in Gleitflugstellung: Geometrische Kenngrossen der Flugprofile und Luftkrafterzeugung. J Orn 120: 30–40
Newman BG, Savage SB, Schouella D (1977) Model tests on a wing section of an *Aeschna* dragonfly. In: Pedley TJ (ed) Scale effects in animal locomotion. Academic Press, London, pp 445–477
Norberg RA (1975) Hovering flight of the dragonfly *Aeschna juncea* L., kinematics and aerodynamics. In: Wu TY, Brokaw CJ, Brennen C (eds) Swimming and Flying in Nature, vol 2. Plenum Press, NY, pp 763–781
Norberg UM (1975) Hovering flight in the pied flycatcher (*Ficedula hypoleuca*). In: Wu TY, Brokaw CJ, Bennen C (eds) Swimming and Flying in Nature, vol 2. Plenum Press, NY, pp 869–881
Norberg UM (1976a) Aerodynamics, kinematics, and energetics of horizontal flapping flight in the long-eared bat *Plecotus auritus*. J Exp Biol 65: 179–212
Norberg UM (1976b) Aerodynamics of hovering flight in the long-eared bat *Plecotus auritus*. J Exp Biol 65: 459–470
Norberg UM (1985) Evolution of vertebrate flight: an aerodynamic model for the transition from gliding to active flight. Am Nat 126: 303–327
Norberg UM (1990) Vertebrate Flight. Springer, Berlin Heidelberg New York Tokyo
Norberg UM, Rayner JMV (1987) Ecological morphology and flight in bats (Mammalia; Chiroptera): wing adaptations, flight performance, foraging strategy and echolocation. Phil Trans R Soc Lond B 316: 335–427
Oehme H, Kitzler U (1975) Untersuchungen zur Flugbiophysik und Flugphysiologie der Vogel. II Zur Geometrie des Vogelflügels. Zool Jb Phys 79: 402–424
Pennycuick CJ (1968) A wind-tunnel study of gliding flight in the pigeon *Columba livia*. J Exp Biol 49: 509–526
Pennycuick CJ (1971a) Gliding flight of the dog-faced bat *Rousettus aegyptiacus* observed in a wind tunnel. J Exp Biol 55: 833–845
Pennycuick CJ (1971b) Gliding flight of the white-backed vulture *Gyps africanus*. J Exp Biol 55: 13–38
Pennycuick CJ (1973) Wing profile shape in a fruit-bat gliding in a wind tunnel, determined by photogrammetry. Period Biol 75: 77–82

Pennycuick CJ (1975) Mechanics of flight. In: Farner DS, King JR (eds) Avian Biology, vol 5. Academic Press, London, pp 1–75
Pennycuick CJ (1982) The flight of petrels and albatrosses (Procellariiformes), observed in South Georgia and its vicinity. Phil Trans R Soc Lond B 300: 75–106
Pennycuick CJ (1983) Thermal soaring compared in three dissimilar tropical bird species, *Fregata magnificens, Pelecanus occidentials* and *Coragyps atratus.* J Exp Biol 102: 307–325
Pennycuick CJ (1986) Mechanical constraints on the evolution of flight. In: Padian K (ed) The Origin of Birds and the Evolution of Flight. Mem Cal Acad Sci 8: 83–98
Pennycuick CJ (1987b) Flight of seabirds. In: Croxall JP (ed) Seabirds: feeling biology and role in marine ecosystems. Cambridge Univ Press, Cambridge, pp 43–62
Pennycuick CJ (1988) On the reconstruction of pterosaurs and their manner of flight, with notes on vortex wakes. Biol Rev 63: 299–331
Pennycuick CJ (1989a) Bird Flight Performance. A Practical Calculation Manual. Oxford Univ Press
Pennycuick CJ (1989b) Span-ratio analysis used to estimate effective lift: drag ratio in the double-crested cormorant *Phalacrocorax auritus* from field observations. J Exp Biol 142: 1–15
Pennycuick CJ, Obrecht HH, Fuller MR (1988) Empirical estimates of body drag of large waterfowl and raptors. J Exp Biol 135: 253–264
Phlips PJ, East RA, Pratt NH (1981) An unsteady lifting line theory of flapping wings with application to the forward flight of birds. J Fluid Mech 112: 97–125
Rayner JMV (1979a) A vortex theory of animal flight. II. The forward flight of birds. J Fluid Mech 91: 731–763
Rayner JMV (1979b) A new approach to animal flight mechanics. J Exp Biol 80: 17–54
Rayner JMV (1979c) A vortex theory of animal flight. I. The vortex wake of a hovering animal. J Fluid Mech 91: 697–730
Rayner JMV (1985a) Mechanical and ecological constraints on flight evolution. In: Hecht MK, Ostrom JH, Viohl G, Wellnhofer P (eds) The Beginnings of Birds. Proc Int Archeopteryx Conf, Eichstatt, pp 279–288
Rayner JMV (1985b) Bounding and undulating flight in birds. J Theor Biol 117: 47–77
Rayner JMV (1986) Vertebrate flapping flight mechanics and aerodynamics, and the evolution of flight in bats. Biona Rep 5: 27–74
Rayner JMV (1988) Form and function in avian flight. In: Johnstone RF (ed) Current Ornithology, vol 5. Plenum Press, NY, pp 1–66
Rayner JMV, Jones G, Thomas A (1986) Vortex flow visualisations reveal change in upstroke function with flight speed in bats. Nature (Lond) 321: 162–164
Reavis MA, Luttges MW (1988) Aerodynamic forces produced by a dragonfly. AIAA-88-0330
Rees CJC (1975a) Form and function in corrugated insect wings. Nature (Lond) 256: 200–203
Rees CJC (1975b) Aerodynamic properties of an insect wing section and a smooth aerofoil compared. Nature (Lond) 258: 141–142
Ruppell G (1989) Kinematic analysis of symmetrical flight manoeuvres of Odonata. J Exp Biol 144: 13–42
Saffman PG (1990) A model of vortex reconnection. J Fluid Mech 212: 395–402
Saharon D, Luttges MW (1989) Dragonfly unsteady aerodynamics: the role of the wing phase relations in controlling the produced flows. AIAA-89-0832
Schmitz FW (1960) Aerodynamik des Flugmodells. Lange, Duisberg
Somps C, Luttges M (1985) Dragonfly flight: novel uses of unsteady separated flows. Science 228: 1326–1329
Spedding GR (1986) The wake of a jackdaw (*Corvus monedula*) in slow flight. J Exp Biol 125: 287–307
Spedding GR (1987a) The wake of a kestrel (*Falco tinnunculus*) in gliding flight. J Exp Biol 127: 45–57
Spedding GR (1987b) The wake of a kestrel (*Falco tinnunculus*) in flapping flight. J Exp Biol 127: 59–78
Spedding GR, Maxworthy T (1986) The generation of circulation and lift in a rigid two-dimensional fling. J Fluid Mech 165: 247–272
Spedding GR, Rayner JMV, Pennycuick CJ (1984) Momentum and energy in the wake of a pigeon (*Columba livia*) in slow flight. J Exp Biol 111: 81–102

Spillman JJ (1987) Wing tip sails; progress to date and future developments. Aero J 91: 445–453
Stolpe M, Zimmer K (1939) Der Schwirrflug des Kolibri im Zeitlupenfilm. J Orn Lpz 87: 136–155
Thomas ALR, Jones G, Rayner JMV, Hughes PM (1990) Intermittent gliding flight in the pipistrelle bat (*Pipistrellus pipistrellus*) (Chiroptera: Vespertilionidae). J Exp Biol 149: 407–416
Tucker VA (1987) Gliding birds: the effect of variable wing span. J Exp Biol 133: 33–58
Tucker VA (1990) Body drag, feather drag and interference drag of the mounting strut in a peregrine falcon, *Falco peregrinus.* J Exp Biol 149: 449–468
Tucker VA, Heine C (1990) Aerodynamics of gliding flight in a Harris' hawk, *Parabuteo unicintus.* J Exp Biol 149: 469–489
Vogel S (1967) Flight in *Drosophila.* III Aerodynamic characteristics of fly wings and wing models. J Exp Biol 46: 431–443
Vogel S (1981) Life in Moving Fluids: the Physical Biology of Flow. Princeton Univ Press, NJ
Van Dyke M (1982) An Album of Fluid Motion. Parabolic, Stanford
Ward-Smith AJ (1984) Biophysical Aerodynamics and the Natural Environment. Wiley, NY
Weis-Fogh T (1956) Biology and physics of locust flight. II. Flight performance of the desert locust (*Schistocerca gregaria*). Phil Trans R Soc Lond B 239: 459–510
Weis-Fogh T (1972) Energetics of hovering flight in hummingbirds and in Drosophila. J Exp Biol 56: 79–104
Weis-Fogh T (1973) Quick estimates of flight fitness in hovering animals, including novel mechanisms for lift production. J Exp Biol 59: 169–230
Weis-Fogh T, Alexander R McN (1977) The sustained power output from striaited muscle. In: Pedley TJ (ed) Scale effects in animal locomotion. Academic Press, London, pp 511–525
Weis-Fogh T, Jensen M (1956) Biology and physics of locust flight. I. Basic principles in insect flight. A critical review. Phil Trans R Soc Lond B 239: 415–458
Wilson JA (1975) Sweeping flight and soaring by albatrosses. Nature (Lond) 257: 307–308
Withers PC (1981) An aerodynamic analysis of bird wings as fixed aerofoils. J Exp Biol 90: 143–162
Wood CJ (1973) The flight of albatrosses (a computer simulation). Ibis 115: 244–256
Yates GT (1986) Optimum pitching axes in flapping wing propulsion. J Theor Biol 120: 255–276

Chapter 4

Principles of Walking and Running

T. McGeer

Contents

Symbols

Roman

c axial position of mass centre
H angular momentum
K spring stiffness
l leg length
m mass
R foot radius
r_{gyr} radius of gyration
SR specific resistance, Eq. (13)
T torque
V linear velocity
W work
w offset from link axis to mass centre

Greek

α_0 leg angle at support transfer
β torso inclination
γ slope, positive downhill
γ_g slope for gravity-powered walking
$\Delta\theta$ rotation from surface normal, positive forward
$\theta_{C,T}$ stance leg angle at take-off
θ_{C*} peak stance leg angle
η coefficient of restitution (Eq. (12))
τ dimensionless time $t\sqrt{g/l}$
τ_0 step period
τ_b flight time
τ_c contact time
Ω angular speed
ω_F swing pendulum frequency

Subscripts

$-$	immediately before support transfer	C	stance leg
$+$	immediately after support transfer	F	swing leg
0	steady cycle conditions	H	at the hip

1 Why Legs?

In the early days of air travel, detractors used to argue that, if God had meant us to fly, he would have given us wings. Had he meant us to roll, he might also have given us wheels – but instead we, along with the great preponderance of land animals great and small, have wound up travelling on legs. Given that situation, it is of course interesting to ask how legs work, and that will be our main concern in this chapter. But throughout we should also wonder, why legs at all? We will make some preliminary observations now, and return to this later question after some study of leg dynamics.

Whatever God's intentions might have been, there is no contesting that the wheel is perfect for travel on smooth terrain. Nature's failure to use it might be explained most easily by the biological impossibility of a continuously rotating joint. But that explanation is not completely satisfactory. It seems that without drastic redesign our bodies could *be* wheels, rolling about via a series of somersaults, or perhaps rotating around our long axes as might a child playing on a hillside. Alternatively, rather than either roll or walk, we could slither like a snake, an approach which, while distasteful from the theological point of view, nevertheless has a certain technical appeal. In particular, rolling and (to a lesser extent) slithering are steady motions, like those of an aeroplane, a boat, a tank, and most other man-made vehicles. To an engineer, steady motion is natural, and so legged locomotion, with its elaborate pattern of eternal disequilibrium, by contrast seems contrived and inelegant. In this light, the question "Why legs?" becomes compelling indeed.

Our first topic will suggest an answer. We will begin by demonstrating that rolling and walking are actually more alike than one might at first think. The demonstration will show that walking is not a contrived product of elaborate motor control, but rather a natural motion which needs no forcing at all. This is the essence of passive dynamic walking, which will be our theme through much of the chapter. We will move from a simple "synthetic wheel" to models having human-like form and gait. Each model will be capable of unforced locomotion, but we will also discuss how the passive motion can be pumped and modulated to produce dextrous behaviour. Later we will turn to running and to quadrupedal locomotion.

2 The Synthetic Wheel

Let us construct an elementary biped. The components are two spokes from a wagon wheel, and their accompanying pieces of rim (Fig. 1). These will be hinged at a hip joint, where we will also put a "payload" mass. The payload will be chosen large enough to put the overall mass centre practically coincident with the hip, just as in the original wagon wheel. Then if one leg is put on the ground and given a push, it will roll forward at constant speed. Meanwhile an infinitesimal shortening of the other leg will keep it clear of the ground, and so leave it dangling freely from the hip.

Now follow this assembly through the motion shown in Fig. 1. The legs begin at equal and opposite angles from the vertical, and with rotational speeds matched (i.e. rotating as if they were fixed in a complete wheel). From that point the stance

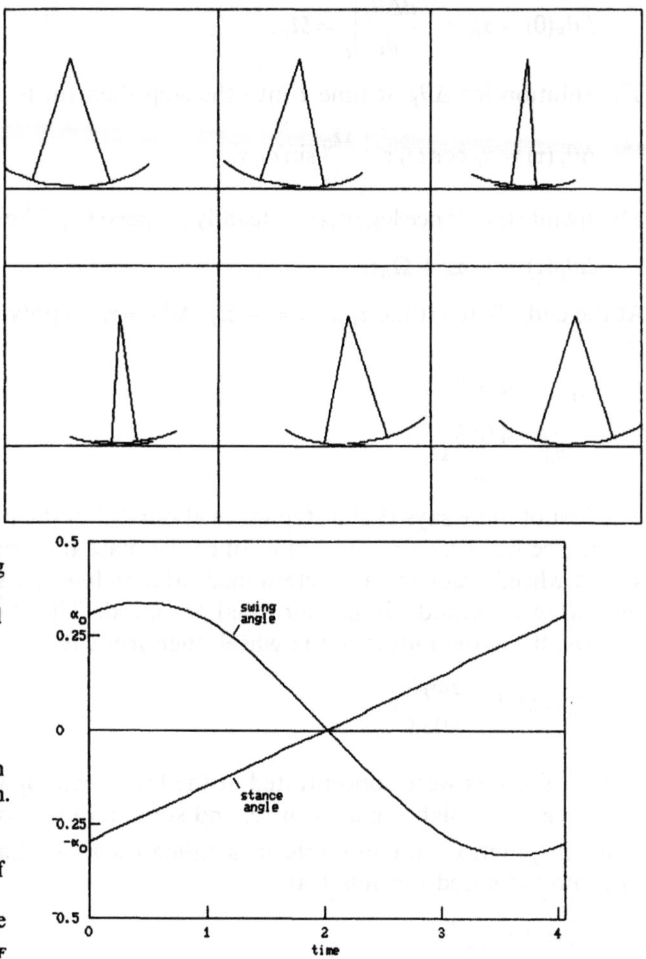

Fig. 1. The simplest of walking models is the synthetic wheel, a biped with straight legs and semicircular feet. The stance leg rolls forward steadily like a spoke in a wheel, while the free leg swings ahead like a pendulum. Support is transferred between legs when their speeds and angles match. The cycle is naturally stable and will repeat continuously, thus synthesising the motion of an ordinary wheel. Here the timescale is normalised by the swing pendulum frequency ω_F

leg continues to roll steadily, but the free leg swings along the sinusoidal trajectory of an unforced pendulum. This leads in time to a very convenient situation: the legs' speeds again become equal just when that their angles are opposite to those at the beginning of the step. If at that point the stance leg is slightly shortened, and the swing leg slightly lengthened, then support will transfer seamlessly from one leg to the next. A new step will begin, just like the last, and so on and on, the model thus rolling (or walking) continuously just like an ordinary wheel.

The synthetic wheel's cadence is determined by the pendulum frequency of its legs, say ω_F. Thus the equation of motion for the swing leg is

$$\frac{d^2\Delta\theta_F}{dt^2} + \omega_F^2\Delta\theta_F = 0, \tag{1}$$

where $\Delta\theta_F$ is the angle from the vertical. We make the approximation that $\sin\Delta\theta_F \approx \Delta\theta_F$, which certainly holds in walking. This equation has to be solved with initial conditions

$$\Delta\theta_F(0) = \alpha_0 \qquad \left.\frac{d\Delta\theta_F}{dt}\right|_0 = \Omega_0. \tag{2}$$

The solution for $\Delta\theta_F$ at time τ into the step then turns out to be

$$\Delta\theta_F(\tau) = \alpha_0 \cos\omega_F\tau + \frac{\Omega_0}{\omega_F}\sin\omega_F\tau. \tag{3}$$

Meanwhile the stance leg rotates steadily at speed Ω_0. Therefore its angle at time τ is

$$\Delta\theta_C(\tau) = -\alpha_0 + \Omega_0\tau. \tag{4}$$

At the end-of-step (time τ_0)$\Delta\theta_F = -\alpha_0$, $\Delta\theta_C = \alpha_0$. Applying these conditions leads to

$$\omega_F\tau_0 = 4.058 \tag{5}$$

$$\alpha_0 = \frac{4.058}{2\omega_F}\Omega_0. \tag{6}$$

The first of these says that a step takes about 2/3 of the period for a full oscillation of the free leg (i.e. $\omega_F\tau = 2\pi$). Thus once the mass distribution is specified, ω_F and so the wheel's cadence are determined. Meanwhile the second formula says that the swing amplitude is proportional to speed. This limits maximum speed as follows. If l is the radius of the wheel, then from Eq. (6)

$$V_0 = \Omega_0 l = \frac{2\omega_F l}{4.058}\alpha_0. \tag{7}$$

If the leg's mass were concentrated at the foot, then ω_F would be $\sqrt{g/l}$. However a real leg has a higher mass centre, and some moment of inertia as well. The net result in practice – for example in a human leg – is that $\omega_F \approx 1.4\sqrt{g/l}$. Putting this into the speed formula leaves

$$V_0 \approx 0.7\alpha_0\sqrt{gl}. \tag{8}$$

Humans seem to find walking uncomfortable with α_0 greater than about 0.4. Taking this as an upper bound implies that

$$V_{0\max} \approx 0.3\sqrt{gl}. \tag{9}$$

Notice that it would be natural in this analysis to normalise speed by $\sqrt{gl}$. Following the long-standing terminology of ship designers, the speed thus normalised is often called a Froude number. On the other hand, it is also common practice to define the Froude number as V^2/gl, so the student must be wary. Leaving the semantic issue aside, however, the essential point is that the speed of a synthetic wheel scales with $\sqrt{gl}$. If similar dynamics hold in nature, then a similar scaling law should be found. Alexander and Jayes (1983) have presented a substantial body of evidence that $\sqrt{l}$ speed scaling does indeed hold, from shrews through to elephants and possibly beyond, up to very large dinosaurs (Alexander 1976). Scaling with gravity is harder to explore, but there is striking evidence from the manned moon missions. Astronauts reported a sensation of slow motion, and indeed since gravity on the moon is $1/6^{\text{th}}$ that on earth, the synthetic-wheel analysis would predict walking at $1/\sqrt{6} \approx 40\%$ normal speed. But rather than walk so slowly, astronauts preferred to hop. The advantage of hopping over running in low-g is discussed in Sect. 4.10.

The synthetic wheel's speed limit is ultimately due to passive recovery of the swing leg, and it might seem that a little muscular intervention might allow accelerated motion. Why, then, did the astronauts not just adopt a higher cadence? A simple experiment suggests the reason. Stand on an elevated platform, and dangle one leg over the edge. Measure its pendulum period while attempting to keep the hip muscles relaxed. Then repeat the measurement while trying to cycle the leg at higher frequency. You will notice some increase in speed, but you will also find it difficult to maintain. The forcing tends to drift into increasing amplitude rather than frequency. Thus the constraints of the synthetic wheel's dynamics can be loosened, but they cannot be let go entirely.

3 The Straight-Legged Biped

One ought not dwell on the issue of speed limits, since the far more important implication of the synthetic wheel analysis is one of convenience rather than constraint. That is, we now have the wonderful result that walking can be generated purely by interaction of gravity and inertia, without muscles, motors, or forcing of any kind. The key question is, do models more anthropomorphic than the synthetic wheel have the same natural talent?

We can pursue the inquiry by relaxing the synthetic wheel's design rules. First, we can allow the foot radius to be less than the leg length. That allows for a natural foot size, while retaining the circular shape which is desirable for mathematical simplicity. Second, we can remove the requirement for a large

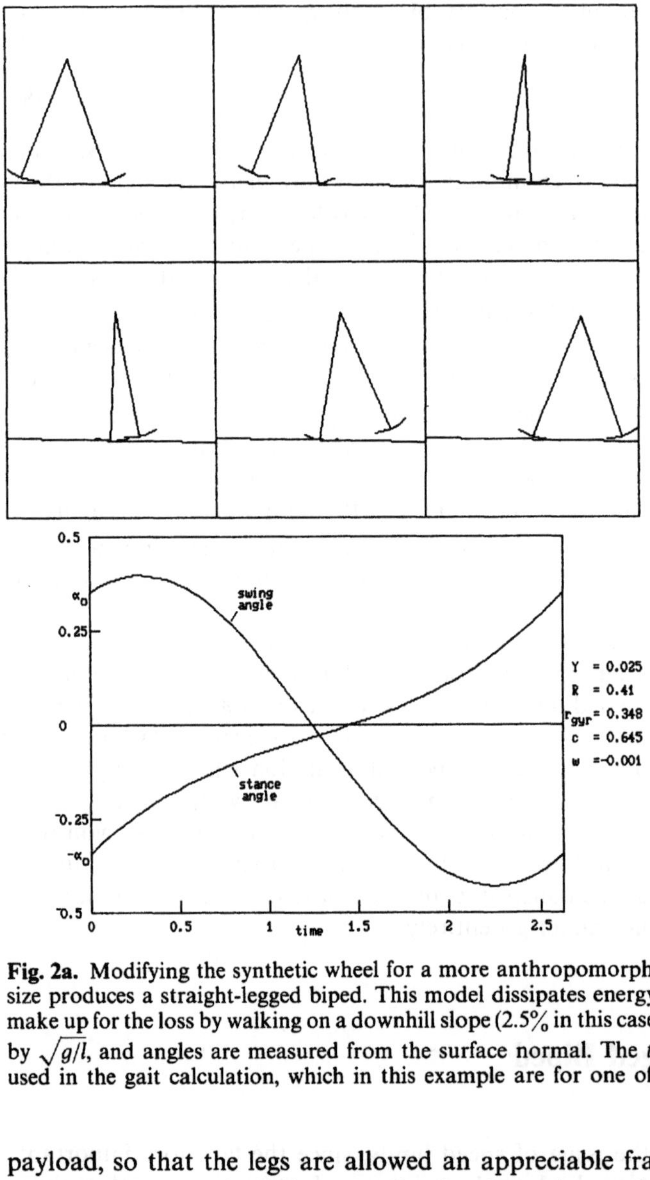

Fig. 2a. Modifying the synthetic wheel for a more anthropomorphic mass distribution and foot size produces a straight-legged biped. This model dissipates energy on each heel strike, but can make up for the loss by walking on a downhill slope (2.5% in this case). The timescale is normalised by $\sqrt{g/l}$, and angles are measured from the surface normal. The *table at right* lists parameters used in the gait calculation, which in this example are for one of our robots (McGeer 1990b)

payload, so that the legs are allowed an appreciable fraction of the total model mass. It turns out that this generalisation from the synthetic wheel, which we call a straight-legged biped, is capable of passive walking over a wide range of parametric variations (McGeer 1990b). Figure 2a shows an example. It has some new dynamics: because of the smaller feet, the stance leg is an inverted pendulum rather than a wheel, and so accelerates and decelerates through the step rather than maintaining a steady rotation. Furthermore, without the large hip mass there is inertial coupling between the legs. These effects complicate calculation of the passive gait (McGeer 1991), but the important point is that despite the complications the gait remains obviously similar to that of the synthetic wheel.

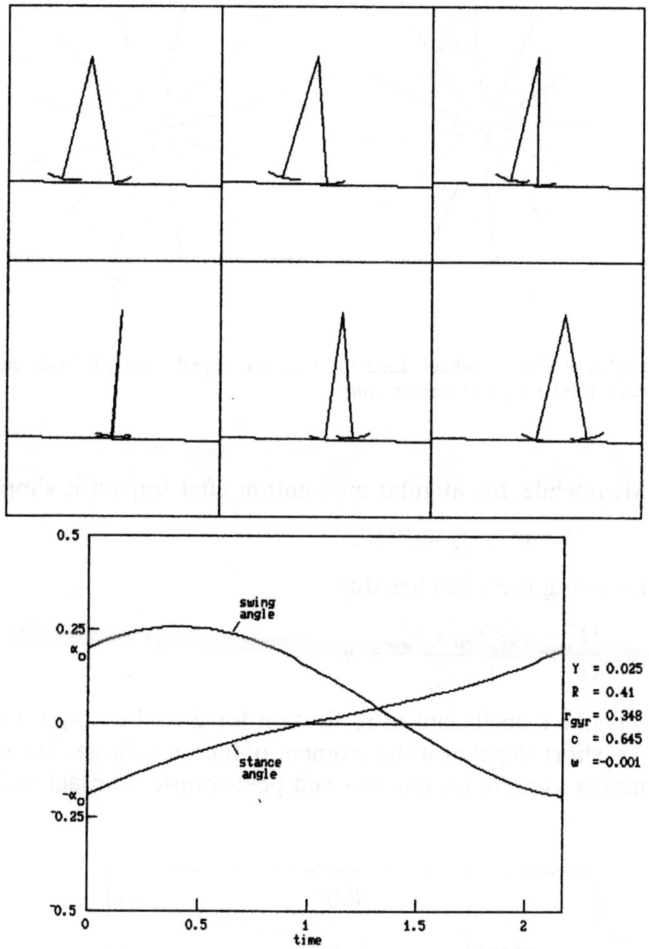

Fig. 2b. A straight-legged biped actually has two passive gaits. On the same 2.5% slope the model could walk in either of the gaits shown; the choice depends upon how the model is started

There is, however, one key difference: the straight-legged biped has to walk downhill to keep its passive cycle going. The energy gained in descent balances energy lost at each heel strike. The loss mechanism is most clearly illustrated by the rimless wheel of Margaria (1976) (Fig. 3). For analytical purposes we suppose that each time one of its legs strikes the ground, the foot is brought to rest instantaneously. No torque can be generated about the point of contact, so angular momentum must be the same before and after impact. This condition implies a drop in rolling speed, as follows. Before impact the rotational speed of the wheel is Ω^-, and the translational speed of the hip has a component $\Omega^- l \cos 2\alpha_0$ normal to the forward leg. The pre-impact angular momentum is therefore

$$
\begin{aligned}
H^- &= ml(l\Omega^- \cos 2\alpha_0) + (ml^2 r_{gyr}^2)\Omega^- \\
&= (\cos 2\alpha_0 + r_{gyr}^2) ml^2 \Omega^- . \qquad (10)
\end{aligned}
$$

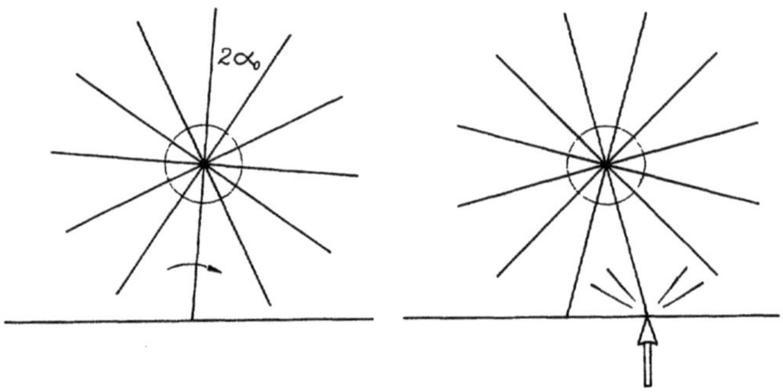

Fig. 3. A rimless wheel. Like the straight-legged biped, it loses energy in a heel strike impulse each time a leg hits the ground

Meanwhile, the angular momentum after impact is simply

$$H^+ = (1 + r_{gyr}^2)ml^2\Omega^+. \tag{11}$$

Equating these implies that

$$\frac{\Omega^+}{\Omega^-} = \frac{\cos 2\alpha_0 + r_{gyr}^2}{1 + r_{gyr}^2} \equiv \eta. \tag{12}$$

η is like a coefficient of restitution for the wheel, approaching unity if α_0 is small (i.e. short steps) or if the moment of inertia is large. The synthetic wheel effectively makes α_0 zero, i.e. the pre- and post-transfer contact points are coincident; hence

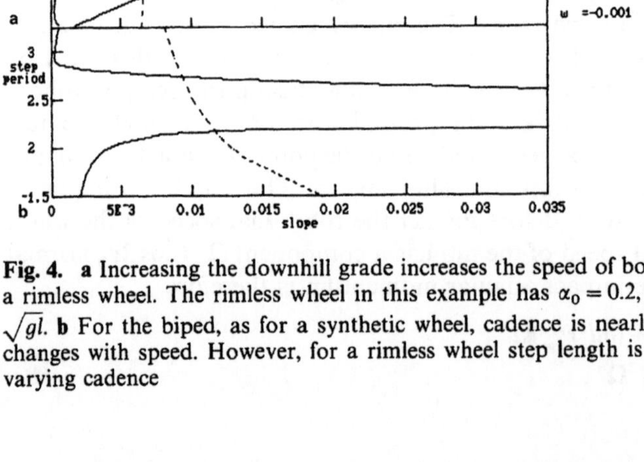

Fig. 4. a Increasing the downhill grade increases the speed of both a straight-legged biped and a rimless wheel. The rimless wheel in this example has $\alpha_0 = 0.2$, $r_{gyr} = 0.5$. Speed is in units of $\sqrt{gl}$. **b** For the biped, as for a synthetic wheel, cadence is nearly constant while stride length changes with speed. However, for a rimless wheel step length is fixed, so speed is changed by varying cadence

it has $\eta = 1$ and can roll steadily on a level surface. Reducing the foot radius moves the contact points apart, and so introduces dissipation. Smaller feet increase the energy loss, and so call for a steeper slope to sustain the motion. The same goes for higher speeds.

Figure 4a shows the speed vs. slope functions for both a rimless wheel and a straight-legged biped. The two are obviously similar, despite the biped's more complicated locomotion cycle. One difference worth noting, however, is that the rimless wheel has a fixed stride length, and so achieves higher speed by accelerating cadence (Fig. 4b). However, the biped has cadence nearly invariant with speed, and so accelerates by lengthening stride like the synthetic wheel. Thus in effect

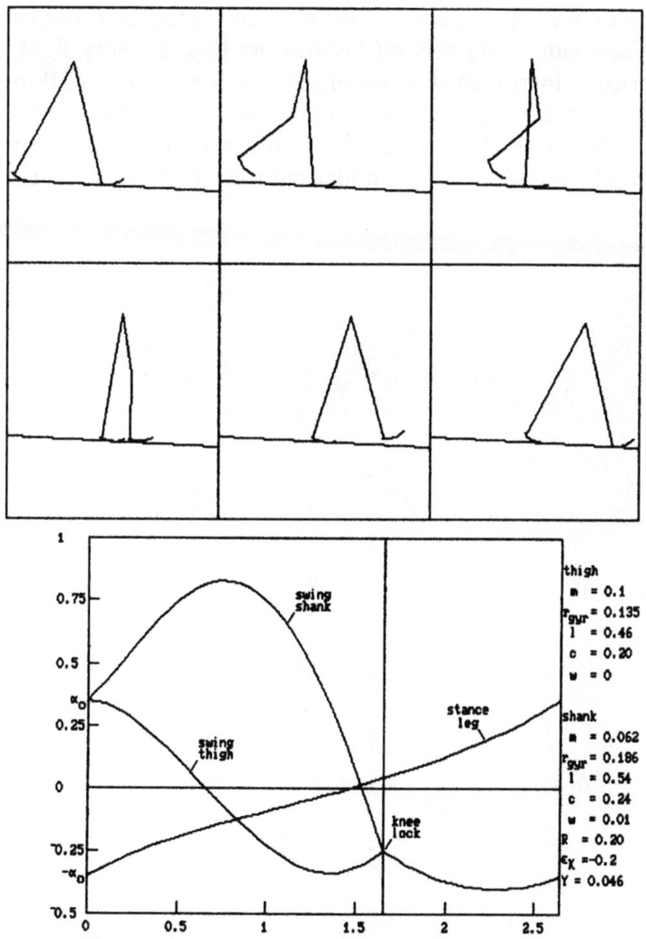

Fig. 5. A biped with knee joints also has passive walking cycles. In this example the leg parameters are roughly anthropomorphic, and the slope is 4.6% downhill. $\varepsilon_K < 0$ indicates that the foot is mounted forward on the leg, as shown in the cartoon

the straight-legged biped combines the synthetic wheel's dynamics with the rimless wheel's energetics.

Figure 4a indicates as well that a straight-legged biped has two possible speeds on any slope. Depending upon the initial conditions, it might adopt the "long period" gait, which we showed in Fig. 2a, or a "short-period" alternative, which is shown in Fig. 2b. Passive walking models seem always to have these paired gaits (unless they have no cyclic gait at all, which is also a possibility). The synthetic wheel provides another example of this rule, and also serves to illustrate that its long-period gait (that shown in Fig. 1) is likely to be preferred. The short-period alternative is realised by starting the step as in Fig. 1, but transferring support when the legs first reach equal and opposite angles (i.e. when $\omega_F \tau_0 = \pi$) rather than waiting until the speeds match as well ($\omega_F \tau_0 = 4.058$). That leaves the foot hitting the ground with substantial forward motion, rather than with no relative motion at all, and a sharp deceleration is required to bring it to rest. The synthetic wheel can shrug this off because its legs are very light, but for a real biped the price is increased dissipation. Hence for a given walking speed, the short-period gait usually requires a steeper slope than the long-period gait.

There is also another vital difference between the two gaits: stability. Given reasonable choices for model parameters, it turns out that if starting conditions

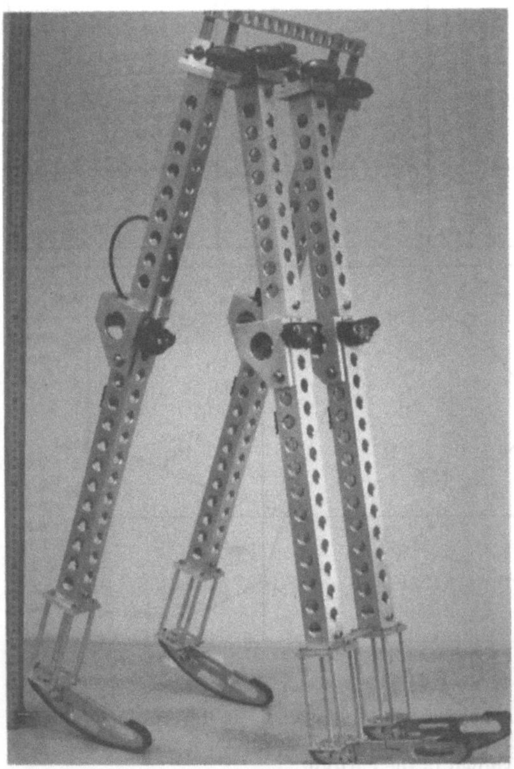

Fig. 6. Dynamite, an experimental passive biped with knees. To keep its motion confined to the longitudinal plane it actually has four legs, which are connected in pairs like crutches. Leg lengths is 80 cm, and mass 6.2 kg

are close to, but not right on, the trajectories of the long-period gait, then over a few steps the model will usually settle into the steady cycle (McGeer 1991). However, if started with conditions only infinitesimally different from those on the short-period gait, the model will almost invariably take a few steps and then collapse. Consequently, in experiments with straight-legged models we have seen only the occasional fleeting glimpse of short-period walking. This gait is not, however, just a curiousity. We will discuss redeeming features below.

4 The Knee-Jointed Biped

Moving on from the straight-legged biped, the next modification to make toward a more natural model is the introduction of knees. We will allow the knees free rotation in the flexural direction, but prevent hyperextention with a mechanical stop. The potential for passive recovery of the swing leg in such a model was noted several years ago by Mochon and McMahon (1980), and recently in exploring the dynamics further I have calculated complete passive gaits (McGeer 1990c). Figure 5 shows an example, which was calculated using human-like model parameters. Passive walking is possible with many other parameter sets, although the designer does not have quite as much freedom as in the straight-legged case.

For experiments on knee-jointed walking we built a model named Dynamite, which is shown in Fig. 6. Design of the hyperextention stops was an important detail. We wanted the knee to reach full extention inelastically after flexing through midstride, as indicated in Fig. 5. There was, however, a tendency to bounce back

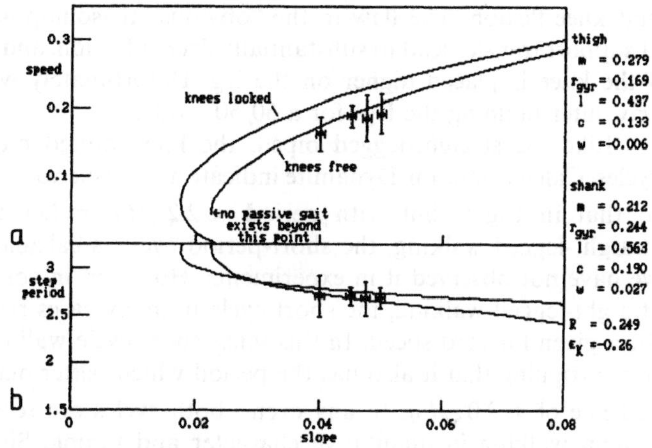

Fig. 7a, b. Speed and step period vs. slope for Dynamite in both knee-jointed and straight-legged gaits. *Continuous curves* show calculations for long-period passive walking. *Circles with uncertainty bars* show experimental measurements of the knee-jointed gait

into flexion; to prevent this, we equipped each knee with a "debouncer" made with a suction cup and a carefully sized leak.

Figure 7a shows walking speeds predicted and measured for Dynamite as a function of slope. The plot also shows performance calculated for the same model in straight-legged walking, i.e. with knees locked. (For straight-legged experiments we put the machine on a chequerboard patern of tiles; these prevented scuffing of the swing foot at midstride, i.e. where $|\Delta\theta_F| \leqq |\Delta\theta_C|$ in Fig. 2a). Slightly steeper slopes are required in knee-jointed walking, because in that case there is energy dissipation not only at support transfer, but also when the swing knee locks at full extention. However, the important point is that despite the new dynamics of the knees-free gait, the motion is quite similar to straight-legged walking, which in turn is quite similar to rolling of the synthetic and rimless wheels. Thus one can be reassured that insight derived from the simpler models is applicable in more complicated circumstances.

Notice that the knee-jointed model, like a human, has its foot placed forward on the leg. The synthetic wheel, by contrast, has symmetric feet, and also (in the long cycle) a symmetric gait. That is, if you were shown a film of the synthetic wheel, you could not tell whether the projector was running backwards or forwards. On the other hand, the asymmetry in the knee-jointed models produces an asymmetric gait. (When the knees are locked the asymmetry manifests itself as very martial goose stepping). The asymmetric gait has a harder heel strike, and so requires generally steeper slopes to sustain the walking cycle. This is undesirable, but the asymmetric foot design remains attractive because it ensures that the contact force vector passes in front of the knee during stance. The contact force thus holds the stance leg locked in the extended position.

A subtler point of anthropomorphic design that becomes important in passive walking is the ratio of thigh and shank lengths. Natural proportions are about $0.46l$ in the thigh and $0.54l$ in the shank (Chapman and Caldwell 1983), as opposed to the "obvious" choice of 50/50 which would maximise swing foot clearance per unit knee flexion. The flaw in the "obvious" reasoning is that rotations induced at support transfer lead to substantially deeper flexion, and so better foot clearance, if the knee is placed higher on the leg. Unfortunately we learned this the hard way, after building the legs for a 50/50 model!

Like the straight-legged biped, the knee-jointed model has paired passive cycles. Calculations for Dynamite indicate a short-period cycle qualitatively similar to that in Fig. 5, but with period $\approx 2.2\sqrt{l/g}$ rather than $\approx 2.6\sqrt{l/g}$. As in straight-legged walking, the short-period cycle is calculated to be unstable, and we have not observed it in experiments. However, in contrast to the situation in straight-legged walking, the short cycle in many cases requires a shallower slope for a given forward speed. In this sense short-cycle walking is more efficient, and it is intriguing that it also has the period which better matches my own preferred cadence of $\approx 2.0\sqrt{l/g}$. In any event, both cycles are reasonably consistent with human walking in qualitative character and timing. Since the human leg also shares the features of foot design and shank/thigh proportions that the passive model indicates are desirable, it seems reasonable to infer that evolution has been sensitive to the potential of passive dynamics.

5 Pumping the Natural Dynamics

Models with passive dynamics are not restricted to walking downhill. Gait can also be sustained on level and uphill grades so long as energy is supplied by an appropriate mechanism. One such mechanism can be appreciated easily by the following argument from symmetry. We mentioned that if you were shown film of the synthetic wheel, you could not tell whether the projector was running forwards or backwards. What about film of a straight-legged biped? In that case perhaps you could tell, because if the project were running backwards then the model would appear to be climbing rather than descending. But on the other hand, coult it not be that the model really was climbing? The necessary energy would be supplied by a series of "toe-off" impulses, each the mirror image of the heel strike impulse in gravity-powered walking. This would be perfectly admissable dynamically; moreover, it is what people actually do through *dorsi-plantarflexion* of the ankle in the latter part of stance.

Of course humans have a knee-jointed rather than straight-legged cycle, and therefore would fail the "projector test" for symmetry. Furthermore, humans prolong the plantarflexion impulse over a finite interval, overlapping into a double-support phase during which both feet are in contact. However, we will restrict attention here to straight-legged models, with true impulses delivered instantaneously at support transfer. The approximation, while liberal, is not too worrisome in view of the similarities between the various walking models, and in any case it will certainly afford conceptual and mathematical simplification.

So let us pursue the symmetry argument further. A remarkable implication is that a "symmetric" climb – i.e. a time-reversed gravity-powered descent – is perfectly efficient. All of the energy supplied is stored in raising the centre of mass. This very happy situation arises because there is no toe-off impulse in a gravity-powered descent; hence there can be no heel strike impulse (and therefore no dissipation) in a symmetric climb. The physical explanation is that the toe-off push from the trailing foot arrests the downward motion of the leading foot, and so makes heel strike perfectly gentle. But sequencing is important: the toe-off push must precede support transfer. If the sequencing is reversed, then there is a hard and dissipative impact (McGeer 1991).

"Symmetric" impulses for climbing thus have special significance, which we shall presently discuss in more detail. But consider first the problem of walking on level ground. Here there is no gravity-powered gait to be run backwards through a projector, and so reveal the right toe-off impulse to apply. Instead we can imagine simply trying impulses with various magnitudes and directions and exploring the consequences. Figure 8 shows a set of example results (McGeer 1991). It turns out that almost any impulse will produce a steady gait! In fact, in most cases it could produce either of two possible gaits; again initial conditions determine the choice between them. These gaits fall into two distinctive sets. One set is symmetric, like long-cycle rolling of the synthetic wheel; energy is supplied by the toe-off impulse, and dissipated in a mirror-image impulse at heel strike. The period of this gait decreases as the toe-off impulse is inclined forward. The other set is asymmetric,

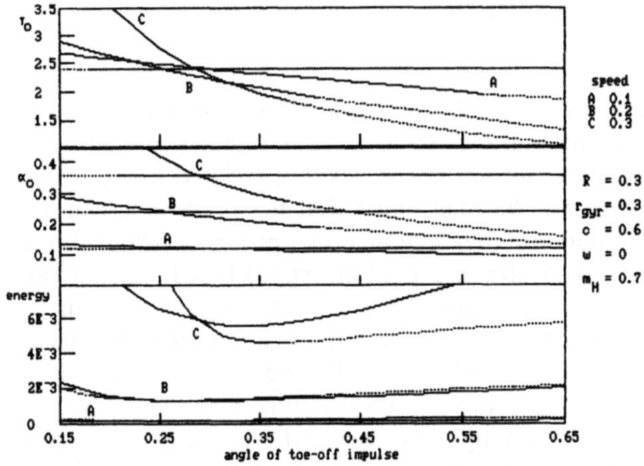

Fig. 8. The natural walking cycle can be pumped by an impulse from the trailing leg each time it leaves the ground. Here the resulting gaits are calculated, as a function of impulse angle, for a straight-legged biped walking on a level surface. Increasing impulse energy (here normalised by mgl) generally increases the steady walking speed. Meanwhile for an impulse of given angle and energy, the mathematics indicate a pair of possible gaits. Where *broken lines* are plotted the shorter-period gait of this pair is physically inadmissable, as explained in the text. Otherwise the choice between them is determined by starting conditions. Notice that one of the gaits has step period independent of impulse angle and energy

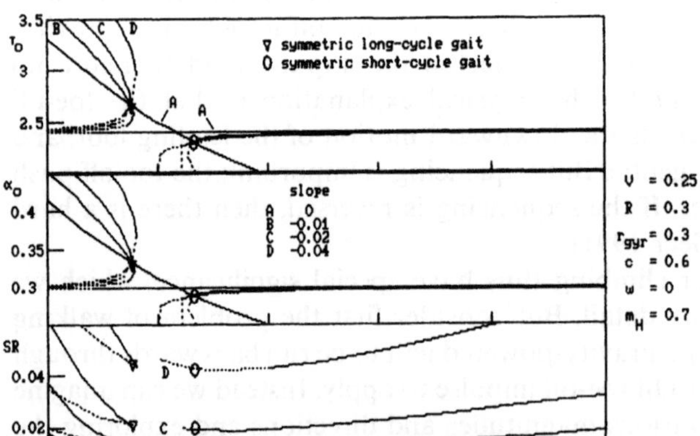

Fig. 9. Another plot of gait parameters as a function of impulse angle, but in this case with speed fixed at $0.25\sqrt{gl}$ and the slope increasing from zero to 4% uphill. The "symmetric" gaits are explained in the text. Energy input here is measured by specific resistance. $SR = -\gamma$ if there is no dissipation. $m_H = 0.7$ indicates that 70% of the model's mass is concentrated in "payload" at the hip

like short-cycle rolling of the synthetic wheel. The period of this gait, as of the corresponding synthetic wheel gait, is exactly half the swing pendulum period, completely independent of impulse magnitude and direction!

Given a good pair of legs, then, it is hard to go wrong. Get them off to a good start, and you can keep them going by pushing in almost any reasonable direction. The harder the push, the faster they go. Analysis does reveal some guidelines for refining the technique; we can observe, for example, that very steep or very shallow pushes will produce gaits with needlessly high dissipation. Stability is also an issue. This depends upon how the toe-off impulses vary during transients, but it is fair to say that, as in gravity-powered walking, the longer-period gaits are typically stable, and the shorter-period gaits unstable. The shorter-period gaits have in addition a more serious problem: often they are not just unstable, but physically inadmissable. Where Fig. 8 shows broken lines, a toe-off impulse strong enough to sustain the gait would make the model jump off the ground (For analytical purposes support transfer is still specified to occur immediately after impulse application, but physically this would require a downward pull on the forward foot.) Notice that selection of the long-period gait would imply symmetric walking if the impulses were nearly vertical, and asymmetric walking if the impulses were inclined further forward.

Let us now return to climbing. Figure 9 shows another exploration of impulse angles, but in this case for climbing various slopes at a specified speed of $0.25\sqrt{gl}$. In gravity-powered walking this speed is achieved on a downhill slope γ_g slightly steeper than 2%, and the angle of the heel strike impulse then turns out to be 0.227 (backward from the surface normal) in the long-period gait, and 0.283 in the short-period. The climbs with mirror-image impulses are indicated on the figure. These have the distinctive feature that their cadences and step lengths are invariant with slope. Climbing could also be pumped by impulses at other angles, but notice that as the slope increases from zero a forbidden zone develops in which either the mathematics produce no solutions for steady walking, or the solutions are physically inadmissable. The symmetric impulses are included in this category on slopes steeper than $-\gamma_g$, as can be appreciated by recalling that on slope $-\gamma_g$ the impulse is just sufficient to arrest the downward motion of the forward foot. The stronger impulse required on a steeper slope would therefore launch the forward foot upward. It follow that the impulse must be applied with increasing care on steep slopes, and moreover that some dissipation must be accepted. The latter point is indicated on the plot by the specific resistance

$$SR \equiv \frac{\text{mechanical work done}}{\text{weight} \times \text{distance travelled}}. \tag{13}$$

If all the work goes into raising the mass centre, then $SR = -\gamma$. This value is reached in a symmetric climbing gait, but otherwise dissipation at heel strike causes SR to rise above the minimum value.

These limitations are inconsequential so long as the slope is shallow, but for steep slopes on is obliged to find better techniques for pumping the walking cycle. Some of these take advantage of the torso.

6 Role of the Torso

We can introduce the torso with another argument from symmetry, provided that we first consider how to manage a steep descent. Ordinary passive walking in steep descent is not possible because the leg angles (α_0) would become unreasonably large. Thus the problem is to keep the cycle amplitude under control. One effective method is to put a braking torque on the stance leg, just as you would brake a wheel going downhill. Of course to apply a torque you need something to react against, and a backward-leaning torso is ideal.

The necessary recline is easily estimated for a torso on a synthetic wheel. Imagine holding a mass m_T, centred a distance c_T from the hip, at an angle β from the vertical. This calls for a torque about the hip of

$$T_H = -m_T g c_T \sin\beta. \tag{14}$$

If this torque is held through one step, then the work done by the reaction on the stance leg is

$$W_H = -\int_{-\alpha_0}^{\alpha_0} T_H d\theta_C = 2m_T g c_T \sin\beta\, \alpha_0. \tag{15}$$

Meanwhile, the energy gained in descent is

$$E = 2mgl\alpha_0\gamma. \tag{16}$$

Matching the two implies that the torso should recline backward at

$$\beta = \sin^{-1}\left(\frac{-ml\gamma}{m_T c_T}\right). \tag{17}$$

Notice that β is independent of step length.

This analysis is not exact, since the hip torque has to vary somewhat through the step to compensate for accelerations. (The wheel no longer rotates uniformly, because its mass centre is no longer coincident with the hip). However Eq. (17) remains a good approximation, and holds also for bipeds with dissipative heel strike so long as γ is replaced by $(\gamma - \gamma_g)$ (McGeer 1991).

That being said, let us reverse the projector on a torso-braked descent. The time-mirror then reveals another perfectly efficient climb, with the torso bent purposefully forward, and energy supplied by a combination of symmetric impulses and hip torque. Moreover, as in Fig. 9, this symmetric motion is an opening to a whole spectrum of torso-mediated (but nonsymmetric) climbing gaits. All have the advantage that the major energy input is distributed uniformly through the motion, rather than being concentrated in large cyclic bursts.

Limitations, however, remain. For a human $m_T \approx 0.7m$ and $c_T \approx 0.2l$, so according to Eq. (17) the torso has to lean about seven degrees for each degree of descent. At this rate one soon runs out of inclination, no matter how purposeful one's bearing. Thus while a torso expands the range of travel, it still cannot provide the answer to stairs and other steep grades. Yet another pumping method is required.

7 Leg Length Variation

In addition to difficulties of energy supply, steep slopes present an elementary geometric problem. A biped having legs of equal length, as in Fig. 2*a*, cannot stand statically on a slope exceeding α_0, since that would place its mass centre outside the base of support. The obvious recourse in this situation is to lengthen the downhill leg, and shorten the uphill leg. Similarly it would seem appropriate when walking on steep slopes to vary leg lengths cyclically. Thus in a climb, for example, one could lengthen the stance leg to $l + \Delta l$ while passing through midstride, and simultaneously shorten the swing leg to $l - \Delta l$. But lengthening the stance leg implies raising the centre of mass, and so entails some energy input. Initially this is stored as potential energy, but it is converted to kinetic as the model tips forward into support transfer. This process has the effect of pumping the cycle, so we can add leg length variation to hip torque and toe-off impulses as a third method of energy supply.

An estimate of the length variation required to climb a given slope is given by

$$\underbrace{mg\,2\alpha_0 l(\gamma_g - \gamma)}_{\text{energy required for climbing}} \approx \underbrace{mg\,2\Delta l}_{\text{energy input by raising mass centre}} \Rightarrow \frac{\Delta l}{l} \approx \alpha_0(\gamma_g - \gamma).$$

This ignores a small change in kinetic energy associated with the length adjustment, and also variations in heel strike dissipation. However, the approximation is nevertheless very good for a rimless wheel pumped by length cycling, holding over the full range from level grades to stairs (McGeer 1991). Similar accuracy can be expected for a biped.

Of course, length cycling can be used in combination with the other pumping methods, and in fact our trusty reversible projector reveals that this is the best strategy. Imagine a pair of legs walking downhill, maintaining speed by dissipating some energy in a heel strike impulse, and the rest in shortening of the stance leg at midstride. Now play the motion backward, and again you see a perfectly efficient climb. Energy is supplied by a combination of toe-off pulsing and stance leg lengthening, and it all goes into raising the mass centre since the heel strike is dissipation-free.

8 Running

Gravity-powered walking leads naturally to the study of slope variation, and we have pursued this study to the point that the techniques outlined above are sufficient to negotiate any slope within reason. Now let us talk about speed.

We have indicated that a synthetic wheel, because of constraints on α_0 and swing frequency, is limited to speeds below about $0.3\sqrt{gl}$ [Eq. (9)]. The straight-legged and knee-jointed models have the same limitation. Furthermore, because these models have small feet, their mass centres follow a convex trajectory during stance; this introduces an independent speed limit due to centrifugal effect. The rimless wheel demonstrates the limit most clearly. Its hub follows a circular arc centred at the foot. Centrifugal effect therefore lightens the contact force on the foot, and when the rotational speed reaches $\sqrt{gl}$ the contact force goes to zero. You can actually feel this effect if you try to walk unusually fast: you reach a speed at which pushing is no longer possible.

To get around these speed limits, you run. Compression of the stance leg puts the mass centre on a concave rather than convex trajectory; leaping through the air eliminates the α_0 constraint; and inter-leg torque application accelerates the swing motion. (Actually, the leap can disappear in unusual situations such as running in tight circles, but the concave trajectory remains (McMahon et al. 1987).) With these ideas in mind, consider a straight-legged biped with springs: one spring in each leg, which can compress telescopically, and another at the hip, which provides the torque necessary for accelerated swinging. Two modes of motion are easily visualised, and might be called bouncing and scissoring. To start the bouncing motion, you would hold the legs vertically and drop the model on to a flat floor. In an ideal world tipping disturbances would be excluded and the rebounds would be perfectly elastic (which implies frictionless springs and massless lower legs), so bouncing would go on indefinitely. Meanwhile, to start the scissoring motion, you would first pull the legs apart and then drop the model. Given exquisite hip bearings, the subsequent back-and-forth swinging would also persist indefinitely, or at least until a foot hit the ground. But what would happen then? One possibility is that you would suddenly have a broken pile of formerly exquisite bearings, massless lower legs, and frictionless springs. However, another possibility, which would emerge if the model landed with an appropriate set of speeds and angles, in *passive dynamic running*.

Figure 10 illustrates an example cycle (McGeer 1990a). The motion is essentially bouncing and scissoring in synchrony. In fact, the cadence indicates that bouncing hardly disturbs the scissor action at all. With the hip stiffness and leg inertia used in this example, the scissor period in free all would be $2.57\sqrt{l/g}$, while in running it is $2.52\sqrt{l/g}$ (i.e. twice the step period in Fig. 10). (There would, however, be a catastrophic disturbance to the cycle if the swing foot were allowed to scuff at midstance. To avoid this problem the swing leg must be shortened at least as much as the stance leg compresses).

Now consider speed control and speed limits. Speed is set by the amplitude of the bouncing and scissoring motions. Figure 11 shows an example. Notice that scissor amplitude (indicated by the leg angles at take-off) and bounce height (indicated by flight time) are directly related; in the steady cycle you cannot change one without changing the other. (Of course you could do in a transient, and what happens then is a question of stability. It turns out that high speeds and cadences lead to passively stable cycles, while low speeds and cadences call for active stabilisation (McGeer 1990a).) The step period remains nearly equal to half the free-fall scissor period throughout the speed range, so that as the flight time goes up, the contact time goes down. Long-term vertical equilibrium requires that the

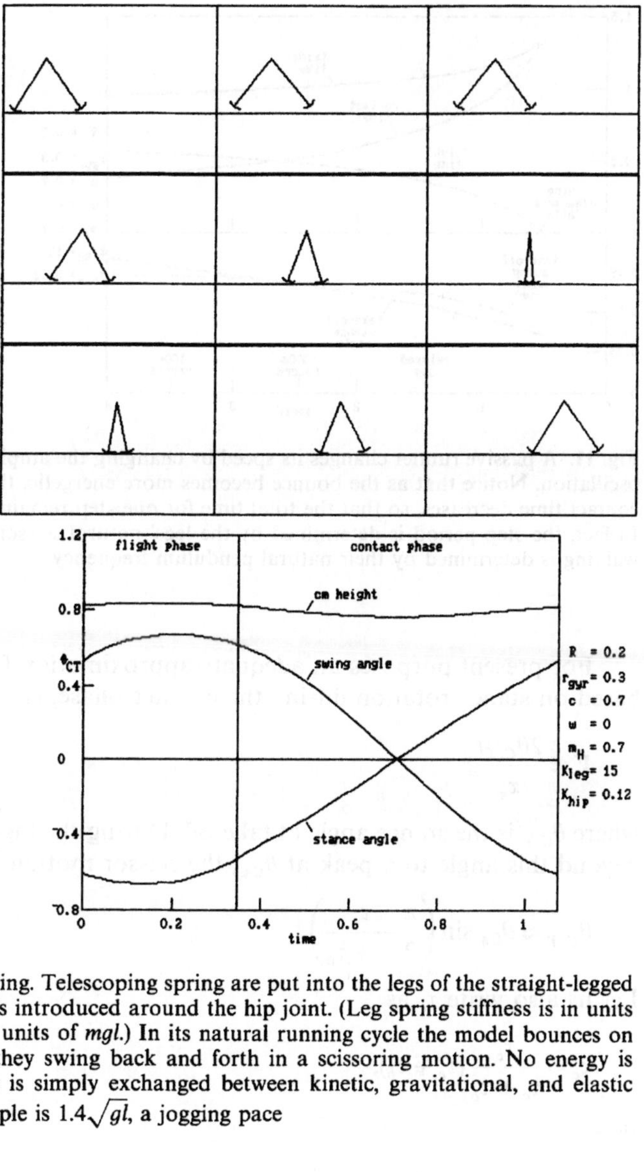

Fig. 10. Passive dynamic running. Telescoping spring are put into the legs of the straight-legged biped, and a torsional spring is introduced around the hip joint. (Leg spring stiffness is in units of *mg/l*; hip spring stiffness in units of *mgl*.) In its natural running cycle the model bounces on the legs in alternation while they swing back and forth in a scissoring motion. No energy is dissipated during the cycle; it is simply exchanged between kinetic, gravitational, and elastic stores. The speed in this example is $1.4\sqrt{gl}$, a jogging pace

upward force averaged over one cycle be *mg*, and hence that the upward force averaged over the contact phase be $mg(\tau_c + \tau_b)/\tau_c$. Declining contact time at fixed cadence therefore implies increasing loads in the stance spring. At the same time, increasing scissor amplitude implies increasing loads on the hip spring. In view of these effects one can see an upper speed limit developing as follows. (Incidentally, the lower speed limit shown in the figure can be eliminated by increasing stance stiffness, which increases flight time.)

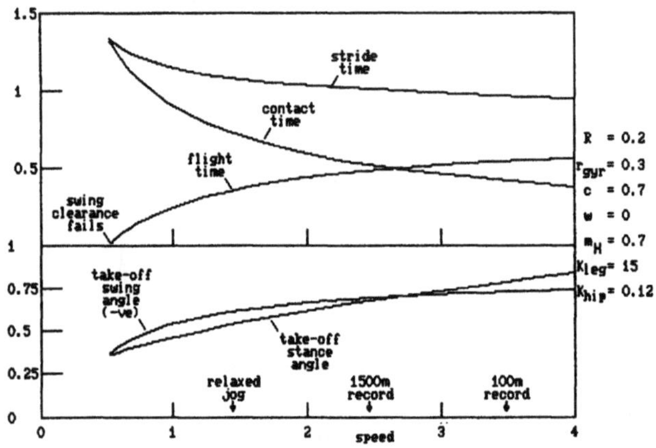

Fig. 11. A passive runner changes its speed by changing the amplitude of its bounce-and-scissor oscillation. Notice that as the bounce becomes more energetic, the flight time increases but the contact time decreases, so that the total time for one step remains nearly independent of speed. In fact, the step period is determined by the legs' natural scissor frequency, just as cadence in walking is determined by their natural pendulum frequency

For present purposes an adequate approximation for average forward speed, based on stance rotation during the contact phase, is

$$V \approx \frac{2\theta_{C,T} l}{\tau_c}, \tag{18}$$

where $\theta_{C,T}$ is the stance angle at take-off. During the flight phase the leg continues beyond this angle to a peak at θ_{C*}; the scissor motion is nearly sinusoidal, so

$$\theta_{C,T} \approx \theta_{C*} \sin\left(\frac{\pi}{2}\frac{\tau_c}{\tau_c + \tau_b}\right). \tag{19}$$

Let us also write τ_c as

$$\tau_c = \frac{\tau_c}{\tau_c + \tau_b}(\tau_c + \tau_b), \tag{20}$$

then

$$V \approx \frac{2l\theta_{C*}}{\tau_c + \tau_b} \frac{\sin\dfrac{\pi}{2}\dfrac{\tau_c}{\tau_c + \tau_b}}{\dfrac{\tau_c}{\tau_c + \tau_b}}. \tag{21}$$

Maximum speed calls for maximum θ_{C*}, which in animals is limited by the construction of the hip joint, and for minimum $\tau_c + \tau_b$, while is limited by hip stiffness. In addition, the second quotient here reaches a maximum value of $\pi/2$ when $\tau_c = 0$; thus τ_c should be made as small as possible, limited as noted above by strength in the stance leg. If we optimistically suppose infinite leg strength and stiffness, and so take the full $\pi/2$ in this factor, then we have

$$V_{\max} \approx \frac{\pi l \theta_{C*\max}}{(\tau_c + \tau_b)_{\min}}. \tag{22}$$

Sprinters achieve step frequencies up to about 4.5 per second, and if we take $\theta_{C*\max} \approx 3\pi/8$ then $V_{\max} \approx 17$ m/s. As expected, this figure is unrealistically fast, but it serves to indicate the nature of the constraints involved. Notice that the limit could be raised somewhat by a nonlinear (stiffening) hip spring, which would flatten the scissor peaks and so allow a larger $\theta_{C,T}$ for a given $\theta_{C*\max}$. Also notice that as in the synthetic wheel we remain limited by maximum leg angles and swing frequency. However, in running, swing frequency has nothing to do with g.

9 Energetics of Running

A remarkable feature of the straight-legged biped runner is that, unlike its walking counterpart, it has zero dissipation in the passive gait. Instead of losing some energy in an impulse at each heel strike, the running model stores the energy elastically, and withdraws it on the rebound. At first glance a similar mechanism would seem feasible for walking: mount the foot on a spring, use it to store the energy that would otherwise be dissipated at heel strike, and later release the energy in a toe-off impulse. However, this idea founders on incompatibility of natural frequencies. The foot spring would have to rebound – i.e. go through half a cycle—in one step period. Using $2.5\sqrt{l/g}$ from Fig. 8 as a typical step period, this implies that

$$\pi\sqrt{\frac{m}{k}} = 2.5\sqrt{\frac{l}{g}} \Rightarrow \frac{mg}{kl} = 0.63. \tag{23}$$

But mg/kl is the fractional spring compression under static load! 63% of leg length seems a bit excessive. A stiffer spring is required to get a more reasonable compression, and a tuned rebound is then possible only with a shorter contact time. That, in turn, implies a running gait.

Why, then, do we not prefer always to run? The reason, of course, is that we are not made of such ideal stuff as the passive model. When running we must continually supply energy to three major consumers: heel strike dissipation, aerodynamic drag, and "active" synthesis of spring action.

The heel strike dissipation occurs because our knees hinge rather than compress telescopically, and consequently allow impulses to propagate up the leg. An estimate of the energy loss can be made if we suppose that a running version of our knee-jointed biped would land with angles and speeds similar to those in the cycle of Fig. 10. Applying conservation of angular momentum about each joint and the point of contact (just as in walking analysis) then determines the heel strike impulse and the drop in kinetic energy. The knee angle at landing is an important parameter; 25° flexure is reasonable, and in that case the impulse turns out to be nearly vertical with magnitude $0.07m\sqrt{gl}$, and the associated energy loss is $0.013mgl$. (Flexing the knee another 10° would cut the loss to $0.005mgl$). Dividing this by the distance travelled between heel strikes gives the contribution to specific resistance, which in this case is about 0.008. In sprinting ($V = 3.5\sqrt{gl}$) the figure would rise to something like 0.03. (Incidentally jogging shoes and fatty foot pads cushion heel strike, so that the landing impulse is distributed over a period of order $0.1\sqrt{l/g}$, but the energy is lost just as irretrievably as in an instantaneous impulse).

Aerodynamic drag has been estimated experimentally and theoretically by various investigators, e.g. (Hill 1927; Pugh 1971). Its contribution to *SR* is about 0.01 in jogging, and rises quadratically with speed to about 0.06 in sprinting.

A further expenditure of energy is required to generate spring action, which while entirely passive in our model, in animals is done by a combination of active muscles and passive tendons. Ker et al. (1987) estimate (from force plate and film records) that during the first part of stance a 70-*kg* jogger removes $\approx 100\,J$ of potential and kinetic energy from the centre of mass, replacing it on the rebound. People seem to store somewhat more than half of this in stance-spring-like elements: $\approx 17\,J$ in bending the arch of the foot; $\approx 35\,J$ in stretching the Achilles tendon, and some significant (but as yet unknown) amounts in tendons of other ankle extensors and, presumably, knee extensors as well. The rest is dissipated in muscle and must be regenerated from chemical reserves. Unfortunately, an estimate of this component of the bouncing action is not available, nor do I know of any data on the balance between passive springs and active muscles in scissoring. It is noteworthy, however, that Alexander (1988a) has found in specialised runners, such as horses, long leg tendons in series with short or even rudimentary muscles. These make spring action almost fully passive. But if for more modestly equipped humans we allow only a 50% passive return, then the input required for the cycle of Fig. 10 would be $\approx 0.14mgl$, and in sprinting $\approx 0.42mgl$. Adding in the work required to balance aerodynamic drag and heel strike losses brings the *SR* up to 0.10 in jogging and 0.5 in sprinting. These figures are much less attractive than for walking (Fig. 9), and the associated power requirements ($SR \times mgV$) are formidable: $0.2mg\sqrt{gl}$ for jogging and $1.8mg\sqrt{gl}$ for sprinting, as compared with $0.0013mg\sqrt{gl}$ for walking at $V = 0.25\sqrt{gl}$. Thus we "discover" that running is hard work – but perhaps we knew that already!

Regardless of how the work is distributed between muscles and tendons, the important point as far as the dynamics are concerned is that both types of tissue behave like springs. Alexander (1988a) has made the case for passive springiness in tendon; McMahon (1990), based on a variety of evidence (Hoffer and Andreassen (1978), Houk (1979), Cavagna (1970), McMahon and Greene (1979)) has argued that muscles appear springy when controlled by spinal reflexes, and that the leg as a whole, when tested dynamically, exhibits characteristic spring properties. The message of the passive model is that these properties alone are sufficient for running; no higher motor control is required. Of course adjustments must be made for variations in speed and terrain, but as in walking these can rely on pumping the underlying oscillation (McGeer 1990a). Again it seems that, for running as for walking, we are well equipped to take advantage of the dynamic possibilities of legs.

10 Why Not Hop?

Bounce-and-scissor running as in Fig. 10 is not limited to bipeds. An analogous cycle in a monoped has been calculated by Thompson and Raibert (1990), with the scissor oscillation between the leg and a torso. Why should we not locomote in this way, using our legs in unison as does a kangaroo? Part of the answer is that by counter-oscillating our legs we prevent annoying bobbing of the torso. Furthermore, passive hopping is unstable, and therefore generally more difficult to control than running. But above all, hopping in practice requires more effort: each step is a full scissor cycle, not just half, and the bounce must be sufficiently energetic to provide the extra flight time.

However, if one's legs are sufficiently strong that bounce height is not a problem, then hopping promises very high speed. Our analysis of maximum running speed (Pugh 1971) suggested that beyond a point extra leg strength could not do you any good; however, this was because we specified a sinusoidal scissor action. As an alternative, imagine hopping with a completely different protocol for hip torque – in fact that used by Raibert (1986) for hopping robots. You would start by pushing forward and upward into a long flight phase, applying (active) torque through the push in such a way that both torso and legs left the ground with zero rotational speed. After take-off you would torque the legs around into the appropriate alignment for landing, and then relax the hip for the duration of the flight phase. Upon landing you would go through a normal rebound, again torquing as necessary to reach the same take-off conditions as on the previous stride. Hip strength is immaterial to such a cycle; maximum speed would be determined only by how high and how long you could jump. Under normal circumstances people cannot jump with enough energy to make this strategy worthwhile, but on the moon, where apparent leg strength increases sixfold, hopping seems to be an attractive proposition.

11 Quadruped Gaits

So far we have concentrated on bipeds, which has been selfishly neglectful of the majority interest. Our only companions in the bipedal camp are birds (and then only when they are not out using their God-given wings). The rest of legged mammals and reptiles are quadrupeds. From our point of view this arrangement needlessly ties up the hands, but its aficionados might point out the merits of a statically stable trunk and reduced loads on the legs. Furthermore, they have the upper hand in speed: a cheetah outpaces an ostrich, despite its shorter legs. But above all, the sheer weight of numbers invites study of quadruped locomotion.

The obvious introductory model might be called the synthetic bicycle, i.e. a pair of synthetic wheels connected by a frame. Each wheel would have exactly the same gait as in the bipedal case, so all the analysis of Sect. 2 holds without modification. In fact, all of our biped models can be connected in this way, again without changing the cyclic gaits provided that fore and hind legs start in-phase. Pairing could be between legs on the same side, in which case the gait would be called a pace, or between opposite sides, which would produce a trot. Of these options, the trot seems more attractive, since it minimises rolling moments when the model is supported on two feet.

Of course, fore and hind legs might not start in-phase. Then new dynamics would develop in pitching of the frame, and in accommodating accelerations of the hips. (The "front biped" might be past midstance and trying to accelerate when the "hind biped" was coming up to midstance and trying to decelerate). The consequences for passive locomotion have not been investigated. It is intriguing, however, that while for running most quadrupeds prefer the trot, for walking the most popular gait is the amble, with front and hind bipeds 90° out-of-phase. Analysis might prove this to be the more stable gait, but one can imagine alternative explanations. While the amble causes pitching of the body, it also reduces vertical motion at the mass centre. Also it may confer some advantage in lateral balance, through coupling from the longitudinal dynamics.

Trotting, pacing, and ambling, then, are the analogues of bipedal walking and running; trotting and pacing quadrupeds are essentially bipeds in formation, while amblers are similar but have some additional dynamics. What about analogues of bipedal hopping? These are the pronk, in which all four legs work in-phase, and the bound, in which the fore and hind bipeds are 180° out-of-phase. Pronking and bounding might be produced by scissor springs acting between paired legs and a rigid frame. In bounding there is also the possibility of scissoring by bending the back; an advantage is that this strategy would effectively lengthen the legs, and so allow a small increase in speed (moderated by some reduction in scissor frequency). It also seems that in practice the back makes a superior passive spring. Alexander (1988b) has identified passive structures in the back of the deer, particularly the aponeurosis (an expanded tendon) of the longissimus muscle, that he calculates would provide most of the spring action necessary for passive scissoring. Similar structures apparently are not present in the hips, or at best have limited effectiveness. Data of Fedak et al. (1982) indicate that the scissor motion in trotting becomes asymmetric and therefore inherently dissipative at high speed. Consequently, bounding should be more efficient than trotting above some characteristic speed, and more efficient than pronking in all conditions. Alexander and Jayes (1983) have observed that virtually all quadrupeds do indeed stop trotting at speeds between $1.4\sqrt{gl}$ and $1.7\sqrt{gl}$, and use a bound-like gait to go faster. Hoyt and Taylor (1981) have verified that the switch minimises oxygen consumption per unit distance travelled.

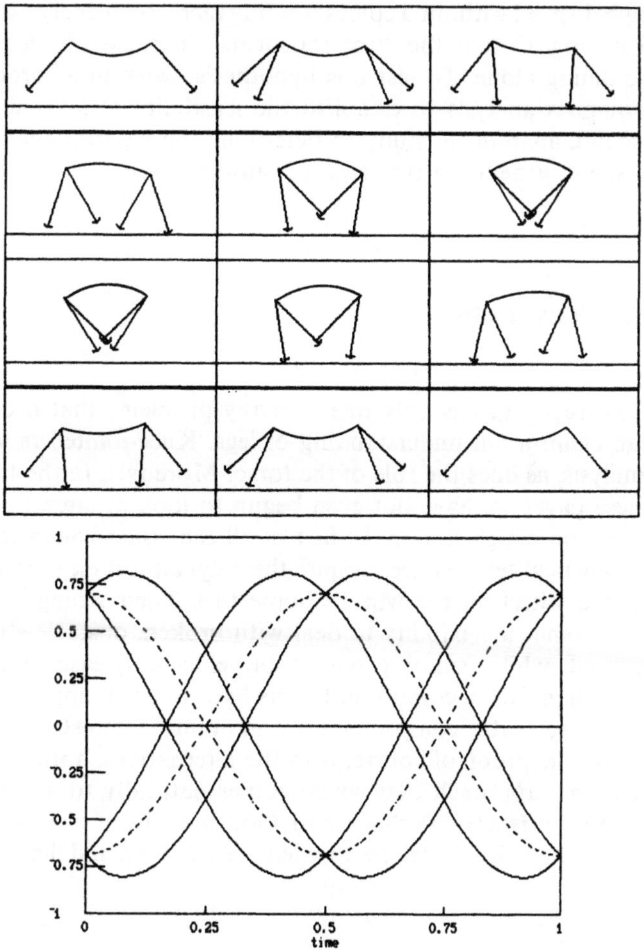

Fig. 12. Almost all quadrupeds gallop at high speeds. A galloping gait might be constructed by superposing two passive scissor motions. The first is scissoring of the front and hind hips, i.e. a sinusoidal variation of their equilibrium positions (*broken lines*). The second is scissoring of the legs about those equilibrium positions. Time here is normalised by the stride period

However, most animals do not abandon the trot entirely at high speed; instead they superpose it upon bounding, as indicated in Fig. 12. Thus each pair of legs scissors out-of-phase as in trotting, but about a median position that itself swings back and forth as in bounding. The result is a gallop. I think it likely that galloping could be sustained by a passive quadruped model, with springs in the hip, back, and legs. However, such a model has not been investigated, so Fig. 12, unlike our other plots, is the result of speculation rather than rigorous analysis. Still, the hypothesis suggests why galloping should be preferred over bounding. Since trotting is apparently efficient at moderate amplitudes (i.e. at speeds up to $\approx 1.4\sqrt{gl}$), it would seem that superposing this motion on bounding should offer higher speed without a corresponding increase in energy consumption. Thus notice from Fig. 12 that the "trot scissoring" increases the leg motion beyond that in bounding alone. To test this hypothesis, work must proceed on two fronts: first, dynamics analysis to establish the feasibility of a passive galloping model, and second, anatomical study to determine the balance between muscles and passive tissue in generating the required spring action.

12 Why Legs?

Quadruped gait is only one of many problems that need more study before we can claim a full understanding of legs. Knee-jointed models also require further analysis, as does the role of the torso. Moreover, in the two-dimensional world of these pages we have not even begun to look at lateral balance and steering, nor travel on rough terrain. In fact of all the issues rough terrain is most significant in practical terms since, despite their dynamical elegance, legs can never compete with a wheel on the wheel's home turf. Their strength is rather that they have the geometric flexibility to deal with broken, obstacle-strewn, and steeply sloped ground, while retaining reasonable efficiency, ease of control, and the wheel's natural gift for locomotion. In this light one can appreciate why legs should offer a very agreeable compromise for an animal interested in travel over terrain of all sorts. The proof, of course, is in the alternatives; nothing in the engineer's garage can even approach that which comes naturally to a horse, or a cat, or for that matter, ourselves. Perhaps had God not given us legs, we should have had to invent them! But instead we are left to discovery, and the effort is only beginning.

References

Alexander R McN (1976) Estimates of the speeds of dinosaurs. Nature (Lond) 261: 129–130

Alexander R McN (1988a) Elastic mechanisms in animal movement. Cambridge Univ Press, Cambridge

Alexander R McN (1988b) Why mammals gallop. Am Zool 28: 237–245

Alexander R McN, Jayes AS (1983) A dynamic similarity hypothesis for the gaits of quadrupedal mammals. J Zool London 201: 135–152

Cavagna GA (1970) Elastic bounce of the body. J Appl Physiol 29: 279–282

Chapman AE, Caldwell G (1983) Factors determining changes in lower limb energy during swing in treadmill running. J Biomech 16(1): 69–77

Fedak MA, Heglund NC, Taylor CR (1982) Energetics and mechanics of terrestrial locomotion II: Kinetic energy changes of the limbs and body as a function of speed and body size in birds and mammals. J Exp Biol 97: 57–66

Hill AV (1927) The air resistance to a runner. Proc R Soc Ser B 102: 380

Hoffer JA, Andreassen S (1978) Factors affecting the gain of the stretch reflex and soleus muscle stiffness in premammillary cats. Soc Neurosci (Abstr) 4: 937

Houk JC (1979) Regulation of stiffness by skeletomotor reflexes. Annu Rev Physiol 41: 99–114
Hoyt DF, Taylor CR (1981) Gait and the energetics of locomotion in horses. Nature (Lond) 292: 239–240
Ker RF, Bennett MB, Bibby SR, Kester RC, Alexander RM (1987) The spring in the arch of the human foot. Nature (Lond) 325(7000): 147–149
Margaria R (1976) Biomechanics and energetics of muscular exercise. Oxford Univ Press, Oxford
McGeer T (1990a) Passive bipedal running. Proc R Soc Ser B 1240(1297): 107–134
McGeer T (1990b) Passive dynamic walking. Int J Robotics Res 9(2): 62–82
McGeer T (1990c) Passive walking with knees. In: Proc 1990 IEEE Robotics and Automation Conference, IEEE Computer Society, Los Alamos, California, pp 1640–1645
McGeer T (1991) Dynamics and control of bipedal locomotion. In: Ho C, Zobrist G (eds) Progress in robotics and intelligent systems, submitted to J Theor Biol
McMahon TA (1990) Spring-like properties of muscles and reflexes in running. In: Winters JM, Woo SL-Y (eds) Multiple muscle systems: biomechanics and movement organisation. Springer, Berlin Heidelberg New York Tokyo, pp 578–590
McMahon TA, Greene PR (1979) The influence of track compliance on running. J Biomech 12: 893–904
McMahon TA, Valiant G, Frederick E (1987) Groucho running. J Appl Physiol 62(6): 2326–2337
Mochon S, McMahon TA (1980) Ballistic walking: an improved model. Math Biosci 52: 241–260
Pugh L (1971) The influence of wind resistance in running and walking and the mechanical efficiency of work against horizontal or vertical forces. J Physiol London 213: 255–276
Raibert M (1986) Legged robots that balance. MIT, Cambridge, Massachusetts
Thompson C, Raibert M (1990) Passive dynamic running. In: Hayward V, Khatib O (eds) Int Symp of Experimental robotics, New York, 1990. Springer, Berlin Heidelberg New York Tokyo

Chapter 5
Empirical Studies of Walking and Running

M.B. Bennett

Contents

1 Introduction

This chapter is about terrestrial locomotion on legs. It concentrates on mammalian locomotion, not because mammals are necessarily the most interesting, but simply because we know less about other groups. The birds, reptiles and invertebrates are considered in places. Whilst the chapter is centered around legged-locomotion, it also examines how locomotion has influenced structure, or how structure has influenced locomotion.

Department of Anatomy, The University of Queensland, Queensland, Australia

2 Investigating Walking and Running

2.1 Analysis Techniques

Cinematography is probably the technique most commonly used in the study of legged-locomotion. Early photographic studies of movement relied on glass-plate cameras which produced a single image. Muybridge (1899) extended this "snapshot" approach by using a number of individual cameras, synchronized to produce a series of photographs of a moving subject. The enormous variety and number of subjects photographed make this a useful reference to this day.

The advent of true cinematography enabled framing rates to be increased. Many studies of terrestrial locomotion use rates of 50–200 frames per second, which is sufficiently fast for animals over about 1 kg, where stride frequencies are generally less than 6 hertz (Alexander et al. 1977; Bennett 1987). Smaller animals may require higher framing rates to allow for their much higher stride frequencies, for example the cockroach (*Periplaneta americana*), that moves with a stride frequency of up to 15 Hz (Full and Tu 1990). High framing rates are needed if accurate resolution of events is to be achieved; for example, the determination of duty factors of small animals. Drawbacks of the technique include the need for sufficient light for high speed cine-filming and the time involved for analysis of the vast amount of information on film.

In many investigations it is desirable or necessary to know the exact position of the skeletal components that lie beneath the skin. It is common practice to use surface marks, either artificial or natural, to identify anatomical landmarks. There are potential errors with this approach due to displacement of the skin in relation to the underlying structures during locomotion. Quantification of these displacements can eliminate the errors (Van Weeren et al. 1988).

Cine X-radiography avoids problems of uncertainty of bone position. The technique has obvious hazards and limitations, but has been applied in small animal locomotion studies with considerable success (Jenkins and Camazine 1977).

Goniometry, the measurement of joint angles, has been used extensively in the analysis of human locomotion. Electrogoniometers, producing continuous records of flexion, extension, abduction, adduction, supination and pronation of the lower limb, are often used in pathological/clinical investigations. They have been used in a few studies of locomotion involving other animals, the horse in particular (Adrian et al. 1977), and their use seems more suited to veterinary investigation than to research.

Electromyography (EMG) is a useful tool in locomotion analysis. Surface electrodes placed directly over a particular muscle, or fine-wire electrodes inserted into the muscle belly are used to monitor the electrical activity produced by the muscle. The EMG signal indicates when the muscles are active during the stride (Goslow et al. 1981) and can provide some information about the relative strength of contraction. Acoustic myography may prove to be useful in the future, but at present there is considerable debate about the source of the sounds recorded during muscle activity.

Animals exert forces against the environment in order to move. Forces that occur between the feet and the ground during locomotion have been investigated in a wide range of animals, from horses (Kingsbury et al. 1978) to cockroaches (Full and Tu 1990). Research involving the mechanics of locomotion of small animals has benefited from custom-built force plates capable of resolving forces down to 0.001 N (Blickhan and Full 1987; Full and Tu 1990). Force plates can operate as ergometers, by using the force-time data to calculate the mechanical energy of the centre of mass of the subject (Cavagna 1975). Other methods for determination of forces that occur between the limb and the ground tend to be more specialized in their applications (Ratzlaff 1989).

2.2 Gait

Animals that move on legs are often referred to as moving with a particular gait. This is normally related to the speed of locomotion. Pennycuick (1975), in this study of the locomotion of wild animals in Africa, defined the term gait as applied to legged animals. In order to encompass other forms of locomotion, Alexander (1989a) produced a modified definition which states that a gait is "a pattern of locomotion characteristic of a limited range of speeds described by quantities of which one or more change discontinuously at transitions to other gaits". The quantities used to discriminate between the gaits of animals that move on legs are commonly, but not necessarily, those measured easily in kinematic analysis, such as stride frequency, stride length, duty factor or the relative phase of the legs.

Bipeds. There are only a few gaits used by bipedal animals in normal locomotion. Humans walk or run with the feet having relative phases of 0 and 0.5. Figure 1 shows how these two gaits can be distinguished by discontinuities in the duty factor or patterns of force generated during ground contact. Macropodid marsupials and some small eutherian mammals hop with a relative phase of zero for both feet (Biewener et al. 1981; Bennett 1987). There are a variety of gaits displayed by birds. Many of the passerines seem to use a bipedal hop with the feet in-phase at all speeds. Other passerines, for example members of the Corvidae, walk at slow speeds but hop at higher speeds with the feet out of phase with each other by 0.14–0.4 cycles (Hayes and Alexander 1983). Ostriches and many other birds walk and run with the feet having the same relative phases as seen in human locomotion.

Quadrupeds. The majority of mammalian species and many reptiles and amphibians are quadrupedal and their gaits have been described comprehensively elsewhere (Gambaryan 1974; Hildebrand 1977; Avery et al. 1987).

Maintenance of equilibrium is of major importance for the majority of animals. Most quadrupeds, when moving slowly, use a stable walk, where one foot is moved at a time, thus ensuring that the centre of mass stays within the triangle of support created by the other three feet still in contact with the ground. However, tortoises use a gait where there are times when only two feet are on the gound, even though the duty factor is high enough for a stable walk. A mathematical model of a tortoise was

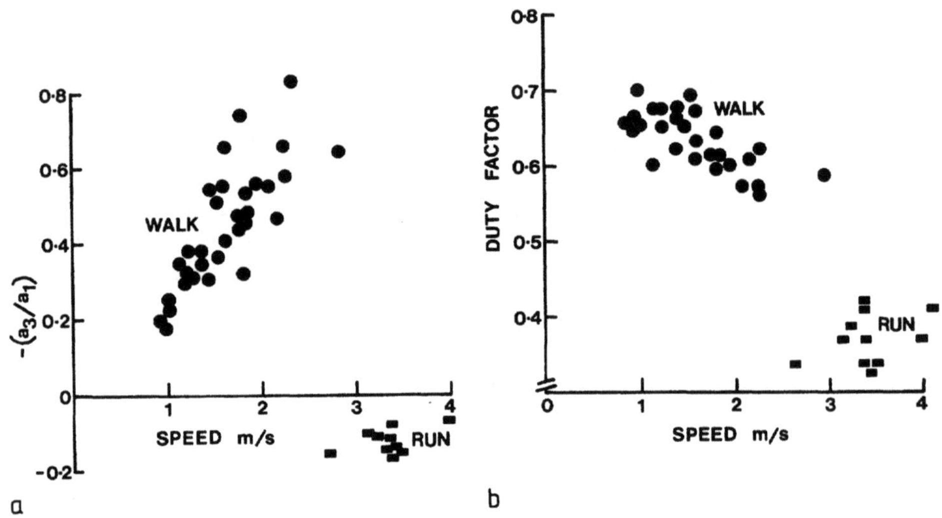

Fig. 1 A, B. Graphs to show how quantities which vary with speed may be used to differentiate between human gaits. **A** The pattern of forces exerted on the ground. The term $-(a_3/a_1)$ is a quantity that describes the shapes of force records. **B** Duty factors of adult males. (Based on data of Alexander and Jayes 1980)

devised (Jayes and Alexander 1980) which indicated that with the slow (energetically economical) muscles of tortoises, this was the only gait that kept unwanted displacements to an acceptable level.

Moving on Six Legs. Stability may be part of the reason for the gaits displayed by insects which, because of their small size, may be more liable to being knocked off-balance by air currents (Alexander 1977).

During normal locomotion on horizontal surfaces, insects will generally use an alternating tripod gait, where the prothoracic and metathoracic limbs of one side of the body are moved simultaneously with the mesothoracic limb of the other side (Delcomyn 1971; Harris and Ghiradella 1980; Full and Tu 1990). To achieve this gait, duty factors have to remain above 0.5 over their complete range of speeds (as seen in cockroaches (*Blaberus discoidalis* and *Periplaneta americana*) and many other insects). Full and Tu (1990) classified the gait of the cockroach (*B. discoidalis*) at "all" speeds as a run, by reference to the absence of pendulum-like energy exchange. The lowest running speed they recorded was 8 cm s^{-1}, whereas a walking gait has been recorded in another cockroach (*P. americana*) at speeds less than 3 cm s^{-1} (Delcomyn 1971). Ground reaction force patterns produced by running cockroaches are virtually indistinguishable from those produced by trotting mammals and running birds.

The land hermit crab (*Coenobita compressus*) only uses three of its five pairs of limbs for locomotion; the last two limb pairs being within the snail shell. It, too, uses an alternating tripod gait in walking (moving forwards rather than sideways, unlike most decapod crustaceans) (Herreid and Full 1986).

Moving on Eight Legs. Animals moving on eight legs tend to use a stepping pattern described as an alternating tetrapod gait (Bowerman 1975). Locomotion is achieved by moving two groups of legs alternately. This may be perfectly in-phase, i.e. four feet leave the ground at exactly the same time, or there may be some phasic differences between the legs within a group. Ghost crabs (*Ocypode*) move sideways and display walking, slow running and fast running gaits (Blickhan and Full 1987). The latter includes a brief aerial phase (Burrows and Hoyle 1973). Once again these gaits have been distinguished primarily using energetic criteria. Footfall patterns within the walk and the slow run in the ghost crab approximated to an alternating tetrapod; similar footfall patterns occurred in the scorpion, but in this instance the gait was referred to as a walk at all speeds of locomotion.

Animals that use more than eight legs for terrestrial locomotion (e.g. annelids and uniramian arthropods) tend to use a gait pattern that involves a "wave" of leg movement that travels along the body. This metachronal wave is sometimes seen in a basic form, superimposed on the alternating tetrapod gaits of some of the crustaceans and chelicerates (see Harris and Ghiradella 1980; Blickhan and Full 1987).

2.3 Effects of Size and Speed

There are a number of theoretical arguments about how running speed should scale with body mass (see McMahon 1975; Gunther 1975). Maximum running speeds have been measured in a wide range of placental mammals, and have been reported in the literature by numerous different authors. Garland (1983) presents data for 107 species of mammal (including three marsupials) and shows that in general, maximum attainable speed increases with body mass up to about 120 kg. Further increases in mass tend to result in a decline in achievable speed (Fig. 2). The data must be regarded with caution because of uncertainties about whether the animals were running at their top speeds and the accuracy of speed estimation. The maximal running speed for all animals scaled to the 0.17 power of body mass in agreement with predictions for dynamic similarity (Gunther 1975), but the exponents for individual groups showed little or no such agreement. Maximum running speeds in marsupial mammals (Bennett 1987; Garland et al. 1988) are similar, on a body weight basis, to placental mammals (Fig. 2).

As an individual increases speed, it alters, in a continuous manner, one or more of the parameters of the gait it is using. At some point it will abruptly change to another gait, marking a discontinuity in a measurable parameter.

In quadrupedal mammals, stride frequency increases linearly with speed within a gait. The rate of increase varies between gait, with large changes of stride frequency occurring in the walk, in contrast to the gallop where frequency is almost constant (Heglund et al. 1974; Heglund and Taylor 1988) (Fig. 3a). Stride length also increases linearly with speed within each gait (Strang and Steudel 1990), and the relatively speed-independent stride frequency seen during galloping can be explained by these increases in stride length. Kangaroos show similar relationships, with stride frequency increasing rapidly when using the low speed pentapedal gait, but only rising 18% when hopping over an 8–25 m s^{-1} speed range. Stride length increases by 180% over the same range of hopping speeds (Dawson and Taylor 1973).

Small animals use higher stride frequencies than do large ones to move at a particular speed. The stride frequency at which a range of animals, from mice to horses, changed from a trot to a gallop was found to vary with mass in a predictable manner (Fig. 3b). This trot-gallop transition has been used as a "physiologically

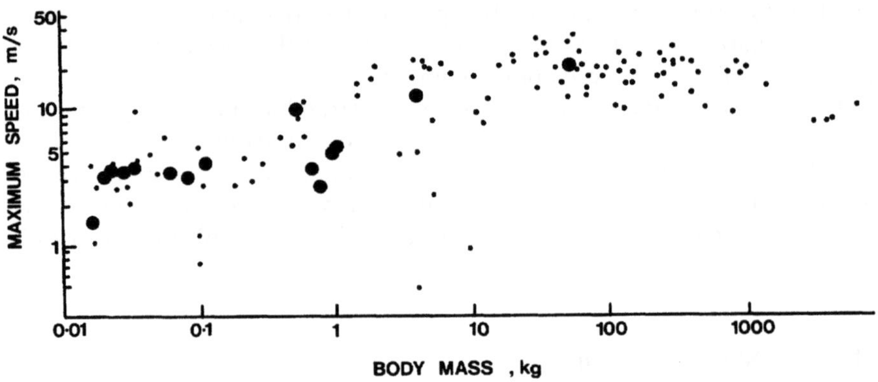

Fig. 2. Graph of maximum running speeds of marsupial ●, and placental mammals •. (After Garland et al. 1988)

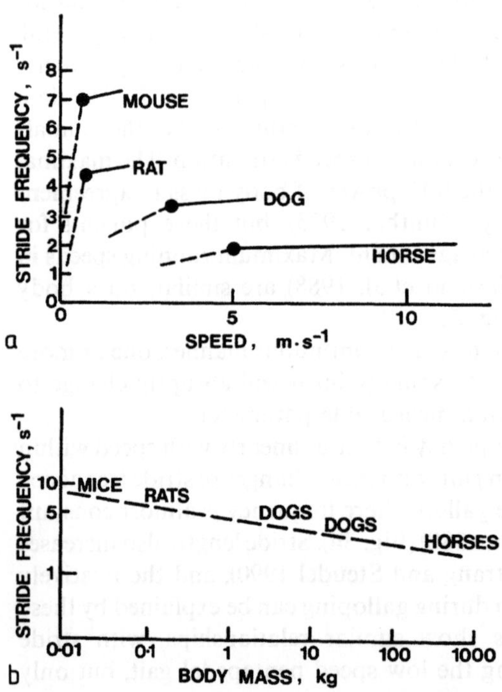

Fig. 3. a Stride frequency as a function of speed for mammals of different sizes. --- = trot, —— = gallop. **b** Stride frequency at the "physiologically equivalent" trot-gallop transition speed. (After Heglund et al. 1974)

equivalent" point for animals of different size, enabling comparisons between animals to be made (Biewener and Taylor 1986; Heglund and Taylor 1988).

One interesting use of the empirical relationships arising from the theory of dynamic similarity (an extension of geometric similarity) has been the investigation of how fast dinosaurs moved. Alexander (1976) showed experimentally that the relative stride length of extant animals when plotted against the Froude number, produced a linear relationship (Fig. 4). This brought fossil trackways, with their preserved dinosaur footprints, "to life". Initial estimates suggested that dinosaurs moved at fairly low speeds – between 1.0 and 3.6 m s^{-1}, although other tracks seem to have been made by medium-sized bipedal theropods running at up to 12 m s^{-1} (Farlow 1981). We are unlikely to find out the accuracy of this method!

Froude numbers have been used to show that quadrupedal primates tend to run at a particular speed by taking longer stride lengths (at a lower stride frequency) than predicted for nonprimates (Alexander and Maloiy 1984). Theory also predicts that different mammals will use the same gait, at any particular Froude number. Experimental evidence shows this to be broadly true. Most quadrupedal mammals (and bipeds) change from a walk to the next-fastest gait at a Froude number of about 0.5, with quadrupeds changing to a gallop at a Froude number of about 2.5 (Hayes and Alexander 1983; Alexander and Jayes 1983).

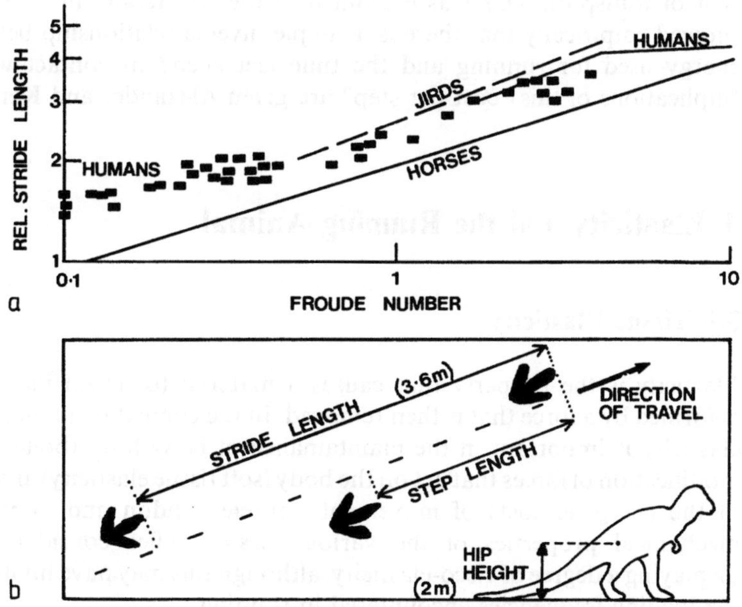

Fig. 4. **a** A graph on logarithmic coordinates of relative stride length, /h, (stride length/hip height) against Froude number (u^2/gh). The *points* represent humans walking and running on a variety of surfaces. The best fit equation is $\lambda/h \simeq 2.3\ (u^2/gh)^{0.3}$ thus $u \simeq 0.25 g^{0.25} \lambda^{1.67} h^{-1.17}$. (After Alexander 1976). **b** Example of a dinosaur trackway; the owner (*bottom right*) of the footprints would have been travelling at about 3 m s^{-1}

One of the major questions in studies of animal locomotion has concerned the energetic cost of transport. The mass-specific cost of locomotion, measured by the rate of oxygen consumption per unit mass per unit time, increases approximately linearly with speed for an individual (kangaroos and lions excepted) (Taylor et al. 1970). The measured energetic cost of locomotion for birds and mammals also varies in a predictable manner with body size, according to the equation:

$$\dot{V}_{O_2}/M_b = 0.533 \quad M_b^{-0.316} \cdot V + 0.300\, M_b^{-0.303}$$

where $\dot{V}_{O_2}$ is the rate of oxygen consumption, in ml $O_2\,s^{-1}$; M_b is body mass, measured in kg; and V is the speed of travel in m s^{-1} (Taylor and Heglund 1982). This means that large animals can move unit mass through unit distance at a lower energetic cost than small animals. Among the parameters that vary with size which may account for this, we see that stride frequency decreases as body mass increases (when measured at the trot-gallop transition or the preferred speed within each gait, which may be regarded as equivalent speeds): it scales to the -0.15 power of body mass. Heglund and Taylor (1988) also showed, by analysis of running animals, that the preferred speeds in each gait increase to the 0.21 power of body weight. Combining these results with information on energy consumption, speed and body mass (Taylor et al. 1982), they found that the mass-specific energy cost of locomotion is proportional to stride frequency at equivalent speeds. Heglund and Taylor (1988) proposed that the time available for limb muscles to generate the forces that occur during the stance phase of running is important in determining the cost of transport. This was examined further by Kram and Taylor (1990), who showed empirically that there is a simple inverse relationship between the rate of energy used for running and the time feet spend in contact with the ground. Implications of this "cost per step" are given Alexander and Ker (1990).

3 Elasticity and the Running Animal

3.1 Tissue Elasticity

Elasticity is the property that causes a material to regain its shape after being deformed by a force that is then removed. In the context of animal locomotion, this elasticity is important in the maintainance of body form (bone elasticity), in the modification of forces that act on the body (soft tissue elasticity) and in the reduction of the energetic costs of movement (muscle, tendon and bone elasticity). The mechanical properties of the various tissues differ considerably, with many displaying a degree of visco-elasticity, although this may have limited importance at the loading frequencies encountered in running.

The limbs of mammals are constructed predominantly of bone, muscle, tendon, ligament, cartilage and fat. The capacity of each tissue for storing energy in the form of elastic strain has been investigated, and it appears that certain mammals exploit tissue elastic properties when running. By using metabolically passive elastic recoil

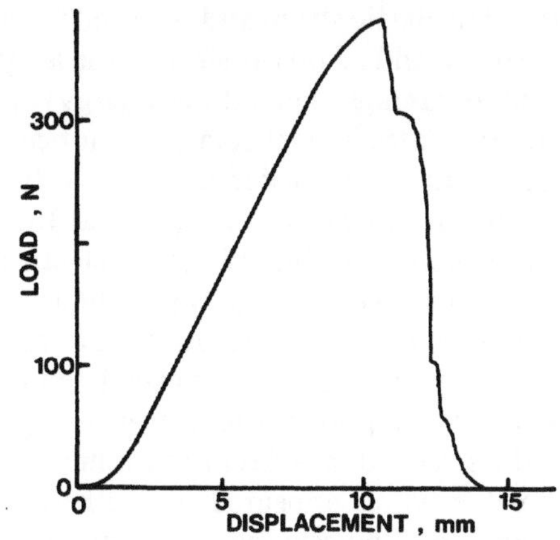

Fig. 5. A force-extension plot for a wallaby tail tendon loaded in tension. Tangent Young's modulus = 1.2 GPa; Ultimate stress = 107 MPa; Ultimate strain = 0.11

to aid locomotion, rather than muscles which require the use of metabolic energy, economies of energy can be made. Of the above-mentioned tissues, tendon and muscle are probably the most important as potential strain energy stores (Alexander 1984a).

In muscle, the compliance of cross-bridges accounts for its ability to store strain energy, with elastic extensions of about 2.5% of fibre length possible before cross-bridges start breaking (Flitney and Hirst 1978). Therefore, fibre length acts as a determinant of the amount of possible energy storage.

Tendons are hierarchical structures, based on collagen, that operate mainly in uniaxial tension as the primary linkage between muscle and bone. Tensile tests performed on tendons from a wide range of species show that a "toe-region", where stiffness increases nonlinearly, leads to a region where the tendon behaves in a near Hookean manner. It is considered that the toe region is produced by the straightening out of collagen fibres which, when unstressed, have a characteristic crimped appearance. Once the waviness has been straightened out (which occurs when the stress is about 30 MPa) any additional load will stretch the collagen fibrils themselves. Tendon has a Young's modulus in the linear region of about 1–1.5 GPa (Bennett et al. 1986) and fails at stresses of at least 100 MPa and strains of about 10–15% (Fig. 5). These values indicate that large amounts of energy can be involved in the stretching of tendon. Ker (1981) demonstrated that up to 93% of the energy used to deform a tendon can be recovered as it recoils, making tendon a good candidate for an important role in energy saving during running.

3.2 Strain Energy Storage in Tendon

The mechanism for elastic strain storage involves the transfer of kinetic energy and potential energy of the body when landing, to strain energy in the tissues. Force-plate records of running mammals show that a braking action is applied to the body as each foot contacts the ground, thus removing kinetic energy. This energy can be stored in elastic strain and restored to the body later in the stride. The use of elastic extension and recoil rather than the lengthening and shortening of muscle fibres can result in savings of metabolic energy (Cavagna et al. 1977). In many large mammals and ungulates in particular, it is noticeable that tendons are long in relation to the fibres of their attached muscles in the distal parts of the limbs. In these circumstances it seems reasonable to suggest that the majority of strain energy will be stored in tendon collagen. Analysis of the elastic properties of the feet of the Donkey (*Equus asinus*) and deer (*Cervus* spp.) (Dimery and Alexander 1985; Dimery et al. 1986a), and of the legs of camels (Alexander et al. 1982) has confirmed that during trotting (or pacing in the case of the camel) tendon elasticity, as opposed to muscle elasticity, is of major importance as a strain energy store. Indeed, in many instances, the fibres of the attached muscles are extremely short or, as in the case of the plantaris of the camel, have disappeared altogether. Energy savings may be maximised by the reduction of muscle, but manoeuvrability may be compromised.

In the pacing camel, where each leg does negative work in the first half of contact time with the ground and positive work in the second half, tendons can do 45% of the required negative and positive work. In a similar situation, tissue components acting passively in donkey limbs may do up to 87% of this work, thus greatly reducing the amount of work the muscles need perform. It is important to note that the generation of force by muscles, rather than the work they perform, may be reflected in the metabolic energy cost of running (Taylor et al. 1982; Kram and Taylor 1990).

The importance of tendon elasticity in horse locomotion has been alluded to via indirect and direct measurements of tendon extensions for walking, trotting and galloping animals (Dimery et al. 1986b; Riemersma et al. 1988). Tendon strains of between about 2 and 6% occur in these gaits.

Large bipedal mammals and birds also appear to utilize their limb tendons for energy storage when running (Alexander et al. 1979a; Ker et al. 1986, 1987). Evidence from physiological experiments, where oxygen consumption measured during activity is considerably less than that which might be expected in the absence of elastic mechanisms, supports the idea that animals do "bounce" along on springs (Dawson and Taylor 1973; Cavagna et al. 1977).

In marked contrast to large animals it seems that strain energy storage in limb tendons is of relatively little importance in small mammals. The data of Biewener and Blickhan (1988) for kangaroo rats (*Dipodomys spectabilis*) shows the tendons of the ankle extensors to be relatively thick in relation to the fibre area of their attached muscles (see Ker et al. 1988). These "thick" tendons undergo small strains (Biewener et al. 1981), resulting in little strain energy storage during normal hopping locomotion. The musculo-skeletal components of the hind limb in this animal show adaptations enabling rapid acceleration of the body to be achieved in preference, it seems, to reducing the cost of locomotion. It is understandable that avoidance of predation may be more important than making elastic energy savings.

The mechanisms described so far will act to reduce the losses of external kinetic energy (energy involved with the movement of the centre of mass) and gravitational potential energy with each step. There are also energy associated with the movement of parts of the body relative to the centre of mass; the internal kinetic energy (Alexander 1983). In the running animal it is the movement of the limbs (particularly

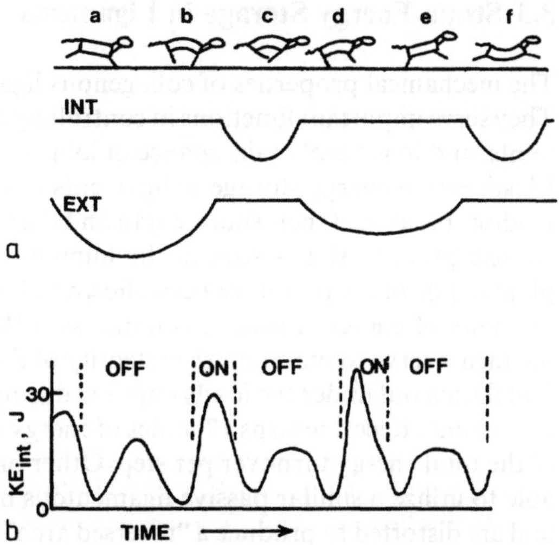

Fig. 6. **a** Diagram showing how internal kinetic energy, *INT*, and (external kinetic energy + gravitational potential energy), *EXT*, vary during a stride. The back acts as an energy store at phases *c* and *f*. (Alexander et al. 1985). **b** Fluctuations in the internal kinetic energy, KE_{int}, of the hind limb of a galloping horse. The times the foot is in contact with the floor are shown. (After Fedak et al. 1982)

the distal parts) that accounts for most of this energy. As animals increase their speed the fluctuations in internal kinetic energy increase more rapidly than the fluctuations in external kinetic energy (Fedak et al. 1982), amounting to much of the additional energetic cost of running (Fig. 6a, b).

Many mammals flex and extend their backs during fast locomotion, which has long been regarded as a method of increasing stride length and thus speed at any given stride frequency. Dawson and Taylor (1973) and Heglund et al. (1982) suggested that elastic structures in the back could produce savings in the external kinetic energy costs, in kangaroos and small mammals. The involvement of the back in reducing the internal kinetic energy costs during galloping has also been demonstrated (Alexander et al. 1985). As the hind limbs approach the end of their forward swing, they convert some of their kinetic energy into strain energy by bending the back into flexion. The forelimbs transfer their energy to the back in a similar manner as they end their backward swing. The stored energy is then available to start the limbs swinging in the opposite directions. Of the 80 J fluctuation of internal kinetic energy for a 50 kg deer (*Dama dama*) galloping at 13 m s^{-1}, it was calculated that about 44 J could be stored in the vertebral column, longissimus muscles and aponeuroses of the back. In the absence of such a system, this energy would have been degraded to heat in the muscles. The mechanism described above will only be applicable for galloping quadrupeds, because of the requirement for the fore and hind limbs to be swinging towards one another during the aerial phase of the stride. The search for structures that could work at the other end of the swing led to the proposal that the fasciae latae of the hind limbs could act in such a role (Bennett 1989). This fascia (composed predominantly of collagen fascicles) was shown to be capable of storing and returning 17–25% of the internal kinetic energy involved with the reversal of the hind limb in a 35-kg dog (*Canis familiaris*). There are, as yet, no candidates for functional "springs" which may act when the forelimb reaches its forward limit of travel.

3.3 Strain Energy Storage in Ligaments

The mechanical properties of collagenous ligaments are similar to those of tendon. They serve important functions in controlling and limiting the range of movement at joints, and in general maintenance of joint integrity (Alexander and Bennett 1987). Elastic strain energy storage in ligaments is generally much less important than in tendons because of their short length and the smaller forces that they are exposed to. An exception to this is seen in the human foot (Fig. 7a), where ligaments on the plantar side of the foot have been shown to be involved in the storage of significant amounts of energy during a running step (Ker et al. 1987). In this instance, the ligaments serve to maintain the integrity of the foot, but stretch as the arches of the foot flatten out under the loads imposed during the first half of the contact time with the ground. Recoil returns 17 joules of energy to the body, amounting to about 17% of the total energy turnover per step. Other primates, while lacking arched feet, are able to utilize a similar passive ligamentous mechanism where the feet begin "flat" and are distorted to produce a "reversed arch" (convex plantar surface) when loaded (Fig. 7b) (Bennett et al. 1989). These mechanisms do not rely on the activation of muscle.

3.4 The Elasticity of Fat Pads

Whether a mammal walks, trots, gallops or hops, each foot makes contact with the ground once in every stride. In many cursorial mammals (excluding the ungulates), the part of the foot that first touches the ground supports a thick pad of fat and

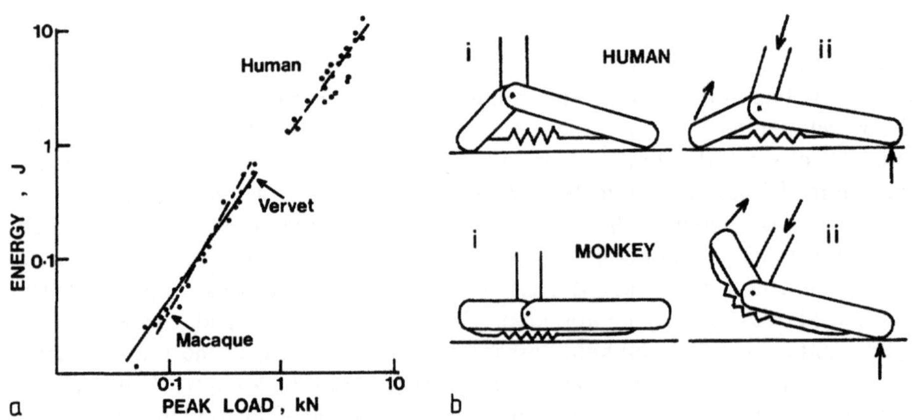

Fig. 7. **a** Graph on logarithmic axes of strain energy against peak load for human and monkey feet loaded in three-point bending. ---human; ——— *Cercopithecus aethiops*; —·—· *Macaca nemestrina*. (After Bennett et al. 1989). **b** Diagrammatic representation of the distortions of human and monkey feet during running. *i* unloaded foot; *ii* loaded foot at midstance. The *arrows* represent (*from left to right*) the force produced by ankle extensor muscles inserting on the calcaneus; the reaction at the talocrural joint and the ground reaction force

connective tissue. The position of these pads in relation to the underlying skeleton varies considerably between species, reflecting the differences in limb posture. For example, humans have a pad on the caudal aspect of the calcaneus, camels have a pad that covers the phalanges and dogs have volar pads situated adjacent to the metacarpophalangeal and metatarsophalangeal joints.

Use of a hydraulic materials testing machine to examine the dynamic compressive properties of pads from a range of mammals has shown that stiffness increases with increasing strain and that energy losses (degraded to heat in each loading-unloading cycle) of 15–43% occur, with little dependence on test frequency from 0.1 Hz to 70 Hz (Alexander et al. 1986). Considerably higher energy losses have been reported for human heel pads, from tests conducted by dropping an instrumented missile onto the pads (Jørgensen and Bojsen-Møller 1989).

The amounts of energy that can be stored as elastic strain in fat pads is small compared to the energy exchange that occurs in a running step (see Ker et al. 1989) and are thus of little importance as "springs" to help make locomotion energetically economical. In addition, the mechanism by which this energy could be returned to the running system is not clear. There is, however, a potential problem in having an elastic pad located beneath the foot, because as a foot hits the ground, kinetic energy is converted into elastic strain energy within the pad. This may be restored to the foot in the form of kinetic energy as the pad recoils and could cause the foot to leave the ground. Such vertical oscillations of the foot, where contact with the floor may be lost one or more times during the course of a step, is referred to as chattering (Alexander et al. 1986) and is presumed to be disadvantageous to the animal. A theoretical model of a leg, foot and paw pad demonstrated the requirements for avoidance of chatter. The experimental results from compression tests, together with additional information derived from films of running animals, suggest that chatter should occur. Animals have to use additional mechanisms to prevent chatter, rather than relying solely on the properties of their pads. They may land on their toes as

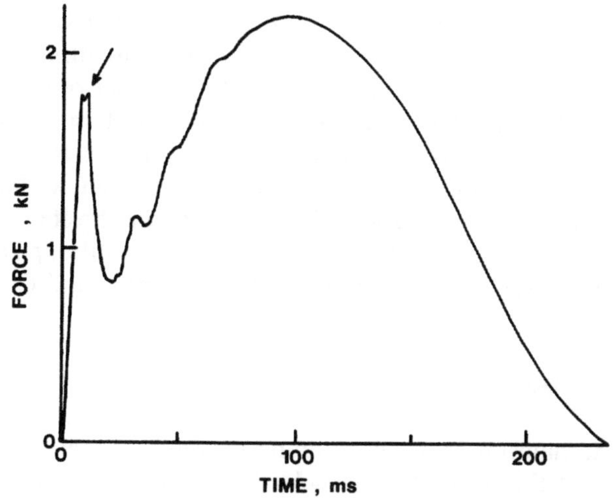

Fig. 8. A force-time plot for human barefoot running illustrating an impact disturbance (*arrowed*) followed by reducing oscillations. (After Dickinson et al. 1985)

opposed to simply landing on the whole of the fat pad (Alexander and Vernon 1975; Jayes and Alexander 1978) or they may change the rate of paw pad loading by the use of digital flexor muscles. Another important adjustment that could be made would be to reduce the velocity of the foot immediately prior to ground contact by muscle action; such rapid decelerations would be difficult to ascertain from film analysis.

Although the presence of pads may induce chatter, they do fulfil an important role in the attenuation of the transient force peak that often occurs shortly after foot-ground contact, and is perceived to be potentially damaging to the body (Paul et al. 1978). Force-plate records for humans running barefoot and for other mammals often show these impact disturbances (Fig. 8) (Dickinson et al. 1985). In these instances the stiffness and thickness of the pads are important in the reduction of the peak force at impact, as a thick pad (of a given stiffness) will decelerate the foot over a greater distance than a thin one. The hysteresis of the material does not affect the magnitude of the impact peak. The "cushioning" provided by fat pads appears to be limited, because of the fairly rapid increase in stiffness displayed when loaded in compression. Paul et al. (1978) have shown that rabbit heel pads can reduce impact forces by up to 28%.

4 Running and Breathing

4.1 Respiratory–Locomotor Coupling

There have been numerous studies where the metabolic cost of locomotion has been assessed by measuring oxygen consumption, but little attention has been paid to how breathing interacts with the mechanics of locomotion.

Bramble and Carrier (1983) studied the relationship between breathing and the half-bound gait in a hare (*Lepus californicus*), and showed that locomotor–respiratory coupling occurs (Fig. 9a). Further studies involving horses and dogs revealed that they too synchronise their breathing with movements of the limbs (Fig. 9b). A coupling ratio of 1:1 (the number of strides per respiratory cycle) was found in trotting and galloping dogs and in cantering and galloping horses, although breathing and gait were less tightly coupled in the trot (Attenburrow, 1982). Data for bipeds shows that the tammar wallaby (*Macropus eugenii*) uses a coupling ratio of 1:1 when hopping (Baudinette et al. 1987), but humans were found to vary the ratio (2:1, 3:1, 4:1, 3:2, 5:2, 7:2) although a 2:1 coupling was used most commonly by experienced runners (Bramble and Carrier 1983).

4.2 Mechanisms

Mechanisms by which running movements may drive breathing movements and assist the respiratory muscles have been suggested (Bramble and Carrier 1983).

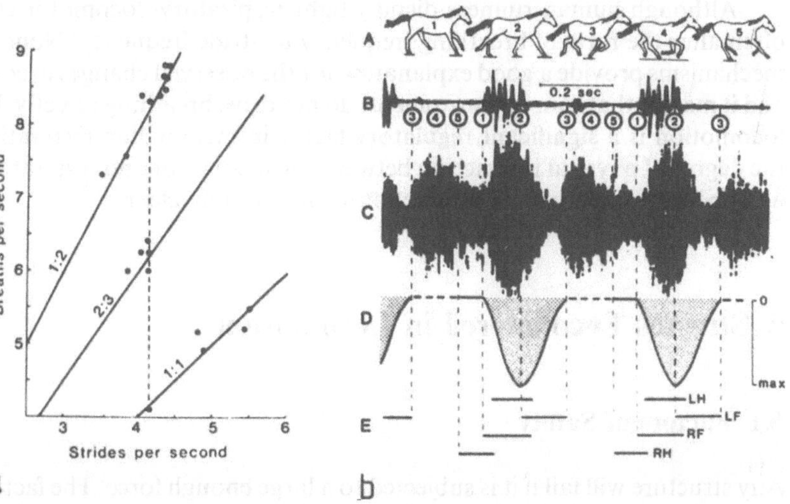

Fig. 9. a Locomotor-respiratory coupling in *Lepus californicus*. **b** Relationship between leg movements and breathing in a galloping horse. *A* body position at marked points of a stride; *B* output from an accelerometer attached to the right forefoot; *C* output from a microphone recording inspiration and expiration; *D* approximate vertical loading on the forelimbs; *E* the *lines* show when the feet were on the ground. *LF* and *RF* left and right forefeet; *LH* and *RH* left and right hindfeet. (Bramble and Carrier (1983)

These have been examined, using an "anatomical approach" (Bramble 1989) and by mathematical modelling (Alexander 1989b). The latter study critically examined the phase relationships between breathing and limb movements.

One mechanism involves the flexion and extension of the vertebral column. Exhalation is driven by components of the lumbo-pelvic region pushing the abdominal viscera and diaphragm cranially during flexion. This is followed by extension of the back, which pulls the viscera caudally and thus drives inhalation. Alexander (1989b) concluded that this system could operate in cantering or galloping horses.

The hopping wallaby probably uses a "visceral piston" mechanism, where the visceral mass, attached to an elastic diaphragm, acts as a damped resonant system. This system will only work effectively if it is tuned to the stride frequency of the animal (Alexander 1989b). Since kangaroos increase speed by taking longer strides, keeping stride frequency fairly constant (Dawson and Taylor 1973), if the visceral piston is tuned to the stride frequency it will operate over the whole range of bipedal hopping speeds. While the piston mechanism may work well for bipedal hopping, its role in quadrupedal locomotion is less certain because of the phase relationships between viscera and legs.

A third mechanism involves the loading of the thorax by forces generated while the forelimbs are on the ground. This cannot apply to bipedal locomotion and is less favoured as a working principle in quadrupeds, because it would probably increase the energetic cost of running to an unacceptable degree (see Alexander 1989b).

Although human runners display tight respiratory–locomotor coupling, they often alter the ratio of breathing frequency to stride frequency. None of the above mechanisms provide a good explanation for the observed changes in coupling ratios, and it may be that running movements do not drive breathing directly. Nevertheless, locomotion is a significant regulatory factor in mammalian respiration, although the degree of physical interaction between the locomotor and respiratory apparatus as opposed to neurogenic interaction is not entirely clear.

5 Stresses Encountered in Locomotion

5.1 Factors of Safety

Any structure will fail if it is subjected to a large enough force. The factor of safety of a given material or structure may be defined as the ratio of the maximum force that it can sustain without failing, to the force it will be required to withstand. This principle is used in engineering in the design of load bearing elements where failure must be avoided. The musculo-skeletal systems of animals are subject to the rigours of natural selection. Consequently, failure of any of the components of the locomotor system may affect the reproductive fitness of the animal by affecting its ability to feed, fight, mate, etc. and would thus be selected against. Indeed, the majority of musculo-skeletal structures appear highly "over-engineered" and seem unlikely to meet forces in normal locomotion that may threaten failure; thus their safety factors are in excess of unity, but how far in excess?

In many studies of locomotion it has been observed that as speed increases, the duty factor (fraction of a stride when a particular foot is in contact with the ground) decreases. A shorter foot-ground contact time results in an increase in the peak ground reaction force, as the mean force exerted by all of the feet still has to match body weight over a stride period. A typical quadrupedal mammal supports 60% of body weight on its forelimbs and 40% on the hind limbs when standing still (Jayes and Alexander 1978). Peak forces that act on the limbs when the animal is running can be approximated by the equation $\pi(1-p)Mg/4\beta$ [where $(1-p)$ is the fraction of body weight, Mg, supported by the pair of legs under consideration, i.e. forelimbs or hind limbs, for an animal running with a duty factor of β] (Alexander et al. 1979a). Peak ground reaction forces (GRF) for animals can thus be determined indirectly by analysis of high speed film where duty factors can be resolved accurately, and this technique has been used to estimate forces for many running quadrupeds and bipeds (e.g. Alexander et al. 1979a). Peak forces have also been determined directly, using force platforms (Riemersma et al. 1988; Biewener et al. 1988). These experiments show that the peak GRF on one foot is typically between two and three times body weight, but occasionally reaches five times body weight for fast-moving mammals. Higher peak forces may result from other activities such as jumping, but these are not considered here. Musculo-skeletal structures must therefore be designed to cope with dynamic loads.

5.2 Bone Safety Factors

To reduce the chances of a limb bone breaking it could be made stronger by making it more massive, but there are obvious disadvantages in this approach. The limb would be clumsy and the energy needed to swing the leg backwards and forwards in running (which is proportional to its moment of inertia about its proximal end) would be large. In consequence, there appear to be at least two conflicting requirements for limb bones, those of strength and lightness, especially for animals that commonly run. There are other considerations about the costs of production and maintenance of tissues.

Stresses in limb bones during locomotion have been estimated from film and force-plate records (Alexander and Vernon 1975; Biewener and Blickhan 1988) and have also been "measured directly" using strain gauges cemented to the bones (Rubin and Lanyon 1982; Biewener et al. 1983). Bone strains can be converted to corresponding stresses using empirical values of Young's modulus for compact bone. The results obtained from the two approaches are generally comparable, but the indirect stress estimate technique appears to be prone to considerable error on occasion (Biewener et al. 1983).

The safety factors of horse limb bones provide an interesting case. The elongation of the limbs, which has occurred presumably to increase stride length and hence running speed, has been accompanied by a reduction in the number of digits. (The third metacarpal and metatarsal bones are very long, whilst the other metacarpals and metatarsals are either absent or have been reduced to splint bones. Horses support themselves on the end of their third digits). The net effect is a reduction in mass, especially in the distal regions, which lowers the inertial costs of swinging a long limb. Unfortunately, the same modifications to limb structure lead to increases in bending and may reduce limb stability. This is coupled with scaling problems associated with large body mass, where body mass has increased more rapidly than the cross-sectional area of the supporting limbs.

Large animals which may suffer these problems of scaling tend to have upright limb postures which result in the bones being well aligned with the ground reaction force. Biewener (1989) argues that this has the effect of increasing the effective mechanical advantage of muscle at joints (Fig. 10a). This reduces the forces that muscles have to generate to balance joint moments during the stance phase of locomotion and results in bones being exposed to lower muscle generated stresses. This is illustrated by the observation that, at the trot-gallop transition [reputedly, a "physiologically equivalent" point for animals of different size (Heglund and Taylor 1988)] the mass-specific force exerted by muscles of a 300-kg horse is only 17% of that exerted by a ground squirrel weighing 0.3 kg (Fig. 10b). This reduction in muscle-generated bone stress, coupled with the restriction of limb excursions to the parasagittal plane, results in peak bone stresses being independent of body size. Large animals have had to sacrifice agility in order to retain reasonable factors of safety.

There are always exceptions to the rule. Dwarf elephants that lived on Mediterranean islands underwent a massive reduction in body size, relative to the continental African elephants. However, this did not result in an equivalent

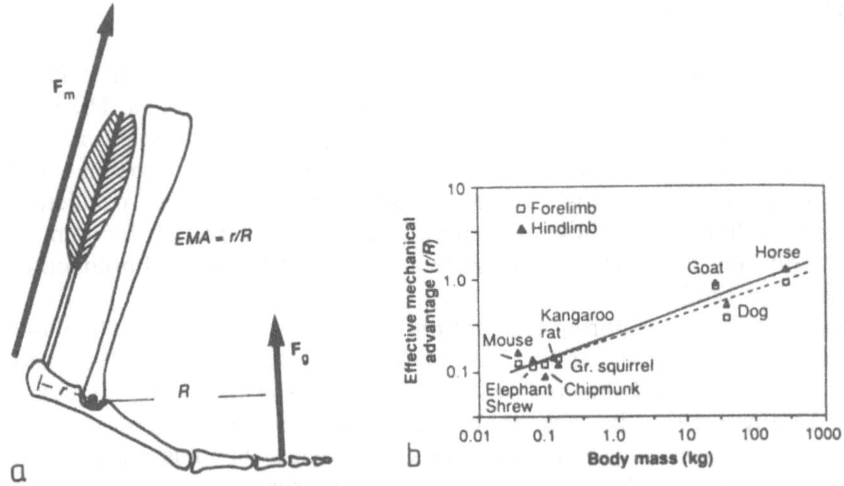

Fig. 10. **a** Schematic illustration of the effective mechanical advantage (*EMA*) of the ankle extensor muscles acting about the ankle joint. F_g is the ground reaction force vector, R is its mechanical advantage about the ankle. F_m is the force the extensor muscles produce, with a mechanical advantage of r, to balance the moments about the ankle. **b** Graphs of forelimb and hindlimb effective mechanical advantages, measured at the trot-gallop transition speed for each of the species shown. (**a, b** Biewener 1989)

reduction in limb bone dimensions. According to Bertram (1989), their limb bones were three times as robust as required, implying they had higher safety factors than shown in Table 1. However, as they are extinct, we cannot examine their agility in order to test this prediction.

Bones can also fail in ways other than by a single catastrophic event. It is likely that limb bones undergo cumulative microdamage (fatigue) and will fail under the repetitive loading regimes found in running. Fatigue-induced failure of horse long bones has been suggested to occur (Currey 1981).

In conclusion, the observed safety factors (of about 2–5) for limb bones (Table 1) may be important in the prevention of bone failure by occasional abnormal loads or by fatigue processes. It must also be remembered that the safety factors reported here are for animals running at high speed. Whilst this exposes limb bones to considerable levels of stress (and strain), the stresses encountered when more strenuous activities are undertaken, for example jumping, are likely to be higher.

5.3 Tendon Safety Factors

As failure of a tendon can severely handicap movement, if seems reasonable to assume that there will be selective pressures acting to minimize the chance of tendon failure.

Table 1. Bone stresses and safety factors for running animals

Animal	Mass (kg)	Bone	Peak stress (MPa) Cranial	Peak stress (MPa) Caudal	Gait	Speed ($m\,s^{-1}$)	Safety Factor	Reference
Chipmunk	0.1	Radius	+65	−86	Bound	4.1	2.6/2.9	Biewener (1983)
Ground Squirrel	0.13	Radius	+44	−76	Bound	4.4	3.9/3.2	Biewener (1983)
Goat	27	Radius	+36	−50	Gallop	4.5	4.7/5.0	Biewener and Taylor (1986)
		Tibia	+38	−48			4.7/5.0	
Kangaroo	42	Tibia	+85	−120	Hop	6.0	2.0/2.1	Alexander and Vernon (1975)
Ostrich	42	Tibio-tarsus	+70	−110	Run	12.0	2.4/2.3	Alexander et al. (1979a)
Horse	280	Radius		−45	Gallop	6.2	/5.6	Biewener et al. (1983)
		Metacarpus		−31			/8.0	
Elephant	2500	Humerus	±47		Run	4.5	2.5–2.9	Alexander et al. (1979b)

The bone stresses are compressive (−) on the caudal sides of these bones and tensile (+) on their cranial sides. Safety factors are calculated assuming cortical bone has a compressive strength of 250 MPa and a tensile strength of 170 MPa; see Currey (1984).

Table 2. Maximum stresses in mammalian limb tendons

Animal	Forelimb tendons Stress (MPa)	Safety Factor	Hindlimb tendons Stress (MPa)	Safety Factor
Oppossum	22	4.5	–	–
Possum	51	2.0	–	–
Dog	49	2.0	–	–
Badger	33	3.0	–	–
Sheep	74	1.4	49	2.0
Cow	69	1.4	–	–
Macaque	22	4.5	30	3.3
Human	17	5.9	67	1.5
Vervet	–	–	22	4.5
Camel	–	–	32	3.1

The maximum stress was calculated for each of the tendons of the wrist, ankle and extrinsic digital flexor and extensor muscles. It was assumed that the muscles would generate a stress of 0.3 MPa. The above values indicate the maximum tendon stress found in each limb. (Data from Ker et al. 1988, Alexander et al. 1982).

At the rates of strain that are likely to be encountered in life, tendons fail under tensile loads of about 100 MPa (Bennett et al. 1986). As the majority of tendons are in series with muscles fibres, the maximum tensile loads they can be exposed to are equal to the highest force that their attached muscle(s) can generate. [The peak isometric stress generated by vertebrate skeletal muscle appears to be about 200–300 kPa (see Biewener et al. 1988), but considerably higher stresses may be generated if a muscle is stretched whilst active]. Maximum tendon stresses have been calculated "indirectly" by using values for peak muscle stress and measurements of muscle fibre cross-sectional area (Ker et al. 1988) or by analysis of limb geometry with or without the simultaneous use of force-plate recordings (Alexander et al. 1982; Biewener and Blickhan 1988) and "directly", by the use of strain gauges attached to the tendon (Riemersma et al. 1988). The stresses to which mammalian limb tendons are exposed during locomotion (Table 2) indicate that safety factors are about 2 to 5, although calculation of maximum stresses in mammalian limb tendons (assuming a maximum muscle stress of 300 kPa) suggests that the range of safety factors is much wider (Ker et al. 1988). The values in Table 2 are for a biased sample, as the digital flexor and ankle extensor tendons are often exposed to high loads (e.g. in locomotion), whereas digital extensor tendons which are rarely, if ever subject to large forces have safety factors of up to 50. A benefit of having "low" safety factor tendons is that when an animal runs, the tendons stretch more under a given load than higher safety factor tendons, and thus store larger amounts of strain energy, which could make useful savings in the metabolic energy cost of running. A disadvantage is the possibility of rupture. This is a real danger, because muscle responds to exercise by hypertrophy more rapidly than its attached tendon, which may lead to a safety factor of less than 1.

6 Optimization

6.1 Muscle and Tendon Mass

The processes of evolution tend to optimize fitness (defined here as the probability of an animal passing on its genes to the next generation) and can act via behavioural or structural modifications. We cannot measure fitness directly, but it is possible to identify factors that will have a predictable effect on it. Using this approach, a number of investigators have attempted to establish how the locomotor system may be optimized.

In their analysis of mammalian limb tendons, Ker et al. (1988) proposed that optimization of tendon thickness is based on minimization of the mass of the muscle-tendon combination. The rationale being, that a reduction in thickness (and hence mass) of a tendon results in an increase in its extensibility. To compensate for this additional stretch, the attached muscle would require longer muscle fibres for it to be able to exert the same forces over the same range of joint movement, thus increasing its mass. A large relative muscle mass implies extra costs in terms of muscle growth and maintenance and in its use. All of these factors favour low area ratios (muscle fibre area/tendon cross-sectional area). Conversely, the disadvantage of having mass and volume located in the distal parts of limbs (which raises internal kinetic energy costs in running) favours high area ratios. The predicted optimum area ratio for a muscle-tendon combination doing positive work is 34, which corresponds to a peak tendon stress of about 10 MPa.

The maximum tendon stresses in limb tendons of a variety of mammals (Fig. 11) indicates that the theoretical optimum stress falls within the main peak, thus giving broad support to the optimization theory. There are also deviations from this value and it is of interest to note that many of the tendons exposed to high stresses have attached muscles that contain short fibres and are the tendons implicated in elastic strain energy storage during fast locomotion.

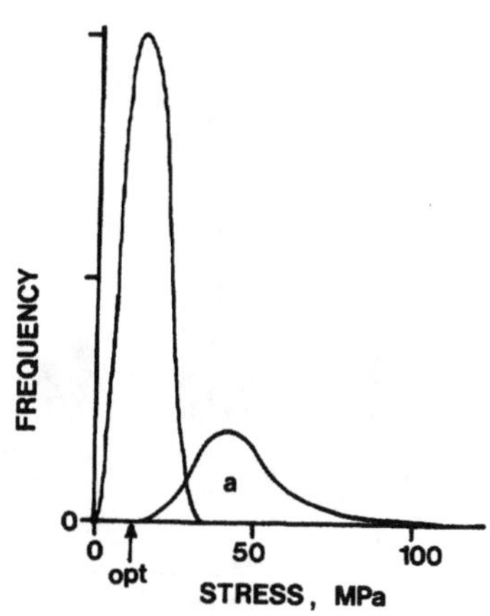

Fig. 11. Schematic illustration of the distribution of maximum tendon stresses among limb tendons of mammals. *opt* the theoretical optimum; a tendons whose muscles have short fibres. (After Ker et al. 1988)

6.2 Bone Mass

Limb bones are also likely to be subject to selective pressures acting to minimize their mass without sacrificing their strength or stability; the tubular nature of many limb bones can be viewed as one such adaptation. An example, given by Currey (1984), shows that a 10% reduction in the mass of distal limb bones of a horse can produce a 5% saving in the energy cost of galloping. Currey and Alexander (1985) used a theoretical approach to predict how the ratio of the internal to the external diameter of tubular bones would vary, whilst fulfilling certain mechanical requirements, assuming that bones are optimized for minimum mass. Data were presented for long bones from a wide variety of vertebrates. Bone dimensions of the 29 terrestrial mammal species included suggest that the majority of limb bones have been optimized for impact strength and, to a lesser degree, ultimate strength. Bone stiffness may be more important in small animals, where limb bones do undergo appreciable bending.

In a further study, the optimizing of bone size (as indicated by its radius) by minimizing the mass of both muscle and bone has been considered (Alexander et al. 1990). The theory suggested that if leg bones were optimized for stiffness then peak mid-shaft bone stresses of about ± 70 MPa would be produced by bending moments during strenuous activities such as fast locomotion. The empirical data (Table 1) shows that measured stresses conform reasonably well with theory. Unfortunately, it is not possible to conclude whether the bones are selected for stiffness, as similar values might be expected if selection was for strength to ensure that the bones have adequate factors of safety (Alexander 1984b).

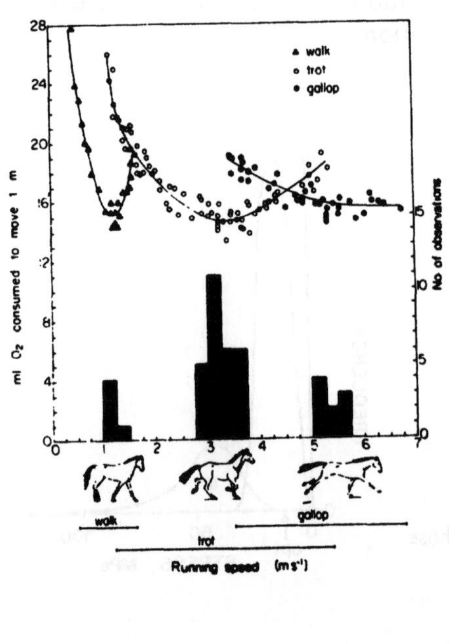

Fig. 12. A graph of the oxygen cost to move unit distance against speed for a horse. The histograms show the speeds that were used when the horse moved freely. (Hoyt and Taylor 1981)

6.3 Gait

It is not only the components of the locomotor system, but locomotion itself, that may be optimized. Animals move with particular gaits at particular speeds. These gaits are used, presumably in preference to others, because they confer some benefit upon the animal. This benefit can be regarded as increasing the fitness of the individual.

Of the many different criteria available. Alexander (1989a) considered that minimising the energetic cost of locomotion would be of general importance to an animal. Evidence to support the hypothesis that gaits are chosen to minimize energy costs comes from measurements of oxygen consumption during locomotion. Hoyt and Taylor (1981) found that although trained ponies could move at a wide range of speeds within each gait on a treadmill, they chose narrow ranges of speed for walking, trotting or galloping when allowed to move freely (Fig. 12). The preferred speeds corresponded to the minimum oxygen consumption to move unit distance, within each gait. Most animals have a preferred speed (see Pennycuick 1975), and it is likely that these coincide with oxygen consumption minima as in the horse.

References

Adrian M, Grant B, Ratzlaff M, Ray J, Boulton C (1977) Electrogoniometric analysis of equine metacarpophalangeal joint lameness. Am J Vet Res 38: 431–435

Alexander R McN (1976) Estimates of speeds of dinosaurs. Nature (Lond) 261: 129–130

Alexander R McN (1977) Terrestrial locomotion. In: Alexander R McN, Goldspink G (eds) Mechanics and energetics of animal locomotion. Chapman and Hall, London, pp 168–203

Alexander R McN (1983) Animal mechanics, 2nd edn. Blackwell, Oxford

Alexander R McN (1984a) Elastic energy stores in running vertebrates. Am Zool 24: 85–94

Alexander R McN (1984b) Optimum strengths for bones liable to fatigue and accidental fracture. J Theor Biol 109: 621–636

Alexander R McN (1989a) Optimization and gaits in the locomotion of vertebrates. Physiol Rev 69: 1199–1227

Alexander R McN (1989b) On the synchronization of breathing with running in wallabies (*Macropus* spp.) and horses (*Equus caballus*). J Zool Lond 218: 69–85

Alexander R McN, Bennett MB (1987) Some principles of ligament function, with examples from the tarsal joints of the sheep (*Ovis aries*). J Zool Lond 211: 487–504

Alexander R McN, Jayes AS (1980) Fourier analysis of forces exerted in walking and running. J Biomech 13: 383–390

Alexander R McN, Jayes AS (1983) A dynamic similarity hypothesis for the gaits of quadrupedal mammals. J Zool Lond. 201: 135–152

Alexander R McN, Ker RF (1990) Running is priced by the step. Nature (Lond) 346: 220–221

Alexander R McN, Maloiy GMO (1984) Stride lengths and stride frequencies of primates. J Zool Lond 202: 577–582

Alexander R McN, Vernon A (1975) Mechanics of hopping by kangaroos (Macropodidae). J Zool Lond 177: 265–303

Alexander R McN, Langman VA, Jayes AS (1977) Fast locomotion of some African ungulates. J Zool Lond 183: 291–300

Alexander R McN, Maloiy GMO, Njau R, Jayes AS (1979a) Mechanics of running of the ostrich (*Struthio camelus*). J Zool Lond 187: 169–178

Alexander R McN, Maloiy GMO, Hunter B, Jayes AS, Nturibi J (1979b) Mechanical stresses in fast locomotion of buffalo (*Syncerus caffer*) and elephant (*Loxodonta africana*). J Zool Lond 189: 135–144
Alexander R McN, Maloiy GMO, Ker RF, Jayes AS, Warui CN (1982) The role of tendon elasticity in the locomotion of the camel (*Camelus dromedarius*). J Zool Lond 198: 293–313
Alexander R McN, Dimery NJ, Ker RF (1985) Elastic structures in the back and their role in galloping in some mammals. J Zool Lond 207: 467–482
Alexander R McN, Bennett MB, Ker RF (1986) Mechanical properties and function of the paw pads of some mammals. J Zool. Lond 209: 405–419
Alexander R McN, Ker RF, Bennett MB (1990) Optimum stiffness for leg bones. J Zool Lond 222: 471–478
Attenburrow DP (1982) Time relationships between the respiratory cycle and the limb cycle in the horse. Equine Vet J 14: 69–72
Avery RA, Mueller CF, Smith JA, Bond DJ (1987) The movement patterns of lacertid lizards: speed gait and pauses in *Lacerta vivipara*. J Zool Lond 211: 47–63
Baudinette RV, Gannon BJ, Runciman WB, Wells S, Love JB (1987) Do cardiorespiratory frequencies show entrainment with hopping in the tammar wallaby? J Exp Biol 129: 251–263
Bennett MB (1987) Fast locomotion of some kangaroos. J Zool Lond 212: 457–464
Bennett MB (1989) A possible energy-saving role for the major fascia of the thigh in running quadrupedal mammals. J Zool Lond 219: 221–230
Bennett MB, Ker RF, Dimery NJ, Alexander R McN (1986) Mechanical properties of various mammalian tendons. J Zool Lond 209: 537–548
Bennett MB, Ker RF, Alexander R McN (1989) Elastic strain energy storage in the feet of running monkeys. J Zool Lond 217: 469–475
Bertram JEA (1989) Size transitions in terrestrial mammals: the interesting case of dwarf elephants. Am Zool 29: 182A, 800
Biewener AA (1983) Locomotor stresses in the limb bones of two small mammals: the ground squirrel and chipmunk. J Exp Biol 103: 131–154
Biewener AA (1989) Scaling body support in mammals: limb posture and muscle mechanics. Science 245: 45–48
Biewener AA, Blickhan R (1988) Kangaroo rat locomotion: design for elastic energy storage or acceleration? J Exp Biol 140: 243–255
Biewener AA, Taylor CR (1986) Bone strain: a determinant of gait and speed? J Exp Biol 123: 383–400
Biewener AA, Alexander R McN, Heglund NC (1981) Elastic energy storage in the hopping of kangaroo rats (*Dipodomys spectabilis*). J Zool Lond 195: 369–383
Biewener AA, Thomason J, Goodship A, Lanyon LE (1983) Bone stress in the horse forelimb during locomotion at different gaits: a comparison of two experimental methods. J Biomech 16: 565–576
Biewener AA, Blickhan R, Perry AK, Heglund NC, Taylor CR (1988) Muscle forces during locomotion in kangaroo rats: force platform and tendon buckle measurements compared. J Exp Biol 137: 191–205
Blickhan R, Full RJ (1987) Locomotion energetics of the ghost crab II. Mechanics of the centre of mass during walking and running. J Exp Biol 130: 155–174
Bowerman RF (1975) The control of walking in the scorpion. I. Leg movements during normal walking. J Comp Physiol 100: 183–196
Bramble DM (1989) Axial-appendicular dynamics and the integration of breathing and gait in mammals. Am Zool 29: 171–186
Bramble DM, Carrier DR (1983) Running and breathing in mammals. Science 219: 251–256
Burrows M, Hoyle G (1973) The mechanism of rapid running in the ghost crab, *Ocypode ceratophthalma*. J Exp Biol 58: 327–349
Cavagna GA (1975) Force platforms as ergometers. J Appl Physiol 39: 174–179
Cavagna GA, Heglund NC, Taylor CR (1977) Mechanical work in terrestrial locomotion: two basic mechanisms for minimizing energy expenditure. Am J Physiol 223: R243–R261
Currey JD (1981) What is bone for? Property-function relationships in bone. Symp Am Soc Mech Eng 45: 13–26
Currey JD (1984) The mechanical adaptations of bones. Univ Press, Princeton

Currey JD, Alexander R McN (1985) The thickness of the walls of tubular bones. J Zool Lond A 206: 453–468
Dawson TJ, Taylor CR (1973) Energetic cost of locomotion in kangaroos. Nature (Lond) 246: 313–314
Delcomyn F (1971) The locomotion of the cockroach *Periplaneta americana*. J Exp Biol 54: 443–452
Dickinson JA, Cook SD, Leinhardt TM (1985) The measurement of shock waves following heel strike during running. J Biomech 18: 415–422
Dimery NJ, Alexander R McN (1985) Elastic properties of the hind foot of the Donkey, *Equus asinus*. J Zool Lond A 207: 9–20
Dimery NJ, Ker RF, Alexander R McN (1986a) Elastic properties of the feet of deer (Cervidae). J Zool Lond A 208: 161–169
Dimery NJ, Alexander R McN, Ker RF (1986b) Elastic extension of leg tendons in the locomotion of horses (*Equus caballus*). J Zool Lond A 210: 415–425
Farlow JO (1981) Estimates of dinosaur speeds from a new trackway site in Texas. Nature (Lond) 294: 747–748
Fedak MA, Heglund NC, Taylor CR (1982) Energetics and mechanics of terrestrial locomotion. J Exp Biol 79: 23–40
Flitney FW, Hirst DG (1978) Cross-bridge detachment and sarcomere "give" during stretch of active frog's muscle. Physiol 276: 449–465
Full RJ, Tu MS (1990) Mechanics of six-legged runners. J Exp Biol 148: 129–146
Gambaryan PP (1974) How mammals run: anatomical adaptations. Wiley, New York
Garland T Jr (1983) The relation between maximal running speed and body mass in terrestrial mammals. J Zool Lond 199: 157–170
Garland T Jr, Geiser F, Baudinette RV (1988) Comparative locomotor performance of marsupial and placental mammals. J Zool Lond 215: 505–522
Goslow GE Jr, Seeherman HJ, Taylor CR, McCutchin MN, Heglund NC (1981) Electrical activity and relative length changes of dog limb muscles as a function of speed and gait. J Exp Biol 94: 15–42
Gunther B (1975) Dimensional analysis and theory of biological similarity. Physiol Rev 55: 659–699
Harris J, Ghiradella H (1980) The forces exerted on the substrate by walking and stationary crickets. J Exp Biol 85: 263–279
Hayes G, Alexander R McN (1983) The hopping gait of crows (Corvidae) and other bipeds. J Zool Lond 200: 205–213
Heglund NC, Taylor CR (1988) Speed, stride frequency and energy cost per stride: how do they change with body size and gait? J Exp Biol 138: 301–318
Heglund NC, Taylor CR, McMahon TA (1974) Scaling stride frequency and gait to animal size: mice to horses. Science 186: 1112–1113
Heglund NC, Cavagna GA, Taylor CR (1982) Energetics and mechanics of terrestrial locomotion. III. Energy changes of the centre of mass as a function of speed and body size in birds and mammals. J Exp Biol 97: 41–56
Herreid CF II, Full RJ (1986) Locomotion of hermit crabs (*Coenobita compressus*) on beach and treadmill. J Exp Biol 120: 283–296
Hildebrand M (1977) Analysis of asymmetrical gaits. J Mammal 58: 131–156
Hoyt DF, Taylor CR (1981) Gait and the energetics of locomotion in horses. Nature (Lond) 292: 239–240
Jayes AS, Alexander R McN (1978) Mechanics of locomotion of dogs (*Canis familiaris*) and sheep (*Ovis aries*). J Zool Lond 185: 289–308
Jayes AS, Alexander R McN (1980) The gaits of chelonians: walking techniques for very low speeds. J Zool Lond 191: 353–378
Jenkins FA Jr, Camazine SM (1977) Hip structure and locomotion in ambulatory and cursorial carnivores. J Zool Lond 181: 351–370
Jørgensen U, Bojsen-Møller F (1989) Shock absorbency of factors in the shoe/heel interaction – with special focus on role of the heel pad. Foot Ankle 9: 294–299
Ker RF (1981) Dynamic tensile properties of the plantaris tendon of sheep (*Ovis aries*). J Exp Biol 93: 283–302

Ker RF, Dimery NJ, Alexander R McN (1986) The role of tendon elasticity in hopping in a wallaby (*Macropus rufogriseus*). J Zool Lond A 208: 417–428
Ker RF, Bennett MB, Bibby SR, Kester RC, Alexander R McN (1987) The spring in the arch of the human foot. Nature (Lond) 325: 147–149
Ker RF, Alexander R McN, Bennett MB (1988) Why are mammalian tendons so thick? J Zool Lond 216: 309–324
Ker RF, Bennett MB, Alexander R McN (1989) Foot strike and the properties of the human heel pad. Proc Instn Mech Eng 203: 191–196
Kingsbury HB, Quddus MA, Rooney JR, Geary JE (1978) A laboratory system for production of flexion rates and forces in the forelimb of the horse. Am J Vet Res 39: 365–369
Kram R, Taylor CR (1990) Energetics of running: a new perspective. Nature (Lond) 346: 265–267
McMahon TA (1975) Using body size to understand the structural design of animals: quadrupedal locomotion. J Appl Physiol 39: 619–627
Muybridge E (1899) Animals in motion. Chapman and Hall, London (Republished 1957)
Paul IL, Munro MB, Abernethy PJ, Simon SR, Radin EL, Rose RM (1978) Musculo-skeletal shock absorption: relative contribution of bone and soft tissues at various frequencies. J Biomech 11: 237–239
Pennycuick CJ (1975) On the running of the gnu (*Connochaetes taurinus*) and other animals. J Exp Biol 63: 775–799
Ratzlaff MH (1989) Quantitative methods for the analysis of equine locomotion and their applications to other species. Am Zool 29: 267–285
Riemersma DJ, Schamhardt HC, Hartman W, Lammertink JLMA (1988) Kinetics and kinematics of the equine hind limb: In vivo tendon loads and force plate measurements in ponies. Am J Vet Res 49: 1344–1352
Rubin CT, Lanyon LE (1982) Limb mechanics as a function of speed and gait: a study of functional strains in the radius and tibia of horse and dog. J Exp Biol 101: 187–211
Strang KT, Steudel K (1990) Explaining the scaling of transport costs: the role of stride frequency and stride length. J Zool Lond 221: 343–358
Taylor CR, Heglund NC (1982) Energetics and mechanics of terrestrial locomotion. Ann Rev Physiol 44: 97–107
Taylor CR, Schmidt-Nielsen K, Raab JL (1970) Scaling of energy cost of running to body size in mammals. Am J Physiol 219: 1104–1107
Taylor CR, Heglund NC, Maloiy GMO (1982) Energetics and mechanics of terrestrial locomotion. I. Metabolic energy consumption as a function of speed and body size in birds and mammals. J Exp Biol 97: 1–21
Van Weeren PR, Van den Bogert AJ, Barneveld A (1988) Quantification of skin displacement near the carpal, tarsal and fetlock joints of the walking horse. Equine Vet J 20: 203–208

Chapter 6

Locomotion of Soft Bodied Animals

M.E. DeMont

Contents

1 Introduction

Many of the basic principles used by soft bodied animals to move have been described in the wonderful books by Trueman (1975) and Alexander (1982). I have tried not to repeat material described previously, although some classic older

Biology Department, St. Francis Xavier University, Antigonish, Nova Scotia, Canada studies have been cited. The chapter is divided into three broad areas; a brief discussion on muscular-hydrostatic skeletons, a section that includes any form of locomotion in, or on a substrate, and a section on swimming.

Soft bodied animals display a remarkable collection of mechanisms to cope with the problem of tissues that do not deform when they should, or do deform when they should not. Movement requires tissue that deforms, so many of the problems of locomotion in soft bodied animals are related to deforming parts of their bodies. Thus, a major portion of this chapter describes the functional morphology of soft bodied animals, and the mechanics of their locomotor tissues. This seems to have been an important area of research in the last decade.

2 Skeletons

Wainwright (1982) described two types of structural support systems in animals; the jointed framework, and the stretched membrane hydrostat. A third support system, the muscular-hydrostat (Kier and Smith 1985), has now been recognized. The function of the first system is not relevant to this chapter, but will be discussed in chapter 7.

The arrangement and function of the musculature in the hydrostatic skeleton has been reviewed by a number of authors (Wainwright 1970; Trueman 1975; Wainwright et al. 1976; Wainwright 1982; Alexander 1983). A hydrostatic skeleton is essentially a membrane-bound container of fluid; the volume of the container is constant. The fluid in the container constitutes a significant proportion of the total mass of the animal. The fluid is incompressible at physiological pressures, so that changes in one dimension result in concomitant changes in at least one other dimension. The muscles powering the system are generally confined to the membrane surrounding the fluid, and consist of two muscle types, circular and longitudinal. If the circular muscles contract, the animal gets long and thin. Contraction of the longitudinal muscles makes the animal short and fat. The muscles are thus antagonistic. The ratio of power output to mass is low, so that the animals are not capable of rapid movements. A hydrostatic skeleton is not intrinsically stiff, indeed organisms with hydrostatic skeletons are very flexible. Trueman and his colleagues have made some recent observations on hydrostatic skeletons in gastropods (Trueman and Brown 1985, 1987), and brachiopods (Trueman and Wong 1987). Gutmann (1981) examined the relationships between invertebrate phyla using a functional analysis of the hydrostatic skeleton, and more recently (Gutmann 1988) used the concept of the hydraulic principle as a basic explanation of the body plans of animals. Wadepuhl and Beyn (1989) used finite element methods to model hydrostatic skeletons.

The muscular-hydrostatic skeleton is essentially a hydrostatic skeleton, except that the fluid is replaced by a three-dimensional array of closely packed muscles. The muscle fibres can be arranged in three different directions, (1) perpendicular to the long axis, (2) parallel to the long axis, and (3) helical or oblique to the

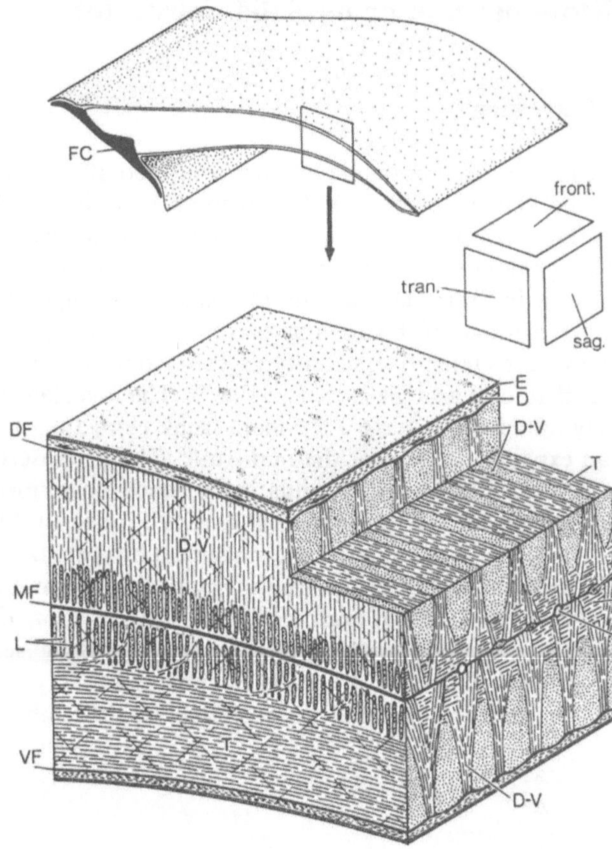

Fig. 1. A muscular-hydrostatic skeleton. The anatomy of the fin of *Sepia officinalis*. The sectional planes (*front.* frontal; *tran.* transverse; *sag.* saggital) are indicated. *D* dermis; *DF* dorsal-ventral muscle; *E* epidermis; *FC* fin cartilage; *L* longitudinal muscle; *MF* median fascia; *N* fin nerve; *T* transverse muscle; *V* blood vessel; *VF* ventral fascia. (Kier 1989)

long axis (Kier and Smith 1985). The power output to mass is large, so muscular-hydrostats are capable of producing rapid movements. Muscular-hydrostats are capable of finer control than fluid-filled hydrostats. For example, localized bending of a muscular-hydrostat is possible if longitudinal muscles (and their antagonists) contract in a restricted area. In a fluid-filled hydrostat, contraction of longitudinal muscles increases pressure in the whole structure, so localized movements are difficult (Kier 1988). Most of the work on muscular-hydrostats has been on muscular organs in molluscs (Kier 1985; Kier and Smith 1985; Kier 1989). Two reviews have been written on the function of muscular-hydrostats; one on squid (Gosline and DeMont 1985), and another on molluscs in general (Kier 1985). Figure 1 shows an example of a muscular-hydrostat, the fin of *Sepia officinalis*.

3 Movements in or on Solid Substrates

3.1 Burrowing

The dynamics and mechanics of burrowing activity in animals has been extensively reviewed (Trueman 1968, 1975; Trueman and Ansell 1969; Trueman and Jones 1977); however, some recent work has appeared on burrowing in *Priapulus caudatus* and *Polyphysia crassa.* Elder and Hunter (1980) and Hammond (1980) described burrowing in *Priapulus caudatus. Priapulus* has a single, undivided coelomic cavity, and its method of burrowing is similar to many soft bodied invertebrates. Burrowing is a repetitive, stereotyped behaviour that can be partitioned into several stages that form a single cycle (see Fig. 2). The first stage starts when the proboscis is fully retracted. In stage 2 the main trunk elongates, pushing anteriorly, resisted by an expanded region of the mid-trunk. The proboscis then starts to protrude, and dilates at the tip. This dilation extends posteriorly. In stage 5 the trunk shortens, resisted by the therminal anchor formed by the dilated proboscis. The later two stages are defined by the movement of a peristaltic wave from the posterior to the anterior region of the trunk. The proboscis simltaneously retracts.

Hunter et al. (1983) examined burrowing of *Priapulus* and *Polyphysia* in an artificial substrate that is nearly translucent in seawater. This permitted photographic records to be made of animals actually moving in the substratum, and thus more detailed analysis was possible, although no substantially new ideas

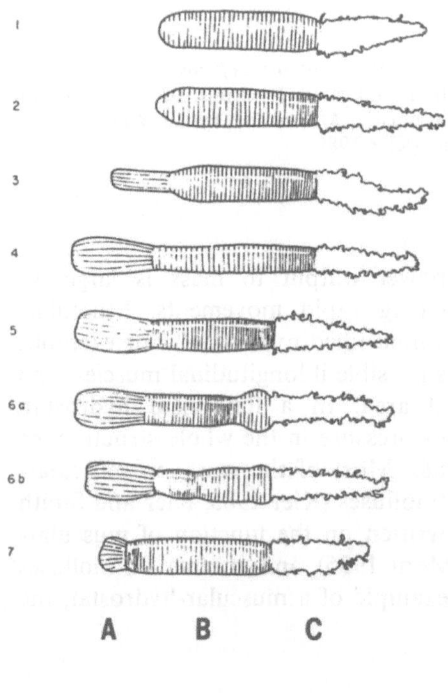

Fig. 2. Burrowing in *Priapulus.* (Elder and Hunter 1980). *A* proboscis; *B* trunk; *C* caudal appendage

were described. Hunter and Elder (1989) quantified the dynamics and mechanical forces generated during burrowing in *Priapulus* and *Polyphysia*. *Priapulus* burrows at 0.76 cm s^{-1}, while *Priapulus* moves more slowly at 0.24 cm s^{-1}. They calculated the net cost of transport (NCT) using the method of Alexander (1975):

$$NCT = \frac{\text{Power during locomotion} - \text{Power at rest}}{\text{body mass} \times \text{velocity}}.$$

Priapulus had a net cost of transport of 314 J kg^{-1} m^{-1}, while the cost of transport for *Polyphysia* was 635 J kg^{-1} m^{-1}. For comparison, the maximum cost of transport for flying or running animals is about 100 J kg^{-1} m^{-1}. The high cost of transport is most likely related to the high substratum resistance.

3.2 Crawling

3.2.1 Muscular Pedal Crawling

The mechanics of muscular pedal crawling has been reviewed (Trueman 1975; Trueman and Jones 1977; Alexander 1982). Movement is caused by pedal waves on the ventral surface of the foot, which is coupled to the substratum by a thin layer of pedal mucus. A variety of wave patterns have been observed, and have been classified with a simple scheme (see Trueman and Jones 1977). Until recently, the mechanism by which these muscular waves results in locomotion has received little attention. Denny (1980, 1981, 1984) and Denny and Gosline (1980) have examined in some detail the mechanical properties of the pedal mucus, and have incorporated the properties in a quantitative model that is able to predict the forces required to propel the animal.

The mucus consists of 96–97% water, with the remaining 3–4% a high molecular weight glycoprotein. Deyrup-Olsen et al. (1983) examined the process in which the mucus is secreted, and suggest that this process differs markedly from the single-step exocytosis of fluid mucus from other mucous membranes. Their results suggest that intact vesicles are extruded, and are disrupted after being freed from the membrane surface.

The pedal mucus exhibits mechanical properties that are extremely dependent on the strain (Denny and Gosline 1980). The mucus behaves as a viscoelastic solid at small deformations, with a shear modulus of about 200 Pa (top graph of Fig. 3). The mechanical properties undergo a remarkable change at a strain of between 5 and 6. The mucus behaves as a viscoelastic solid below strains of about 5 to 6, while at strains above 6 it behaves as viscous liquid. The stress at which the material changes its properties increases as the rate of deformation increases (bottom graph of Fig. 3). The properties the mucus exhibits as a viscoelastic solid return if the material is permitted to heal. The transitions described above are ideally suited to the locomotion of a slug.

Locomotion consists of the movement of muscular waves that start at that posterior end and move toward the head (Denny 1980, 1981). At any one time there are about 15 waves present on the foot. The leading edge of a wave stresses the mucus to its yield point, so that the area of the foot contained in the wave

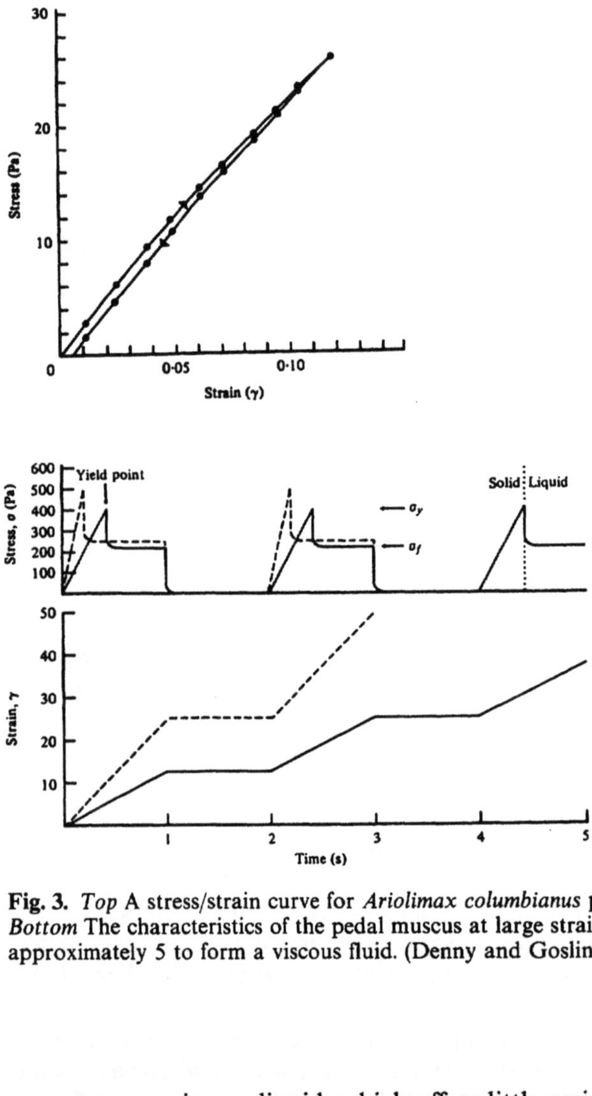

Fig. 3. *Top* A stress/strain curve for *Ariolimax columbianus* pedal muscus at a small strain rate. *Bottom* The characteristics of the pedal muscus at large strains. The muscus yields at a strain of approximately 5 to form a viscous fluid. (Denny and Gosline 1980)

moves over a viscous liquid, which offers little resistance. The mucus then heals, to become a viscoelastic solid. Thus the area of the foot behind the waves passes over a solid surface. The mucus is able to resist the force generated by the waves, so that the slug is propelled forward.

The cost of transport for several species of crawling gastropods has been measured. Denny (1980) measured the metabolic cost of crawling in the terrestrial slug *Ariolimax columbianus*, and found a value of about 900 J kg^{-1} m^{-1}, which is considerably higher than other forms of locomotion. Houlihan and Innes (1982) measured the metabolic cost of transport in four species of Mediterranean trochids (Mollusca, Archeogastropoda): *Monodonta turbinata*, *Monodonta articulata*, *Gibbula richardi* and *Gibbula rarilineata*, and found values lower than reported by Denny, but still higher than the cost of transport for flying and swimming. Innes and Houlihan (1985) measured the metabolic cost of transport for the temperate intertidal gastropod *Littorina littorea*, and found values similiar to those for Mediterranean gastropods. Both of the latter studies measured the cost of locomotion for animals moving in air and water, and showed that the cost of transport was considerably higher in air. Crawling speeds are in the order of 0.3 mm s^{-1}.

3.2.2 Crawling in Worms

Most of the work on crawling in worms that has appeared since the reviews by Trueman and Packard (1977) and Alexander (1982) has examined the importance of the cylindrical trellis of collagen fibres that wrap around the bodies of many worms. Seymour (1983) examined the constraints imposed by the helical fibres. Alexander (1987) developed a model that describes what happens when cylindrical animals with helical fibres in their skin bend, and later reviewed his ideas (Alexander 1988). In his model, the body is a right circular cylinder, with fibres running uninterrupted along its length. The bent curvature is uniform, with a circular cross-section. His model predicts that the animals are unstable when they are straight, so will tend to bend one way. Pabst and Healy (1989) have suggested that previous models do not account for the noncircular cross-sections of many cylindrical animals, which may be using other load-bearing elements to maintain their noncircular cross-sections. Keller and Falkovitz (1983) describe a quantitative dynamical theory of crawling in worms. Dobrolyubov (1986) models travelling waves of deformation in elongate bodies, the local deformations being transverse or longitudinal. Wainwright (1988) has examined the importance of the cylindrical shape of worms.

3.3 Crawling in Other Animals

Dimock (1985) examined the ciliary-powered locomotion of the marine snail *Ilyanassa obsoleta*, and found maximum speeds of about 3 mm s^{-1}.

Bond and Harris (1988) examined locomotion in several species of freshwater and marine sponges. Sponges can crawl at speeds of about 4 mm day^{-1}. There was no evidence of travelling waves of surface contractions. Locomotion is achieved by the cumulative locomotion of the cells that compose the lower surface of the sponge.

Stern-Tomlinson et al. (1986) completed a kinematic study of crawling in the leech *Hirudo medicinalis*. The leech crawls by repeated extensions and contractions of the body, with alternating attachments of the head and posterior sucker. They described two types of crawling, inchworm crawling and vermiform crawling. In the former variant, the body forms a loop at the end of the cycle, so that the two suckers are close to each other. In vermiform crawling, the body remains close to the substrate throughout the cycle. The velocity for each type of crawling was similar, and reached a maximum of about 1.3 cm s^{-1}.

4 Swimming

Soft bodied animals use a vast collection of mechanisms to swim. The best swimmers use jet propulsion, although this is inherently inefficient (Alexander 1977). Fin-based swimming has appeared in the cephalopods, and may be a mechanism to overcome the inherent inefficiency.

4.1 Coelenterates

4.1.1 Sea Anemones

Swimming in the Cnidaria is restricted to the sea anemones and jellyfish. Some aspects of locomotion were reviewed by Mackie (1974). Swimming is rare in the sea anemones, but has been recorded in several species. Some species (*Stomphia*) swim with sharp flexions of the column (Sund 1958), while others (*Gonactinia*) swim with synchronous oscillations of the tentacles (Robson 1971). The mechanical properties of the mesoglea in sea anemones have been examined extensively, but have not been correlated with swimming behaviour. I summarize the properties here, however, since some are important to the response of sea anenomes to fluid flow. Chapman (1953b) made simple mechanical measurements that lasted in the order of minutes, but Alexander (1962) showed that the important time-dependent behaviour lasted in the order of hours. Gosline (1971a, b) extended Alexander's observations and was able to formulate a macromolecular model that described the behaviour. The mesoglea consists of collagen fibres embedded in a matrix. The mechanical behaviour of the intact mesoglea is dominated by the behaviour of

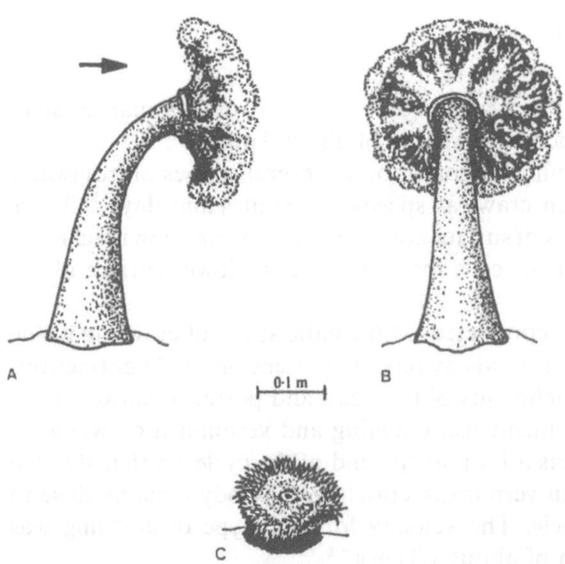

Fig. 4 A–C. A large *Metridium senile* exposed to currents bends at the oral disc (*arrow* indicates direction of flow. **A** Side view. **B** View from upstream. **C** An *Anthopleura xanthogrammica*. (Koehl 1977a)

the matrix, which exhibits rubber elasticity. The collagen acts as a filler that can reinforce the matrix, thus providing some short-term rigidity to the material. Koehl (1977a, b, c) measured mechanical and structural properties of the mesoglea of two species of anemones: *Metridium senile* and *Anthopleura xanthogrammica. Metridium* occur in calm areas, but are comparatively tall, so experience the free stream velocity. *Anthopleura* live where wave action exists, but the animals are short, so experience little free stream velocity. Koehl demonstrated several mechanical and morphological features that allowed these anemones to accommodate to the very different flow regimes that each encounters. The flexural stiffness, for example, of the upper column of *Metridium* is about two orders of magnitude less than the flexural stiffness of the body wall of *Anthopleura. Metridium* can easily bend to expose the greatest area of its oral disc for filter feeding, and *Anthopleura* remains upright to catch mussels that fall onto its oral discs (see Fig. 4). Some of these ideas were reviewed elsewhere (Koehl 1982, 1984). Denny (1989) has extensively examined the biology of animals living in wave swept environments.

4.1.2 Jellyfish

General Biology. Medusae are another example of jet-propelled swimmers. Gladfelter (1973) made a broad survey of the locomotor systems of both hydro- and scypho-medusae, examining the locomotor structures and function. His study attempted to draw general conclusions about fundamental characteristics of swimming in these animals. Two general features distinguish the hydromedusan and scyphomedusan locomotor systems: Scyphomedusae are generally much larger, and lack a velum. It is clear that the locomotor systems in general can have a broad range of structures, all of which can effect the function of the system. In all animals, radial mesogleal fibres were found in regions of the bell that deformed during contraction. The control of locomotion in both scyphomedusan (Satterlie and Spencer 1979) and Hydromedusan (Mackie 1990; Spencer 1978, 1979, 1982; Spencer and Satterlie 1981) jellyfish have been extensively examined. Leonard (1980, 1982) examined the temporal organization of swimming activity in the hydrozoan *Sarsia tubulosa.* She found that the variability of timing of the spontaneous swimming bouts is due to the wide range of rhythms that can occur, and also due to the animal's ability to change from one activity to the other. The study of complex dynamics, such as those seen in the swimming behaviour of *Sarsia,* is currently a major field of study in many areas of science (Glass and Mackey 1988). The application of nonlinear mathematics to the organization of behaviour in jellyfish may provide an important application of nonlinear dynamics to a biological system, especially since the physiology of the nervous system is well known.

Functional Morphology. The most detailed studies have been made on the hydrozoan, *Polyorchis* (DeMont and Gosline 1988a–c; Gladfelter 1972). The anatomy of the locomotor structure has been described in detail, and is summarized here. The bell cavity is approximately cylindrical, and in cross-section the exumbrellar and subumbrellar outlines are essentially circular. The subumbrellar mesoglea is scalloped into eight longitudinal adradial ridges, or "joints" that extend from above the subumbrellar summit nearly to the margin of the bell. Most of the bell consists of transparent, noncellular mesoglea that is traversed by numerous radially arranged fibres.

Circular muscles line the subumbrellar surface and power the jet cycle. The cycle is initiated by contraction of these swimming muscles, which reduces the diameter of the bell. The bell does not change length during this process. Deformation of the bell is not uniform around the circumference, since the mesoglea folds around the adradial joints. Recoil of the bell to resting dimensions is a passive process, powered by the potential energy stored during the deformation of the bell (see Fig. 5).

Some confusion seemed to have existed as to the exact nature of the radial mesogleal fibres (Gladfelter 1972; Mackie and Mackie 1967; Chapman 1953a, 1959); however, recent work (DeMont and Gosline 1988a) confirms that the fibres are "elastic" in nature. These materials are typically highly extensible (Gosline 1980). DeMont and Gosline (1988a) measured strains of about 0.57 in the radial fibres of *Polyorchis* mesoglea. For comparison, collagen fibres can be reversibly extended to strains of no more than 0.1 (Wainwright et al. 1976).

Mechanical Properties of the Mesoglea. Alexander (1964) performed some creep-recovery experiments on isolated samples of jellyfish mesoglea. He found that jellyfish mesoglea is much more compliant than sea anemone mesoglea. It had a broad distribution of retardation times, indicating that the mesoglea resembles a simple polymeric gel. These results show that the viscoelastic properties of jellyfish mesoglea and sea anemone are very different.

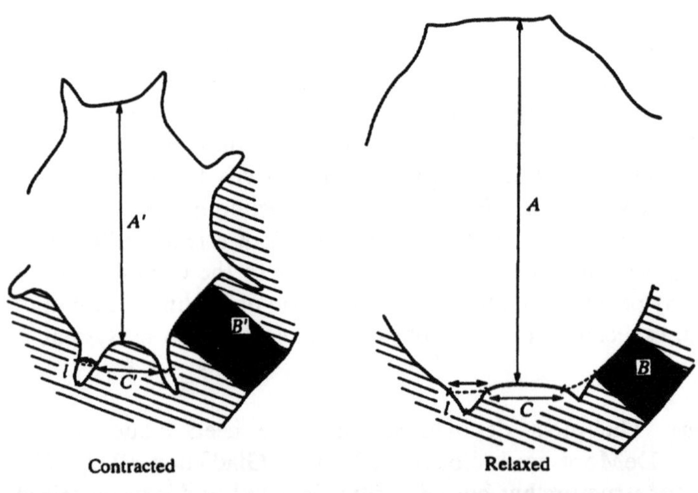

Fig. 5. An outline of the changes in the bell geometry of the hydromedusan jellyfish *Polyorchis penicillatus* during a single jet. *Letters* with and without primes indicate corresponding dimensions in the contracted and relaxed states. Notice the change in the bell thickness (*B*). (DeMont and Gosline 1988a)

DeMont and Gosline (1988a) measured the mechanical properties of an intact bell in the hydromedusan *Polyorchis*. This stress–strain curve for the intact bell showed a nonlinear relationship. The initial low modulus of the curve probably results from deformation of the nonfibrous joints, while the stiffer, high modulus region at higher strains could result from the loading of the mesogleal fibres themselves. The nonlinearity has important implications for jet-propelled swimming, and is discussed below. DeMont and Gosline (1988a) made some rough calculations and showed that most of the energy stored in the mesoglea during the deformation of the bell can be stored in the radial mesogleal fibres.

Mechanics and Energetics of Locomotion. Maximum velocities for swimming medusa are low: *Stomolophus* reaches a maximum velocity of about 13 cm s^{-1} (bell diameter of 6–10 cm) (Larson 1987), *Polyorchis* reaches a maximum of 7 cm s^{-1} (bell height 2.8 cm) (Gladfelter 1972), and *Gonionemus* reaches a maximum of 11.8 cm s^{-1} (bell height 2 cm) (Gladfelter 1973). Daniel (1985) showed that the metabolic cost of locomotion for *Gonionemus* and *Stomotoca* is about an order of magnitude larger than a vertebrate swimmer with the same mass, and argued that the larger cost of transport is due to the energy expended in accelerating and decelerating the animal's mass and added mass. Larson (1987) calculated the metabolic cost of transport for the much larger *Stomolophus*, and found that the cost of transport is about the same as that for fish of equivalent size. He thought the cost-efficient transport of these jet-propelled animals was due to their low specific metabolic rate. The swimming muscles of *Stomolophus* account for only 5% of the total body mass, while this value can exceed 50% for some fishes.

DeMont and Gosline (1988a–c) completed a detailed analysis of the mechanics of the jet cycle in the hydrozoan jellyfish, *Polyorchis*. They measured the mechanical energy generated by the contraction of the locomotor muscle, and were able to partition the energy into three components (percentage of total): the energy generating the pressure in the subumbrellar cavity (39–60%), the energy associated with the inertia of the bell (19–32%), and the energy stored in the deformation of the mesoglea (20–29%). They also showed that the bell, when forced to oscillate in rhythmic contractions, is working at, or near, the natural frequency of the structure. The animal can thus take advantage of the phenomenon of resonance, and reduce the energetic requirement of the cycle about 24–37% of the total cost. Daniel (1983) developed a theoretical model of medusan jet propulsion, that is discussed in his Chap. 2, this volume. Bone and Trueman (1982) examined jet propulsion in the siphonophores *Chelophyes* and *Abylopsis*.

4.2 Bivalves

4.2.1 General Biology

Swimming in bivalves appears to be restricted to the Pectinacea, although some reports have been made of a primitive form of jet propulsion in the jacknife clam *Ensis* (McMahon and McMahon 1983), and other species described in that paper.

Scallop swimming has been described in several species: *Amusium* (Morton 1980), *Argopecten* (Winter and Hamilton 1985) *Chlamys* (Gruffydd 1976; Moore and Trueman 1971; Stephens and Boyle 1978), *Pecten* (Yonge 1936), *Placopecten* (Caddy 1968), and *Propeamussium* (Morton and Thurston 1989). Most workers have described it as an escape response, but some suggest that swimming might be associated with seasonal migrations (Dadswell and Weihs 1990; Morton 1980), or efforts to locate suitable habitats (Gruffydd 1976; Winter and Hamilton 1985). The neural control of swimming was originally thought to be a reflex dependent on feedback from stretch receptors in the adductor muscles (Mellon 1969; Stephens and Boyle 1978), but more recent work suggests that the neural basis of swimming is a motor circuit located in the central nervous system (Wilkens 1981). Thomas (1988) discussed the evolutionary convergence of bivalve shells.

4.2.2 Functional Morphology

The locomotor structure of swimming scallops is powered by a single striated adductor muscle. Contraction of the muscle reduces the gape between the two valves, and simultaneously compresses a rubbery material called abductin, which is contained in the ligament at the base of the hinge between the two valves. The energy stored in the abductin is released to increase the gape of the valves, and to stretch the adductor muscle.

The adductor muscle has many structural and mechanical similarities to vertebrate skeletal muscle (Rall 1981); for example, both exhibit similar isometric twitch kinetics. Rall also found some significant differences between the scallop and frog muscle. The scallop muscle demonstrated a 30% lower isometric stress and a poor ability to maintain force during a tetanus. Marsh (1990) measured thermal effects on muscle kinetics in the scallop *Argopecten irradians*. He showed that all parts of the locomotor cycle were not equally affected by temperature. The rate of valve closure and maximum rate of opening were relatively thermally insensitive. The largest effect of temperature appeared to be on the initial phase of the valve opening. He argued that this indicated some thermal dependence on deactivation of the muscle. However, thermal dependence on the physical properties of the abductin should be taken into account, since the energetics of the initial phase of the opening is dominated by the release of energy stored in the abductin. Indeed, for some scallops, most of the energy associated with swimming may be fluctuating between the kinetic energy of the valves, and the strain energy stored in the abductin (DeMont 1990).

The morphology of the hinge ligament has been described in several species of bivalves: *Mytilus* (Trueman 1950), *Ostrea* (Trueman 1951), *Pecten* (Trueman 1953b) and *Spondylus* and *Plicatula* (Yonge 1973). *Pecten* is the only swimmer in this group, so further discussion is confined to that species. The ligament consists of two layers. The outer layer is found along the dorsal margin of the valves. The inner layer is found between the two valves, and divided into three parts, a large noncalcified structure, the abductin, and two lateral calcified regions attaching the abductin to the valves. The abductin functions mechanically as a spring during locomotion, so its physical properties are functionally important.

4.2.3 Mechanical Properties of Abductin

Trueman (1953a) appears to be the first to have measured the mechanical properties of the abductin. His mechanical tests may have taken several minutes to perform, so his conclusions are difficult to interpret. He showed that the hysteresis loop for the loading and unloading curves of the intact ligament of *Pecten* or *Chlamys* was considerably less than for other bivalves. This is clearly functionally important for the refilling process. Alexander (1966) was able to measure the resilence of the inner hinge ligament for both *Pecten* and *Chlamys* at the physiological frequency (about 3 Hz), and found a resilence of about 90%. Alexander also performed some thermodynamic experiments on the abductin, and showed that the elastic restoring force is mainly associated with entropy changes. Thus abductin exhibits true rubber-like elasticity. Kelly and Rice (1967) compared some properties of abductin to elastin and resilin, and found that it exhibited similar physical properties, but found that its amino acid composition was distinct in that it contained high concentrations of the amino acids glycine and methionine. Ono et al. (1990) recently showed that the mechanical anisotropy of the ligament of the clam *Pseudocardium sachalinensis* can be attributed to the uniaxially oriented structure of the aragonite crystals embedded in the matrix. Gosline (1980) and Shadwick and Gosline (1983) have reviewed the molecular biomechanics of abductin. Surprisingly, there have been no measurements of the dynamic mechanical properties of abductin, or any other components of the ligaments of bivalves.

4.2.4 Mechanics and Energetics of Locomotion

Swimming speeds for scallops are low: Morton (1980) found a maximum swimming speed of $73\,\mathrm{cm\,s^{-1}}$ for a 70 mm *Amusium*, Dadswell and Weihs (1990) report a maximum speed of $79.4\,\mathrm{cm\,s^{-1}}$ for a 60-mm *Placopecten*, at a Reynolds number of about 4×10^4. Gould (1971) argued that swimming ability should decline during the ontogeny of scallops. Morton (1980) found no evidence for this in *Amusium*, but Dadswell and Weihs (1990) found that midsized *Placopecten* are the best swimmers. They partially based their arguments on measurements of the fineness and aspect ratios of the shells; however, the differnces in these quantities were not large. It would have been useful to estimate the differences in the total drag that these values produced. Their arguments were based on conventional beliefs that high aspect ratio hydrofoils produce the best performance. Ennos (1989) showed that the performance of small gliders is optimized by the use of low aspect ratio wings. DeMont (1990) mathematically modelled a swimming scallop, and showed that animals swim at, or near the resonant frequency of the shell-hinge system. This may reduce the cost of locomotion, since at resonance little work is required to keep the shells oscillating, so that most of the work done by the muscles can be used to generate thrust. Vogel (1985, 1988) showed that there is a hydrodynamically generated force that can assist in the reopening of the shells.

4.3 Cephalopods

The cephalopods exhibit a wide range of locomotor mechanisms, all based to some extent on an inherently inefficient jet-propelled system. Much has been written on comparisons of cephalopods and fish (Packard 1972; O'Dor and Webber 1986), and comparisons within the cephalopods themselves (Wells 1990). The best swimmers of the cephalopods are the squids. Fin-based locomotion exists in many cephalopods, and is an attempt to overcome the inefficiency of jet propulsion.

4.3.1 Squid

General Biology. The squids are the fastest swimmers of all the aquatic invertebrates; their performance is comparable to fish of roughly the same size (O'Dor and Webber 1986). Packard (1972), in his extensive and important review, found that at all levels of analysis, there were striking resemblances between the fish and squid, indeed suggesting that squid functionally are fish. O'Dor and Webber (1986) reviewed the more recent literature and argued that there are a number of physiological and biochemical contraints that prevent squid from competing directly with fish. The squids seem to overcome the inherent inefficiency of jet propulsion by maximizing their metabolic rates, growing fast, and having a short life span (compared to fish of equivalent size). Much of the recent work on locomotion has been by O'Dor and his collagues (O'Dor 1982, 1988a, b, 1989; O'Dor et al. 1990a; O'Dor and Webber 1986; Webber and O'Dor 1985, 1986). Squid make extensive vertical and horizontal migrations, sometimes exceeding 2000 km/year (Dawe et al. 1981). Most pelagic squid are found in schools, which appear similiar to those of pelagic fish (Hurley 1978; Mather and O'Dor 1984).

The squid are well known for their giant axons (Young 1938), which have been used extensively to study the conduction mechanisms of nervous impulses (Gilpin-Brown 1977). For years, it has been assumed that the giant axons functioned to generate rapid escape responses (Wilson 1960), but recent work suggests that the small axon pathways are used to generate both the slow respiratory jets, and rapid escape jets (Otis and Gilly 1990; Mackie 1990). The giant axon system is not utilized until after the nongiant axon system is stimulated; the system is timed to augument the rise in the mantle pressure. Squid can continuously vary their swimming velocity (O'Dor and Webber 1986; O'Dor 1988a).

Functional Morphology. The mechanical organization of the squid mantle has recently been reviewed (Gosline and DeMont 1985). Ward and Wainwright (1972) and Ward (1972) were the first to describe locomotory aspects of the mantle. They showed that it consists of circular and radial muscles that are arranged in alternate rings along the mantle. The muscles are obliquely striated. Ward (1972) proposed that refilling of the mantle might be effected by contraction of the radial fibres, or elasticity of the mantle itself. Packard and Trueman (1974) showed that contraction of the circular muscles causes the mantle to thicken, while contraction of the radial muscles causes the mantle to thin. Gosline et al. (1983) made electromyographic recordings, while measuring changes in mantle diameter. This allowed them to distinguish activity in the radial and circular muscles. They observed several patterns of muscle activity for both escape jets and respiratory movements. For example, one pattern of respiratory movements was powered by the radial muscles alone, while another was powered by the circular muscles alone. Both were antagonized by elastic mechanisms. The animals were restrained during the measurements, so continuous variations of swimming activity, that have recently been described (O'Dor and Webber 1986; O'Dor 1988a), were not observed. L-glutamate is the excitatory transmitter for the circular muscles, while the radial muscles contract in response to acetylcholine (Bone et al. 1982).

Moon and Hulbert (1975) described the mantle muscle ultrastructure of the squid *Symplectoteuthis oualaniensis*, and demonstrated that the muscle has the characteristics of fast muscle types. Bone et al. (1981) were able to describe the muscle in more detail. The circular muscles are of two types: a poorly vascularized central zone containing few mitochondria, and a well-vascularized inner and outer zone rich in mitochondria. The radial fibers are similiar to the inner zone. Mommsen et al. (1981) extended these observations to show that the inner zone is specialized for anaerobic metabolism, and the outer zones are specialized for a highly oxidative metabolism.

Connective tissue fibres are embedded in the muscles, and extend from the inner to the outer tunics (Ward and Wainwright 1972). The intermuscular fibres form a complex network that run in several orientations at more or less regular angles to each other (Bone et al. 1981) (see Fig. 6). Gosline and Shadwick (1983a, b) examined in detail the role of the connective tissue fibres in the mechanics of jetting. They showed that as the mantle wall thickens during the contraction of the muscles the connective tissue fibres are put in tension, and store strain energy. This strain energy is released to partially power the refilling. Their calculations

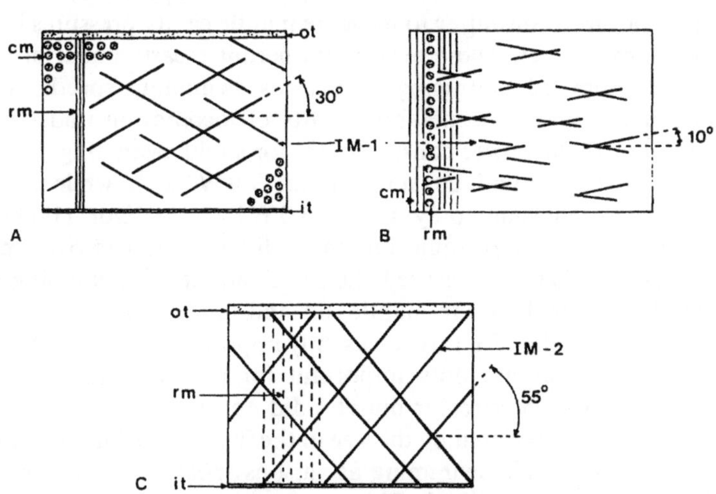

Fig. 6 A–C. The organization of the intermuscular fibers of the squid mantle. **A** Longitudinal section. **B** Section cut tangentially to the surface of the mantle. **C** Transverse section. *ot* outer tunic; *it* inner tunic; *cm* circular muscle; *rm* radial muscle. (Gosline and Shadwick 1983b)

suggest that the fibres are collagenous. The spring appears to be constructed so that the energy is stored at a time when the muscles are unable to apply their mechanical output to generate hydrodynamic thrust. This would increase the utilization of the mantle muscles. This idea was also applied to the mechanics of jetting in the jellyfish *Polyorchis* (DeMont and Gosline 1988b). The unusual long-duration action potential seen in the locomotor muscle of *Polyorchis* may function to insure that sufficient energy is stored in the elastic structures to power the refilling. Recent work (Robinson et al. 1986) has compared the mechanics of refilling in human hearts to that described for squid.

Kier (1989) has examined the morphology and mechanics of the fin musculature in squid, and has also examined the musculature of both the arms and tentacles (Kier 1985). The fins consist of a three-dimensional array of tightly packed muscles, which function both to generate force, and provide structural support. He noted some similarities between the mantle musculature and fin musculature. The fins contain connective tissues that may function as an elastic mechanism. Some of the muscles are divided into oxidative and anaerobic glycolytic muscle fibres. These may have functions similar to those of the mantle muscles.

Mechanics and Energetics of Locomotion. Squid can swim almost as fast as fish of the same size; a 350-g *Loligo vulgarius* can reach a burst speed of $2.0\,\mathrm{m\,s^{-1}}$, while a 250-g *Salmo gairdneri* can reach a burst speed of $2.7\,\mathrm{m\,s^{-1}}$ (O'Dor and Webber 1986). O'Dor (1982) measured the swimming performance of the squid *Loligo opalescens*, and found that the metabolic cost of transport for squid is considerably higher than for fish. He calculates a net cost of transport for *Loligo* to be $7.8\,\mathrm{J\,kg^{-1}\,m^{-1}}$, and compares this to the sockeys salmon, which has a cost of transport of $2.0\,\mathrm{J\,kg^{-1}\,m^{-1}}$. The squid were also shown to have a much higher standard and active metabolic rates.

Webber and O'Dor (1985, 1986) have used pressure transducers connected to an ultrasonic transmitter to measure mantle cavity pressures in animals swimming in a respirometer. These techniques permit measurement of power requirements for jet-propelled swimming at various swimming speeds, and will allow field measurements of the metabolic costs of locomotion under a variety of natural conditions. The power consumptions for squid swimming are considerably higher than for fish. Squid consume about $7.0\,\mathrm{W\,kg^{-1}}$, while fish under equivalent conditions consume only about $3.0\,\mathrm{W\,kg^{-1}}$. O'Dor (1988b) argues that the circulatory system of squid ultimately limits sustained performance.

O'Dor (1988a) measured the forces acting on swimming squid. His analysis clearly showed how the mantle, fin and funnel interact to produce a variety of swimming modes. Fin thrust was only important for speeds below $0.2\,\mathrm{m\,s^{-1}}$; above that speed they are only important for attitude control. The intensity of the contraction of the circular muscles did not vary much with the swimming speed; jet thrust was regulated by the size of the funnel aperture. He calculated the Froude efficiency for squid swimming at various speeds. A squid swimming at $0.1\,\mathrm{m\,s^{-1}}$ had an efficiency of 0.12. This confirms Alexander's (1977) prediction that the Froude efficiency of jet-propelled swimmers is low. Vogel (1987, 1988) showed that there may be some flow-assisted mantle cavity refilling at high swimming speeds.

4.3.2 Octopus

Octopus are not active swimmers, and spend most of their time on the sea bottom, using their arms to move over the substrate. The arrangement and function of octopus arms has been reviewed by Kier (1988). The arms are capable of a range of movements, including elongation, shortening, torsional movements, and bending in any plane. Jet-propelled swimming is used on occasion, mostly for escape from predators. The nervous control of the mantle was described by Wilson (1960), and has been reviewed by Wells (1978). The mechanics of the mantle muscle has been reviewed by Gosline and DeMont (1985). The mantle consists of three sets of orthogonal muscles. Simultaneous contraction of the three sets of muscles stiffens the wall. A more forceful contraction of one set will cause the mantle to change shape. Contraction of circular muscles decreases the diameter of the mantle, the simultaneously increases the thickness. Refinning is not passive, as in the squid, but powered by contraction of radial muscles. Contraction of longitudinal muscles prevent the mantle from lengthening.

4.3.3 Nautilus

General Biology. Nautilus live sedentary lives at moderate depths of between 150 and 500 m (Chamberlain 1990). They are of interest to zoologists and palaeontologists, since *Nautilus* are the most primitive of all living cephalopods, and unusual compared to all of the swimming molluscs in having an external shell. Shelled cephalopods have small mantle cavities, little propulsive muscle, and thus have low swimming velocities. Bidder (1962) was one of the first to examine swimming in *Nautilus*, and more recently, Packard et al. (1980) examined breathing and swimming movements. Most of our current knowledge, however, can be attributed to Chamberlain (1976a, b, 1980a, b, c, 1987, 1990, 1991), who has documented in considerable detail the basic principles of *Nautilus* locomotion, especially in relation to their evolutionary history.

Functional Morphology. Chamberlain (1987, 1990) has described the mechanism of thrust generation in *Nautilus*. The mantle is not muscular, as in the squids, so mantle cavity pressure (and thus thrust) is not increased by contraction of mantle muscles. Instead, mantle cavity pressure is increased by contraction of the paired cephalic muscles and the funnel muscles. The paired cephalic muscles, which contain bundles of longitudinal and radial muscles (Hochachka et al. 1978), couple the body to the shell wall; contraction of the longitudinal muscles reduces the volume of the mantle cavity. Muscular antagonism is elicited by contraction of the radial muscles in the cephalic muscles. The funnel muscles overlap ventrally, and control the size and shape of the funnel opening. This effectively modulates locomotion, since flow, and thus thrust, varies as the size of the orifice changes. The propulsive muscles comprise only 9% of the total weight of the animal (Chamberlain 1987). For comparison, fish (*Salmo giardneri*) and squid (*Loligo vulgaris*) have mass ratios of 60% and 35%, respectively. Kier (1987) has examined the functional morphology of the tentacle musculature of *Nautilus*.

Mechanics and Energetics of Locomotion. Nautilus do not swim fast; maximum velocities are about 25 cm s^{-1} (Chamberlain 1987, 1990). Chamberlain (1990) compared the metabolic cost of transport for *Nautilus* (1.9 J kg^{-1} m^{-1}), squid (9.6 J kg^{-1} m^{-1}) and salmon (1.9 J kg^{-1} m^{-1}). Several reasons exist for the large differences between *Nautilus* and squid; the former are neutrally buoyant, and the latter have a life-history strategy that demands high energy consumptions (O'Dor and Webber 1986). O'Dor et al. (1990b) showed that at a typical speed of 0.5 m s^{-1} *Nautilus* have a lower cost of transport than fish. Froude efficiencies for *Nautilus* are likely to be in the order of 0.29, which makes it comparable to other cephalopods (Chamberlain 1990).

4.3.4 Cuttlefish

The cuttlefish use both jet propulsion and fins to swim. Maximum mantle cavity pressures generated during a jet are about half that of squid (Trueman and Packard 1968), although the mechanics of the mantle muscle appear similar (Packard and Trueman 1974). The fins extend the length of the mantle, and can produce waves running forward and backward, or in different directions on either side of the animal (Packard 1972). Kier (1989) described the fin musculature, which is similar to that described above for squid. Kier et al. (1989) were able to measure electrical activity in the fin muscles, and showed that oxidative muscles power gentle fin movements, while anaerobic glycolytic muscles power vigorous fin movements. Kier et al. (1985) found mechanoreceptors that may help to control fin movements. Surprisingly, there seems to be little work done on thrust generation from the fins.

4.4 Tunicates

The doliolids and salps swim with jet propulsion, although the mechanism is different than all other jet-propelled swimmers. Water is inhaled into a jet chamber through an opening at the anterior end, and exhaled in another opening at the posterior end. Doliolids spend most of the time motionless in the water, but on occasion produce a jet to maintain position in the water column (Bone 1989). Bone and Trueman (1984) have examined jet propulsion in *Dolioum.* Animals as small as 4–5 mm long can reach instantaneous velocities as high as 25 cm s^{-1}. A large *Doliolum* has a mechanical cost of transport of 5.6 J kg^{-1} m^{-1}, while the mechanical cost of transport for a large salp is only 0.55 J kg^{-1} m^{-1}. The differences are not surprising since chamber pressure and jet velocities in *Doliolum* are considerably higher than those measured in salps (Bone and Trueman 1983). The Froude efficiency of jet propulsion in salps is high. Bone and Trueman (1983) give values as high as 0.36, and indicate that these measurements are taken from maximum performances. The efficiency should be even higher at cruising speeds. The high efficiency of salp jet propulsion is due to the large exhalant apertures and relatively slow jet velocities. These observations have been recently extended by Madin (1990). The locomotor muscles of *Doliolum* are obliquely striated and lack a sarcoplasmic reticulum (Bone and Ryan 1974); external calcium is required for contraction (Bone 1989).

4.5 Swimming in Other Animals

Shaw and Fontaine (1990) recently describe swimming in the crinoid *Florometra serratissima*. Thrust is generated by repetitive arm strokes.

This chapter described some of the basic principles used by soft bodied animals to propel themselves. The arrangement and function of the basic structural support systems were described. Some of the mechanisms used to move on or on solid substrates were described, but most of this work has been reviewed previously. Soft bodied animals use a vast collection of mechanisms to swim. The functional morphology and the mechanics and energetics of swimming were reviewed, with some attention given to swimming in the coelenterates and molluscs.

References

Alexander R McN (1962) Visco-elastic properties of the body-wall of sea anemones. J Exp Biol 39: 373–386

Alexander R McN (1964) Visco-elastic properties of the mesoglea of jellyfish. J Exp Biol 41: 363–369

Alexander R McN (1966) Rubber-like properties of the inner hinge-ligament of Pectinidae. J Exp Biol 44: 119–130

Alexander R McN (1975) Biomechanics. Chapman and Hall, London

Alexander R McN (1977) Swimming. In: Alexander R McN, Goldspink G (eds) Mechanics and energetics of animal locomotion. Chapman and Hall, London, pp 222–248

Alexander R McN (1982) Locomotion of animals. Blackie, London

Alexander R McN (1983) Animal mechanics, 2nd edn. Blackwell, Oxford

Alexander R McN (1987) Bending of cylindrical animals with helical fibers in their skin or cuticle. J Theor Biol 124: 97–110

Alexander R McN (1988) Elastic mechanisms in animal movement. Cambridge Univ Press, Cambridge

Bidder AM (1962) Use of the tentacles, swimming and buoyancy control in the pearly *Nautilus*. Nature (Lond) 4853: 451–454

Bond C, Harris AK (1988) Locomotion of sponges and its physical mechanism. J Exp Zool 246: 271–284

Bone Q (1989) On the muscle fibres and locomotor activity of Doliolids (Tunicata: Thaliacea). J Mar Biol Ass UK 69: 597–607

Bone Q, Ryan KP (1974) On the structure and innervation of the muscle bands of *Doliolum* (Tunicata: Cyclomyaria). Proc R Soc Lond B 187: 315–327

Bone Q, Trueman ER (1982) Jet propulsion of the Calycophoran Siphonophores *Chelophyes* and *Abylopsis*. J Mar Biol Ass UK 62: 263–276

Bone Q, Trueman ER (1983) Jet propulsion in salps (Tunicata: Thaliacea). J Zool Lond 201: 481–506

Bone Q, Trueman ER (1984) Jet propulsion in *Doliolum* (Tunicata: Thaliacea) J Exp Mar Biol Ecol 76: 105–118

Bone Q, Pulsford A, Chubb AD (1981) Squid mantle muscle. J Mar Biol Ass UK 61: 327–342

Bone Q, Packard A, Pulsford AL (1982) Cholinergic innervation of muscle fibres in squid. J Mar Biol Ass UK 62: 193–199

Caddy JF (1968) Underwater observations on scallop (*Placopecten magellanicus*) behavior and drag efficiency. J Fish Res Bd Can 25(10): 2123–2141

Chamberlain JA (1976a) Flow patterns and drag coefficients of cephalopod shells. Palaeotology 19(3): 539–563

Chamberlain JA (1976b) Hydrodynamic properties of cephalopod shell ornament. Paleobiology 2(4): 316–331
Chamberlain JA (1980a) The role of body extension in cephalopod locomotion. Palaeotology 23(2): 445–461
Chamberlain JA (1980b) Hydromechanical design of fossil cephalopods. Syst Assoc Spec Vol 18: 289–336
Chamberlain JA (1980c) Motor performance and jet propulsion in *Nautilus*: implications for cephalopod paleobiology and evolution. Am Malacol Union Inc Bull 123–127
Chamberlain JA (1987) Locomotion in *Nautilus*. In: Saunders WB, Landman NH (eds) *Nautilus*: the biology and paleobiology of a living fossil. Plenum, New York, pp 489–525
Chamberlain JA (1990) Jet propulsion of *Nautilus*: a surviving example of early Paleozoic cephalopod locomotor design. Can J Zool 68: 806–814
Chamberlain JA (1991) Cephalopod locomotor design and evolution: the constraints of jet propulsion. In: Rayner JMV and Wootton, RJ (eds) Biomechanics in evolution. Cambridge Univ Press, London (in press)
Chapman G (1953a) Studies of the mesoglea of coelenterates. I. Histology and chemical properties. J Micro Sci 94(2): 155–176
Chapman G (1953b) Studies on the mesoglea of coelenterates II. Physical properties. J Exp Biol 30: 440–451
Chapman G (1959) The mesoglea of *Pelagia noctiluca*. J Micro Sci 100(4): 599–610
Dadswell MJ, Weihs D (1990) Size-related hydrodynamic characteristics of the giant scallop, *Placopecten magellanicus* (Bivalvia: Pectinidae). Can J Zool 68: 778–785
Daniel TL (1983) Mechanics and energetics of medusan jet propulsion. Can J Zool 61: 1406–1429
Daniel TL (1985) Cost of locomotion: unsteady medusan swimming. J Exp Biol 119: 149–164
Dawe EG, Beck P, Drew HJ, Winters GH (1981) Long distance migration of a short-finned squid, *Illex illecebrosus*. J Northw Atl Fish Sci 2: 75–76
DeMont ME (1990) Tuned oscillations in the swimming scallop *Pecten maximus*. Can J Zool 68: 786–791
DeMont ME, Gosline JM (1988a) Mechanics of jet propulsion in the hydromedusan jellyfish, *Polyorchis penicillatus*. I. Mechanical properties of the locomotor structure. J Exp Biol 134: 313–332
DeMont ME, Gosline JM (1988b) Mechanics of jet propulsion in the hydromedusan jellyfish, *Polyorchis penicillatus*. II. Energetics of the jet cycle. J Exp Biol 134: 333–345
DeMont ME, Gosline JM (1988c) Mechanics of jet propulsion in the hydromedusan jellyfish, *Polyorchis penicillatus*. III. A natural resonating bell; the presence and importance of a resonant phenomenon in the locomotor structure. J Exp Biol 134: 347–361
Denny M (1980) Locomotion: the cost of gastropod crawling. Science 208: 1288–1290
Denny MW (1981) A quantitative model for the adhesive locomotion of the terrestrial slug, *Ariolimax columbianus*. J Exp Biol 91: 195–217
Denny MW (1984) Mechanical properties of pedal mucus and their consequences for gastropod structure and performance. Am Zool 24: 23–36
Denny M (1988) Biology and the mechanics of the wave-swept environment. Princeton Univ Press, Princeton, New Jersey
Denny MW, Gosline JM (1980) The physical properties of the pedal mucus of the terrestrial slug, *Ariolimax columbianus*. J Exp Biol 88: 375–393
Deyrup-Olsen I, Luchtel DL, Martin AW (1983) Components of mucus of terrestrial slugs (Gastropoda). Am J Physiol 245: R448–R452
Dimock RV (1985) Quantitative aspects of locomotion by the mud snail *Ilyanassa obsoleta*. Malacologia 26(1–2): 165–172
Dobrolyubov AI (1986) The mechanism of locomotion of some terrestrial animals by travelling waves of deformation. J Theor Biol 119: 457–466
Elder HY, Hunter RD (1980) Burrowing of *Priapulus caudatus* (Vermes) and the significance of the direct peristaltic wave. J Zool Lond 191: 333–351
Ennos AR (1989) The effect of size on the optimal shapes of gliding insects and seeds. J Zool Lond 219: 61–69
Gilpin-Brown JB (1977) The squid and its giant nerve fibre. Symp Zool Soc Lond 38: 233–241
Gladfelter WB (1972) Structure and function of the locomotory system of *Polyorchis montereyensis* (Cnidaria, Hydrozoa). Helgol Wiss Meeresunters 23: 38–79

Gladfelter WG (1973) A comparative analysis of the locomotory system of medusoid Cnidaria. Helgol Wiss Meeresunters 25: 228–272

Glass L, Mackey MC (1988) From clocks to chaos – the rhythms of life. Princeton Univ Press, Princeton, New Jersey

Gosline JM (1971a) Connective tissue mechanics of *Metridium senile* I. Structural and compositional aspects. J Exp Biol 55: 763–774

Gosline JM (1971b) Connective tissue mechanics of *Metridium senile* II. Visco-elastic properties and macromolecular model. J Exp Biol 55: 775–795

Gosline JM (1980) The elastic properties of rubber-like proteins and highly extensible tissues. In: Vincent JFV, Currey JD (eds) The mechanical properties of biological materials. Princeton Univ Press, Princeton

Gosline JM, DeMont ME (1985) Jet-propelled swimming in squids. Sci Am 252(1): 96–103

Gosline JM, Shadwick RE (1983a) The role of elastic energy storage mechanisms in swimming: an analysis of mantle elasticity in escape jetting in the squid, *Loligo opalescens*. Can J Zool 61(6): 1421–1431

Gosline JM, Shadwick RE (1983b) Molluscan collagen and its mechanical organization in squid mantle. In: Hochachka PW (ed) The Mollusca, vol 1. Academic Press, London, pp 371–399

Gosline JM, Steeves JD, Harman AD, DeMont ME (1983) Patterns of circular and radial mantle muscle activity in respiration and jetting of the squid *Loligo opalescens*. J Exp Biol 104: 97–109

Gould SJ (1971) Muscular mechanics and the ontogeny of swimming in scallops. Palaeontology 14(1): 61–94

Gruffydd LlD (1976) Swimming in *Chlamys islandica* in relation to current speed and an investigation of hydrodynamic lift in this and other scallops. Norw J Zool 24: 365–378

Gutmann WF (1981) Relationships between phyla based on functional-mechanical analysis of the hydrostatic skeleton. Am Zool 21: 63–81

Gutmann WF (1988) The hydraulic principle. Am Zool 28: 257–266

Hammond RA (1980) The hydraulic system in a burrowing worm, *Priapulus caudatus*. J Zool Lond 192: 489–499

Hochachka PW, French CJ, Meredith J (1978) Metabolic and ultrastructural organization in *Nautilus* muscles. J Exp Zool 205: 51–62

Houlihan DF, Innes AJ (1982) Oxygen consumption, crawling speeds, and cost of transport in four Mediterranean intertidal gastropods. J Comp Physiol B 147: 113–121

Hunter RD, Elder HY (1989) Burrowing dynamics and energy cost of transport in the soft-bodied marine invertebrates *Polyphysia crassa* and *Priapulus caudatus*. J Zool Lond 218: 209–222

Hunter RD, Moss VA, Elder HY (1983) Image analysis of the burrowing mechanism of *Polyphysia crassa* (Annelida: Polychaeta) and *Priapulus caudatus* (Priapulida). J Zool Lond 199: 305–323

Hurley AC (1978) School structure of the squid *Loligo opalescens*. Fish Bull 76: 433–442

Innes AJ, Houlihan DF (1985) Aerobic capacity and cost of locomotion of a cool temperate gastropod: a comparison with some Mediterranean species. Comp Biochem Physiol 80A: 487–493

Keller JB, Falkovitz MS (1983) Crawling in worms. J Theor Biol 104: 417–442

Kelly RE, Rice RV (1967) Abductin: a rubber-like protein from the internal triangular hinge ligament of *Pecten*. Science 155: 208–210

Kier WM (1985) The musculature of squid arms and tentacles: ultrastructural evidence for functional differences. J Morph 185: 223–239

Kier WM (1987) The functional morphology of the tentacle musclulature of *Nautilus pompilius*. In: Saunders WB, Landman NH (eds) *Nautilus*: the biology and paleobiology of a living fossil. Plenum, New York, pp 257–269

Kier WM (1988) The arrangement and function of molluscan muscle. In: Trueman ER, Clarke MR (eds) The Mollusca, vol II. Academic Press, New York, pp 211–252

Kier WM (1989) The fin musculature of cuttlefish and squid (Mollusca, Cephalopoda): morphology and mechanics. J Zool Lond 217: 23–38

Kier WM, Smith KK (1985) Tongues, tentacles and trunks: the biomechanics of movement in muscular-hydrostats. Zool J Linn Soc 83: 307–324

Kier WM, Messenger JB, Miyan JA (1985) Mechanoreceptors in the fins of the cuttlefish, *Sepia officinalis*. J Exp Biol 119: 369–373

Kier WM, Smith KK, Miyan JA (1989) Electromyography of the fin musculature of the cuttlefish *Sepia officinalis*. J Exp Biol 143: 17–31

Koehl MAR (1977a) Effects of sea anemones on the flow forces they encounter. J Exp Biol 69: 87–105
Koehl MAR (1977b) Mechanical organization of cantilever-like sessile organisms: sea anemones. J Exp Biol 69: 127–142
Koehl MAR (1977c) Mechanical diversity of connective tissue of the body wall of sea anemones. J Exp Biol 69: 107–125
Koehl MAR (1982) The interaction of moving water and sessile organisms. Sci Am 247: 124–134
Koehl MAR (1984) How benthic organisms withstand moving water. Am Zool 24(1): 57–71
Larson RJ (1987) Costs of transport for the scyphomedusa *Stomolophus melagris* L. Agassiz. Can J Zool 65: 2690–2695
Leonard JL (1980) Temporal organization of swimming activity in *Sarsia tubulosa* M. Sars (Hydrozoa). J Comp Physiol 136: 219–225
Leonard JL (1982) Transient rhythms in the swimming activity of *Sarsia tubulosa* (Hydrozoa). J Exp Biol 96: 181–193
Mackie GO (1974) Locomotion, flotation and dispersal. In: Muscatine L, Lenhoff HM (eds) Coelenterate biology: reviews and new perspectives. Academic Press, New York, pp 313–357
Mackie GO (1990) Giant axons and control of jetting in the squid *Loligo* and the jellyfish *Aglantha*. Can J Zool 68(4): 799–805
Mackie GO, Mackie GV (1967) Mesogloeal ultrastructure and reversible opacity in a transparent siphonophore. Vie Milieu Ser A 18(A): 47–71
Madin LP (1990) Aspects of jet propulsion in salps. Can J Zool 68: 765–777
Marsh RL (1990) Deactivation rate and shortening velocity as determinants of contractile frequency. Am J Physiol 259: R223–R230
Mather JA, O'Dor RK (1984) Spatial organization of schools of the squid *Illex illecebrosus*. Mar Behav Physiol 10: 259–271
McMahon RF, McMahon CO (1983) Leaping and swimming as predator escape responses in the jacknife clam, *Ensis minor* Dall (Bivalvia: pharellidae). Nautilus 93(3): 55–58
Mellons DF Jr (1969) The reflex control of rhythmic motor output during swimming in the scallop. Z Vergl Physiol 62: 318–336
Mommsen TP, Ballantyne J, MacDonald D, Gosline J, Hochachka PW (1981) Analogues of red and white muscle in squid mantle. Proc Natl Acad Sci USA 78(5): 3274–3278
Moon TW, Hulbert WC (1975) The ultrastructure of the mantle musculature of the squid *Symplectoteuthis oualaniensis*. Comp Biochem Physiol 52B: 145–149
Moore JD, Trueman ER (1971) Swimming of the scallop, *Chlamys opercularis* (L.). J Exp Mar Biol Ecol 6: 179–185
Morton B (1980) Swimming in *Amusium pleuronectes* (Bivalvia: Pectinidae). J Zool Lond 190: 375–404
Morton B, Thurston MH (1989) The functional morphology of *Propeamussium lucidum* (Bivalvia: Pectinacea), a deep-sea predatory scallop. J Zool Lond 218: 471–496
O'Dor RK (1982) Respiratory metabolism and swimming performance of the squid, *Loligo opalescens*. Can J Fish Aquat Sci 39(4): 580–587
O'Dor RK (1988a) The force acting on swimming squid. J Exp Biol 137: 421–442
O'Dor RK (1988b) Limitations on locomotor performance in squid. J Appl Physiol 64(1): 128–134
O'Dor RK, Shadwick RE (1989) Squid, the olympian cephalopods. J Cephalopod Biol 1(1): 33–55
O'Dor RK, Webber DM (1986) The constraints on cephalopods: why squid aren't fish. Can J Zool 64: 1591–1605
O'Dor RK, Portner HO, Shadwick RE (1990a) Squid as elite athletes: locomotory, respiratory and circulatory integration. In: Gilbert DL, Adelman WJ, Arnold JM (eds) Squid as experimental animals. Plenum Press, New York, pp 481–503
O'Dor RK, Wells Jm Wells MJ (1990b) Speed, jet pressure and oxygen consumption relationships in free-swimming *Nautilus*. J Exp Biol 154: 383–396
Ono K, Kikuch Y, Higashi K, Tamiya N, Yasuoka N (1990) Elastic anisotropy of bivalve hinge-ligament. J Biomech 23(4): 307–312
Otis TS, Gilly WF (1990) Jet-propelled escape in the squid *Loligo opalescens*: concerted control by giant and non-giant motor axon pathways. Proc Natl Acad Sci USA 87: 2911–2915
Pabst DA, Healy MD (1989) Worms and whale tails: flattening the circular cylinder. Am Zool Abstr 29(4): 182A
Packard A (1972) Cephalopods and fish: the limits of convergence. Biol Rev 47: 241–307

Packard A, Trueman ER (1974) Muscular activity of the mantle of *Sepia* and *Loligo* (Cephalopoda) during respiratory movements and jetting, and its physiological interpretation. J Exp Biol 61: 411–419

Packard A, Bone Q, Hignette M (1980) Breathing movements in a captive *Nautilus*. J Mar Biol Assoc UK 60: 313–327

Rall JA (1981) Mechanics and energetics of contraction in striated muscle of the sea scallop, *Placopecten magellanicus*. J Physiol 321: 287–295

Robinson TF, Factor SM, Sonnenblick H (1986) The heart as a suction pump. Sci Am 254(6): 83–91

Robson EA (1971) The behaviour and neuromuscular system of *Gonactinia prolifera*, a swimming sea-anemone. J Exp Biol 55: 611–640

Satterlie RA, Spencer AN (1979) Swimming control in a cubomedusan jellyfish. Nature (Lond) 281(5727): 141–142

Seymour MK (1983) Some implications of helical fibers in worm cuticles. J Zool Lond 199: 287–295

Shadwick RE, Gosline JM (1983) Molecular biomechanics of protein rubbers in molluscs. In: Hochachka PW (ed) The Mollusca, vol 1, pp 399–430, Academic Press New York

Shaw GD, Fontaine AR (1990) The locomotion of the comatulid *Florometra serratissima* (Echinodermata: Crinoidea) and its adaptive significance. Can J Zool 68: 942–950

Spencer AN (1978) Neurobiology of *Polyorchis*. I. Function of effector systems. J Neurobiol 9(2): 143–157

Spencer AN (1979) Neurobiology of *Polyorchis*. II. Structure of effector systems. J Neurobiol 10(2): 95–117

Spencer AN (1982) The physiology of a coelenterate neuromuscular synapse. J Comp Physiol 148: 353–363

Spencer AN, Satterlie RA (1981) The action potential and contraction in subumbrellar swimming muscle of *Polyorchis penicillatus* (Hydromedusae). J Comp Physiol 144: 401–407

Stephens PJ, Boyle PR (1978) Escape responses of the queen scallop *Chlamys opercularis* (L.) (Mollusca: Bivalvia). Behav Physiol 5: 103–113

Stern-Tomlinson W, Nusbaum MP, Perez LE, Kristan WB (1986) A kinematic study of crawling behaviour in the leech, *Hirudo medicinalis*. J Comp Physiol A 158: 593–603

Sund PN (1958) A study of the muscular anatomy and swimming behaviour of the sea anemone, *Stomphia coccinea*. Q J Microsc Sci 99(3): 401–420

Thomas RDK (1988) Evolutionary convergence of bivalve shells: a comparative analysis of constructional constraints on their morphology. Am Zool 28: 267–276

Thorburn IW, Gruffydd LlD (1979) Studies of the behavior of the scallop *Chlamys opercularis* (L.) and its shell in flowing sea water. J Mar Biol Ass UK 59: 1003–1023

Trueman ER (1950) Observations on the ligament of *Mytilus edulis*. Q J Microsc Sci 91(3): 225–235

Trueman ER (1951) The structure, development, and operation of the hinge ligament of *Ostrea edulis*. Q J Microsc Sci 92(2): 129–140

Trueman ER (1953a) Observations on certain mechanical properties of the ligament of *Pecten*. J Exp Biol 30(4): 453–467

Trueman ER (1953b) The ligament of *Pecten*. Q J Microsc Sci 94(2): 193–202

Trueman ER (1968) The burrowing activities of bivalves: Symp Zool Soc Lond 22: 167–186

Trueman ER (1975) The locomotion of soft-bodied animals. Arnold, London

Trueman ER, Ansell AD (1969) The mechanisms of burrowing into soft substrata by marine animals. Oceanogr Mar Biol Ann Rev 7: 315–366

Trueman ER, Brown AC (1985) The mechanism of shell elevation in Haliotis (Mollusca: Gastropoda) and a consideration of the evolution of the hydrostatic skeleton in Mollusca. J Zool Lond 205: 585–594

Trueman ER, Brown AC (1987) Proboscis extrusion in *Bullia* (Nassariidae): a study of fluid skeletons in Gastropoda. J Zool Lond 211: 505–513

Trueman ER, Jones HD (1977) Crawling and burrowing. In: Alexander R McN, Goldspink G (eds) Mechanics and energetics and animal locomotion. Chapman and Hall, London, pp 204–221

Trueman ER, Packard A (1968) Motor performances of some cephalopods. J Exp Biol 49: 495–507

Trueman ER, Wong TM (1987) The role of the coelom as a hydrostatic skeleton in lingulid brachiopods. J Zool Lond 213: 221–232
Vovel S (1985) Flow-assisted shell reopening in swimming scallops. Biol Bull 169: 624–630
Vogel S (1987) Flow-assisted mantle cavity refilling in jetting squid. Biol Bull 172: 61–68
Vogel S (1988) How organisms use flow-induced pressures. Am Sci 76: 28–34
Wadepuhl M, Beyn W-J (1989) Computer simulation of the hydrostatic skeleton. The physical equivalent, mathematics and application to worm-like forms. J Theor Biol 136: 379–402
Wainwright SA (1970) Design in hydraulic organisms. Naturwissenschaften 57: 321–326
Wainwright SA (1982) Structural systems: hydrostats and frameworks. In: Taylor CR, Johansen K, Bolis L (eds) A companion to animal physiology. Cambridge Univ Press, Cambridge, pp 325–338
Wainwright SA (1988) Axis and circumference – the cylindrical shape of plants and animals. Harvard Univ Press, Cambridge
Wainwright SA, Biggs WD, Currey JD, Gosline JM (1976) Mechanical design in organisms. Arnold, London
Ward DV (1972) Locomotory function of the squid mantle. J Zool Lond 167: 487–499
Ward DV, Wainwright SA (1972) Locomotory aspects of squid mantle structure. J Zool Lond 167: 437–449
Webber DM, O'Dor RK (1985) Respiration and swimming performance of short-finned squid (*Illex illecebrosus*). NAFO Sci Coun Stud 9: 133–138
Webber DM, O'Dor RK (1986) Monitoring the metabolic rate and activity of free-swimming squid with telemetered jet pressure. J Exp Biol 126: 205–224
Wells MJ (1978) Octopus. Chapman and Hall, London
Wells MJ (1990) Oxygen extraction and jet propulsion in cephalopods. Can J Zool 68: 815–824
Wilkens LA (1981) Neurobiology of the scallop. I. Starfish-mediated escape behaviors. Proc R Soc Lond B 211: 341–372
Wilson DM (1960) Nervous control of movement in cephalopods. J Exp Biol 37: 57–72
Winter MA, Hamilton PV (1985) Factors influencing swimming in bay scallops, *Argopecten irradians* (Lamarck, 1819). J Exp Mar Biol Ecol 88: 227–242
Yonge CM (1936) The evolution of the swimming habit in the Lamellibranchia. Mem Mus Hist Nat Belg (2): 77–99
Yonge CM (1973) Functional morphology with particular reference to hinge and ligament in *Spondylus* and *Plicatula* and a discussion on relations within the superfamily Pectinacea (Mollusca: Bivalvia). Phil Trans R Soc Lond B 267: 171–208
Young JZ (1938) The functioning of the giant nerve fibres of the squid. J Exp Biol 15: 170–185

Chapter 7

Muscle Function in Locomotion

J.L. van Leeuwen

Contents

1 Introduction

Most animal locomotor systems are driven by cross-striated muscles under control of a nervous system. Apart from their mechanical function, muscles may also have important sensory functions through the presence of muscle spindles. Much is known about the physiological and structural properties of muscles, owing to a

Neuroregulation Group, Department of Physiology, University of Leiden, P.O. Box 9604, 2300 RC, Leiden, The Netherlands great many studies on isolated muscles or muscle fibres. Only recently, our knowledge of muscle performance in vivo (the main topic of this chapter) has increased considerably.

This chapter starts with a general discussion of muscle properties, with emphasis on mechanical aspects. Special attention will be paid to the dynamics of the muscle–tendon complex. Thereafter, techniques for in vivo monitoring of muscular performance will be reviewed. The major part considers the performance and design of muscles in swimming fish, in flying insects, and in running and jumping mammals. Other aspects of muscle function are discussed in Chap. 5 (mechanical stresses in muscles), Chap. 6 (muscle function in soft bodied animals), and Chap. 8 (muscle energetics). The word contraction is used synonymously with shortening in this chapter, and is quite distinct from the term activation (Sect. 2.2). A list of symbols and abbreviations is given in Appendix A.

2 Muscle Properties

Muscle and muscle fibre structure is discussed in many recent textbooks on physiology, cell biology and histology. Here, only a brief summary is given. The various levels of structural organization of a cross-striated muscle are schematized in Fig. 1. The muscle belly – composed of muscle fibres, connective tissue, blood vessels and nerves – is connected to the skeleton via tendons, with collagen fibres as the dominant component. Muscle fibres are elongated multi-nucleated cells, and contain myo-filaments (arranged in sarcomeres: the contractile units), mitochondria (for energy supply) and endoplasmic (sarcoplasmic) reticulum (with an important function in the activation process). Muscle fibres are connected to their tendon(s) via tendinous sheets. In a musculous attachment, muscle fibres insert via very short collagen fibres to the skeleton. Tendons and tendinous sheets act as force transmitters and elastic energy stores (Sect. 2.7). To investigate the function of muscles in locomotor systems, quantitative models of the mechanical behaviour of sarcomeres, muscle fibres and muscle–tendon complexes are required. In the following, such models will be developed, based on quantified phenomenological descriptions of the fundamental properties of underlying structural and functional levels.

2.1 Sarcomeres

The force and mechanical power production of sarcomeres depends on (1) sarcomere length and geometry, (2) shortening velocity, and (3) the activation of the sarcomere. This section deals with the first two aspects; activation dynamics is discussed in Sect. 2.2.

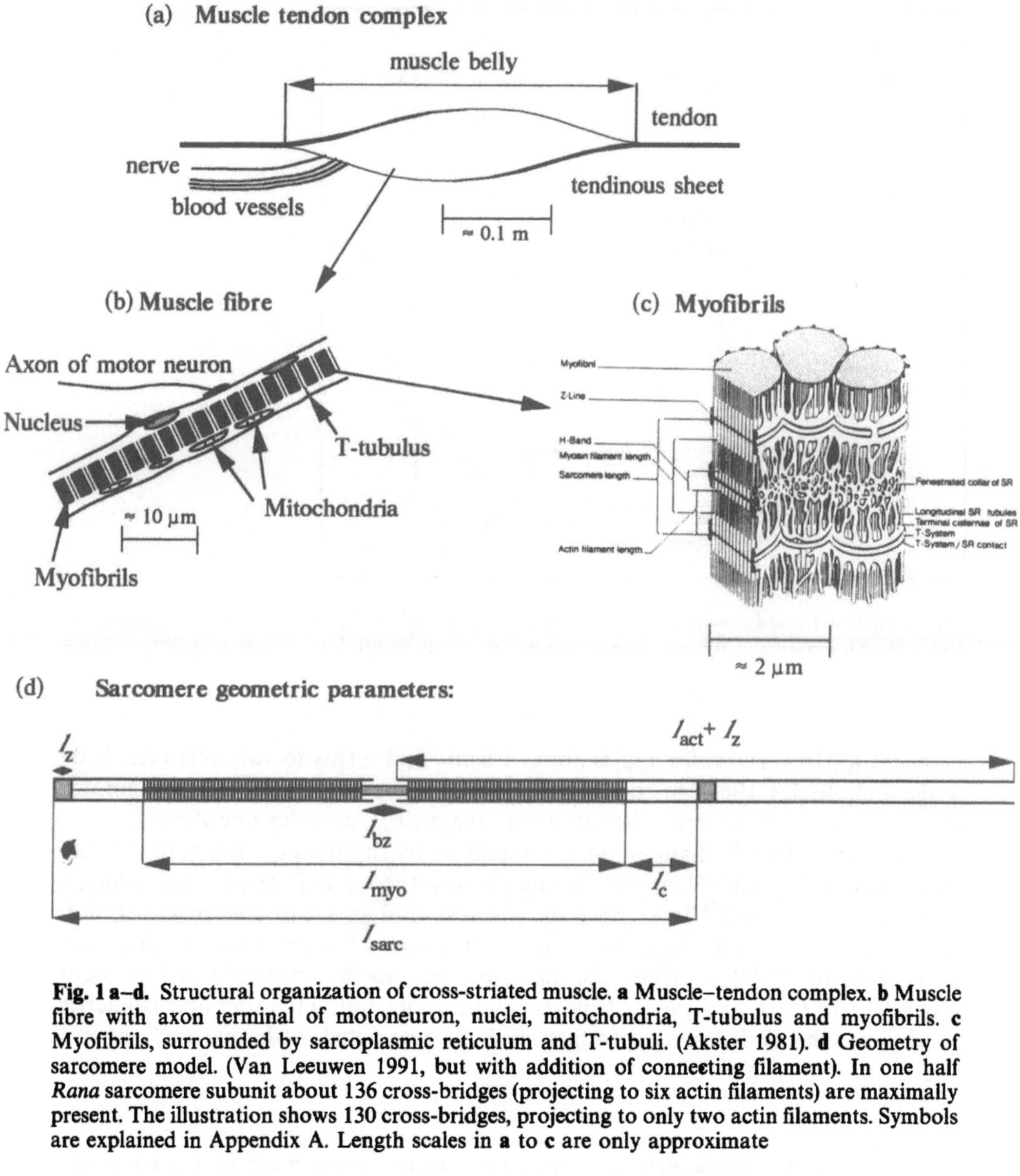

Fig. 1 a–d. Structural organization of cross-striated muscle. **a** Muscle–tendon complex. **b** Muscle fibre with axon terminal of motoneuron, nuclei, mitochondria, T-tubulus and myofibrils. **c** Myofibrils, surrounded by sarcoplasmic reticulum and T-tubuli. (Akster 1981). **d** Geometry of sarcomere model. (Van Leeuwen 1991, but with addition of conneeting filament). In one half *Rana* sarcomere subunit about 136 cross-bridges (projecting to six actin filaments) are maximally present. The illustration shows 130 cross-bridges, projecting to only two actin filaments. Symbols are explained in Appendix A. Length scales in **a** to **c** are only approximate

2.1.1 Force Production and Sarcomere Length and Geometry; Filamentary Overlap Function

The main structural elements of the sarcomere (Fig. 1d) are the myosin (thick) filament (with length l_{myo}), the actin (thin) filament (with length l_{act}) and the Z-disc (with thickness l_z). The myosin filament is connected to the Z-disc via the connecting filament (denoted also as titin filament, Wang 1985) of length l_c. From the myosin filament, myosin heads project outwards, except in a central part, the "bare zone",

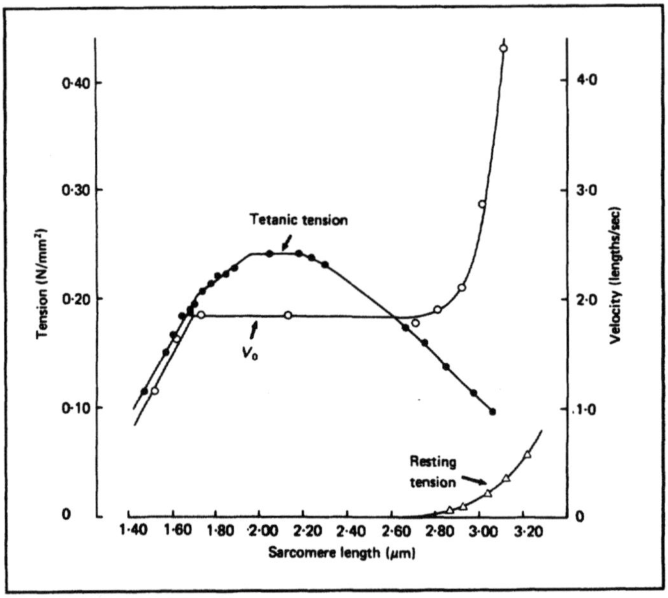

Fig. 2. Tension (*closed symbols*), maximum externally unloaded shortening velocity (*open circles*), and passive forces ($F_{pas\,2}$, *open triangles*) against sarcomere length for a frog muscle fibre. Edman (1979)

of length l_{bz}. In vertebrates, l_{myo} is about 1.6 μm, and seems to vary relatively little (Page and Huxley 1963), in contrast to l_{act} which is highly variable in vertebrates (from about 1.7 to 2.8 μm). The sarcomere length l_{sarc} includes one Z-disc.

Sarcomere length changes are accompanied by sliding movements of the thin filaments past the thick filaments (Sliding filament theory: H.E. Huxley and Hanson 1954; A.F. Huxley and Niedergerke 1954). Cross-bridges are formed by interaction of the myosin heads with neighbouring actin filaments. Experiments (Gordon et al. 1966; Edman 1979) indicate the cross-bridges to be mutually independent generators of force and power (Fig. 2). Thus, under full activation and isometric conditions, the active force production F_{cu0} associated with one half myosin

Fig. 3 a–f. Illustration of fundamental mechanical properties of a muscle system. **a** Characteristic lengths (indicated above each pictogram) in a full contraction of a sarcomere are numbered 1 to 5. **b** Force–length relationship of sarcomere with full activation and zero interfilamentary speed. Pictogram numbers of **a** correspond to those given at each characteristic length. Roman numbers denote curves of active force (*I*), $F_{pas\,1}$ (*II*), active force plus $F_{pas\,1}$ (*III*), and F_{pas2} (*IV*). **c** Velocity dependence function. **d** Normalized mean cross-bridge power as a function of u^*. **e** Normalized metabolic cost for stress production $P^*_\sigma(u^*)$. The left part is an estimate based on a discussion by Woledge et al. (1985). The right part is based on observations by Heglund and Cavagna (1987). **f** Stress–strain relationship of tendinous tissue, based on observations by Ker (1981) and Bennett et al. (1986). A plot is also shown of the strain energy density function, which shows the elastic energy stored per unit volume. Viscous effects are neglected

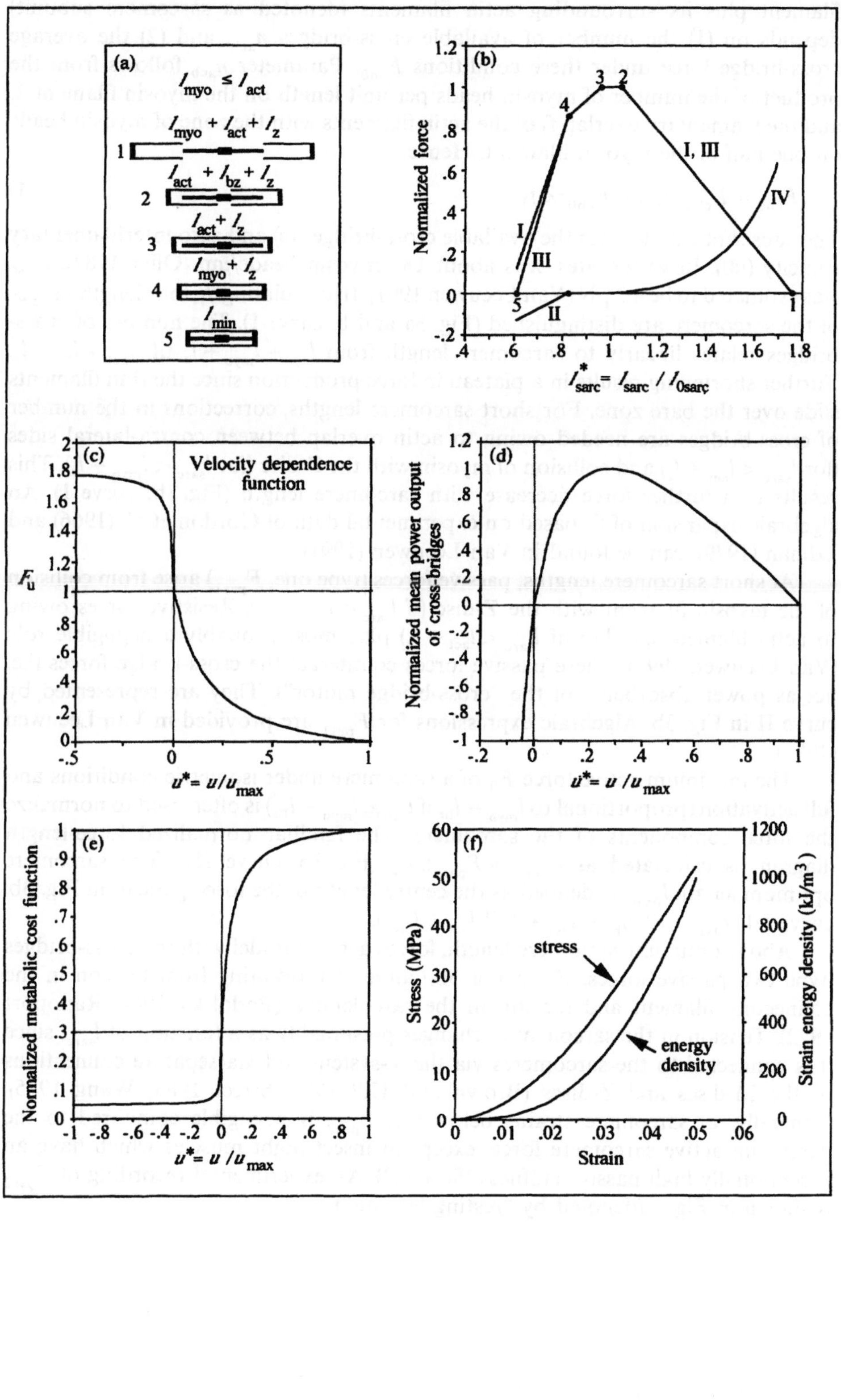
(a)
$l_{myo} \leq l_{act}$
$l_{myo} + l_{act} + l_z$
1
$l_{act} + l_{bz} + l_z$
2
$l_{act} + l_z$
3
$l_{myo} + l_z$
4
l_{min}
5
(b)
Normalized force
I, III
IV
I
III
II
$l^*_{sarc} = l_{sarc} / l_{0sarc}$
(c)
Velocity dependence function
F_u
$u^* = u / u_{max}$
(d)
Normalized mean power output of crossbridges
$u^* = u / u_{max}$
(e)
Normalized metabolic cost function
$u^* = u / u_{max}$
(f)
Stress (MPa)
Strain energy density (kJ/m^3)
stress
energy density
Strain

filament plus its surrounding actin filaments (denoted as sarcomere subunit) depends on (1) the number of available cross-bridges n_{acb}, and (2) the average cross-bridge force under these conditions F_{cb0}. Parameter n_{acb} follows from the product of the number of myosin heads per unit length on the myosin filament λ, and the filamentary overlap f_l of the actin filaments with the zone of myosin heads on one half of the myosin filament. Hence:

$$F_{cu0} = F_{cb0} \cdot n_{acb} = F_{cb0} \cdot \lambda \cdot f_l \tag{1}$$

The subscript cu0 denotes the available cross-bridges (c) and zero interfilamentary velocity (u0). In vertebrates, λ is about 187 myosin heads/μm (Offer 1987). F_{cb0} was estimated to be 4.3 pN (Van Leeuwen 1991). In calculating f_l, four length ranges of the sarcomere are distinguished (Fig. 3a and b, curve I). The number of cross-bridges relates linearly to sarcomere length from $l_{act} + l_{myo} + l_z$ till $l_{act} + l_{bz} + l_z$. Further shortening results in a plateau in force production since the thin filaments slide over the bare zone. For short sarcomere lengths, corrections in the number of cross-bridges are needed owing to actin overlap between contra-lateral sides (for $l_{sarc} < l_{act} + l_z$) and collision of myosin with the Z-disc (for $l_{sarc} < l_{myo} + l_z$). This results in a further force decrease with sarcomere length (Fig. 3b, curve I). An algebraic expansion of f_l, based on experimental data of Gordon et al. (1966) and Edman (1979), can be found in Van Leeuwen (1991).

At short sarcomere lengths, passive forces (type one, F_{pas1}) arise from collision of the myosin filament with the Z-disc (if $l_{sarc} < l_{myo} + l_z$). Resistive forces owing to actin filament overlap (if $l_{sarc} < l_{act} + l_z$) play most probably a negligible rôle (Van Leeuwen 1991). These passive forces counteract the cross-bridge forces (i.e. act as power absorbents of the "cross-bridge motor"). They are represented by curve II in Fig. 3b. Algebraic expressions for F_{pas1} are provided in Van Leeuwen (1991).

The maximum active force F_0 of a sarcomere under isometric conditions and full activation (proportional to $l_{myo} - l_{bz}$ if $l_{act} \geqq l_{myo} - l_{bz}$) is often used to normalize the force components of the sarcomere. The familiar normalized force-length diagram is calculated as $(F_{cu0} + F_{pas1})/F_0$ (Fig. 3b, curve III). The sarcomere optimum length l_{0sarc} is defined as the centre length of the force plateau in Fig. 3b, curve III ($l_{0sarc} = 0.5 l_{bz} + l_{act} + l_z$ if $l_{act} \geqq l_{mvo}$).

Above optimum sarcomere length, forces act in parallel with the cross-bridges (type two passive forces, F_{pas2}: Fig. 3b, curve IV), resulting from tension in the connecting filament and tension in the sarcolemma (Podolsky 1964; Rapoport 1972). Tension in the sarcolemma changes presumably as a function of l_{sarc}, since it is connected to the sarcomeres via the T-system and via separate connections to the M-discs and Z-discs (Brown and Hill 1982; Street 1983; Wang 1985). Generally, at sarcomere strains below 0.1, F_{pas2} is negligible compared to the maximum active sarcomere force, except in insect flight muscles which have an exceptionally high passive stiffness (Sect. 4.2). An experimental recording of F_{pas2} is shown in Fig. 2 (denoted by "resting tension").

2.1.2 Force Production and Interfilamentary Velocity; Velocity Dependence Function

The interfilamentary velocity u is the velocity of the myosin filament with respect to its corresponding actin filaments. To follow the conventions from the physiological literature, u is chosen to be positive if the sarcomere shortens. Hence, the absolute contraction velocity of the sarcomere u_{sarc} is equal to $2u$. The relationship between the normalized average cross-bridge force and u is defined as:

$$F_u = (1 - u^*)/(1 + u^*/k) \quad \text{for } u < 0 \qquad (2a)$$

$$F_u = 1.8 - 0.8(1 + u^*)/(1 - 7.56u^*/k) \quad \text{for } u > 0, \qquad (2b)$$

where $u^* = u/u_{max}$, u_{max} is the maximum interfilamentary velocity for which force can be generated, and k is a constant which depends on the muscle fibre type (Otten 1987b). Formula (2) will be denoted as the velocity dependence function, a term introduced by Hatze (1981). This function depends only on u, whereas the force–velocity relationship, a term often used in muscle physiology and mechanics, depends not only on u, but also on sarcomere length and active state. Thus, a single force–velocity relationship for a muscle fibre does not exist. Edman (1979) showed that the externally unloaded maximum specific shortening velocity $\hat{v}_{max}$ decreases linearly with sarcomere shortening for $l_{sarc} < l_{myo} + l_z$ (Fig. 2, V_0). Most probably, this decrease is caused by internal resistive forces owing to collision of myosin with the Z-discs. He showed also that $\hat{v}_{max}$ increases above optimum sarcomere length. This is probably caused by a contribution of passive elastic forces (F_{pas2}). Nevertheless, formulae (2a) and (2b) are equal to formulae for the normalized force–velocity relationship from the literature, since at fibre optimum length the normalized force–velocity relationship should have a shape close to the velocity dependence function. Formula (2a) corresponds to the normalized Hill (1938) equation, whereas formula (2b) is based on Aubert (1956).

From the velocity dependence function, the mean normalized cross-bridge power can be obtained as a function of u, shown in Fig. 3d. With the selected k of 0.17, the maximum power output is found at $u = 0.276u_{max}$.

The usage of the velocity dependence function (although sufficient for the present purposes) has some limitations since measurements indicate that a long-term (hundreds of ms) force enhancement occurs after pre-stretch (Edman et al. 1978), while, during a forcible stretch, force is increased with sarcomere length (Granzier et al. 1989), probably caused by filament compression. The velocity dependence function cannot be used if very rapid step changes of length are applied to a muscle fibre as in the experiments by Huxley and Simmons (1971).

2.2 Muscle Fibre Activation

Muscle tissue is generally activated through nervous input, although mechanical activation occurs also in fibrillar flight muscles. Different types of innervation are found. Fast vertebrate muscle fibres are generally innervated by just one nerve fibre with one end-plate (see e.g. Vrbová et al. 1978). The nerve fibre may also terminate with an array of end-plates (multiterminal innervation). Alternatively, muscle fibres may be innervated by several nerve fibres, each with several end-plates (polyneural and multiterminal innervation). This last type is found in many invertebrate muscles (e.g. Usherwood 1967), but is also one of the possible innervations in fish muscles (Bone 1978). It is thought that this type of innervation allows a more precise control of tension development in a muscle fibre. This seems to be especially important in very small muscles with a low number of muscle fibres. In mammals, a motor unit is the collection of muscle fibres being under control of one motoneuron. The nerve actionpotential is transmitted to the muscle fibre through release of a transmitter in the neuromuscular junction or synapse. Transmitter release causes a local depolarization of the sarcolemma (end-plate potential), which may lead to a muscle impulse transmitted along the membrane. The membrane generally has transverse invaginations known as the T-tubuli or T-system, which link the membrane to the sarcoplasmic reticulum (SR). Depolarization of the membrane causes a release of Ca^{2+} from the SR into the sarcoplasm which initiates contraction.

The mechanical response of a mucle fibre to a single nerve impulse is known as a twitch. Characteristics of the twitch are often used to make a functional distinction between different muscle fibre types. If nerve impulses follows each other with enough delay, a series of twitches is generated. If the delay shortens, twitches may partly or completely fuse to give a tetanus (Fig. 5a, c).

For the discussion of locomotor function it is convenient to have a mathematical description of the excitation-contraction coupling, enabling a simulation of twitch and tetanus. With some additions and alterations, the simple description formulated by Otten (1987b) is followed here. Quantitative details are provided in Appendix B. Following Ebashi and Endo (1968), the active state F_a is defined as the relative amount of Ca^{2+} bound to troponin. $F_a = 1$ if all interactive sites on the actin filament are exposed through the action of Ca^{2+}. In a resting muscle fibre, $F_a = F_{a0}$, so that $F_{a0} \leqq F_a \leqq 1$. Active force development is directly proportional to F_a.

2.3 Calculation of Force, Power and Work of a Muscle Fibre Under Dynamic Conditions

The average cross-bridge force of an activated sarcomere can now be described as the product of average cross-bridge force under isometric conditions and full activation, velocity dependence function, and active state: $F_{cb0} \cdot F_u \cdot F_a$. The dynamic force of the sarcomere F_{sarc} is equal to the sum of the cross-bridge forces and the passive forces:

$$F_{sarc} = n_{acb} \cdot F_{cb0} \cdot F_u \cdot F_a + F_{pas1} + F_{pas2} = F_{cu0} \cdot F_u \cdot F_a + F_{pas1} + F_{pas2}. \tag{33}$$

The work output W_{sarc} for a particular contraction from time t_a to t_b is described by:

$$W_{sarc} = \int_{l_{sarc}(t_a)}^{l_{sarc}(t_b)} F_{sarc} \, dl_{sarc}. \tag{4}$$

A solution of this integral is most conveniently obtained by numerical integration on a digital computer, since, during locomotor activities, both F_{sarc} and l_{sarc} vary in a complex way. Sarcomere power output P_{sarc}, the product of sarcomere force and velocity, is positive if force output and length change are in the same direction. Power output is negative when a muscle fibre produces force while it is stretched. In this case, energy is absorbed by the fibre.

For full activation ($F_a = 1$) and given values of l_{act}, l_{myo}, l_{bz}, l_z, a "force surface" is defined with the phase plane (l^*_{sarc}, u^*) as its base (Fig. 4a). Under all circumstances, force production of the muscle fibre is below this surface or just in this surface. Therefore, this surface may be called the "normalized force envelope" of the muscle fibre. Analogously, Fig. 4b shows a normalized power envelope. If $u^* > 0$, then power output is always below or just in the surface of maximum power. If $u^* < 0$, then power output is always above or just inside the surface. For both surfaces, F_{pas2} was neglected.

Figure 5a shows schematized twitch and tetanus responses of slow and fast muscle fibres of carp (Granzier et al. 1983). The above formulae may serve to simulate the twitch and the tetanus of a muscle fibre, as well as (more important) power output during dynamic circumstances as in locomotion. Figure 5c, left graph, shows twitch and tetanus simulations of a fast muscle fibre of the carp. Neither a "sag" phenomenon (i.e. a decline in tension during a tetanus, Fig. 5a) nor fatigue are incorporated in the model.

To investigate muscle fibre function in locomotion, it is very rewarding to measure the force output of isolated muscle fibres or small fibre bundles during length changes imitating in vivo strain variations. This technique was first developed by Machin and Pringle (1959, 1960) in their classical studies on the mechanics of asynchronous flight muscle. Quite recently, the technique has also been applied to synchronous insect flight muscle (Josephson 1985; Stevenson and Josephson 1990), frog sartorius muscle (Stevens 1988) and fish swimming muscles (Altringham and Johnston 1990). An example of a recording for fast twitch fibres of swimming muscles from a bullrout with a sinusoidal length change is shown in Fig. 5b. The work done in one cycle is proportional to the surface of the loop

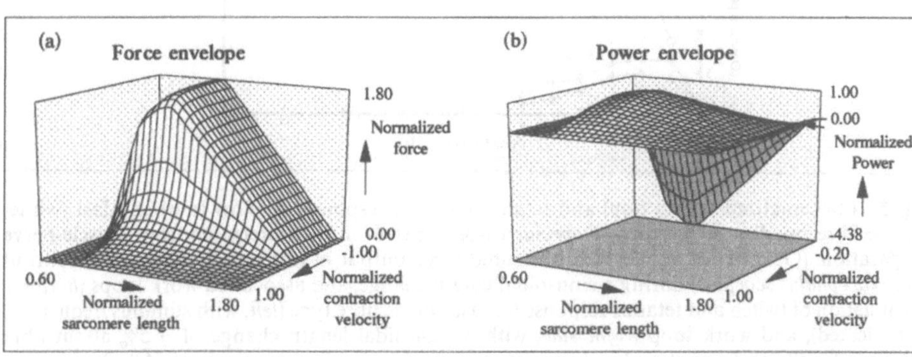

Fig. 4 a, b. Force (**a**) and power (**b**) envelopes as discussed in Sect. 2.3. *Small arrow* in **b**, *right side*, points to zero-power line through surface (along this line, the interfilamentary speed is zero)

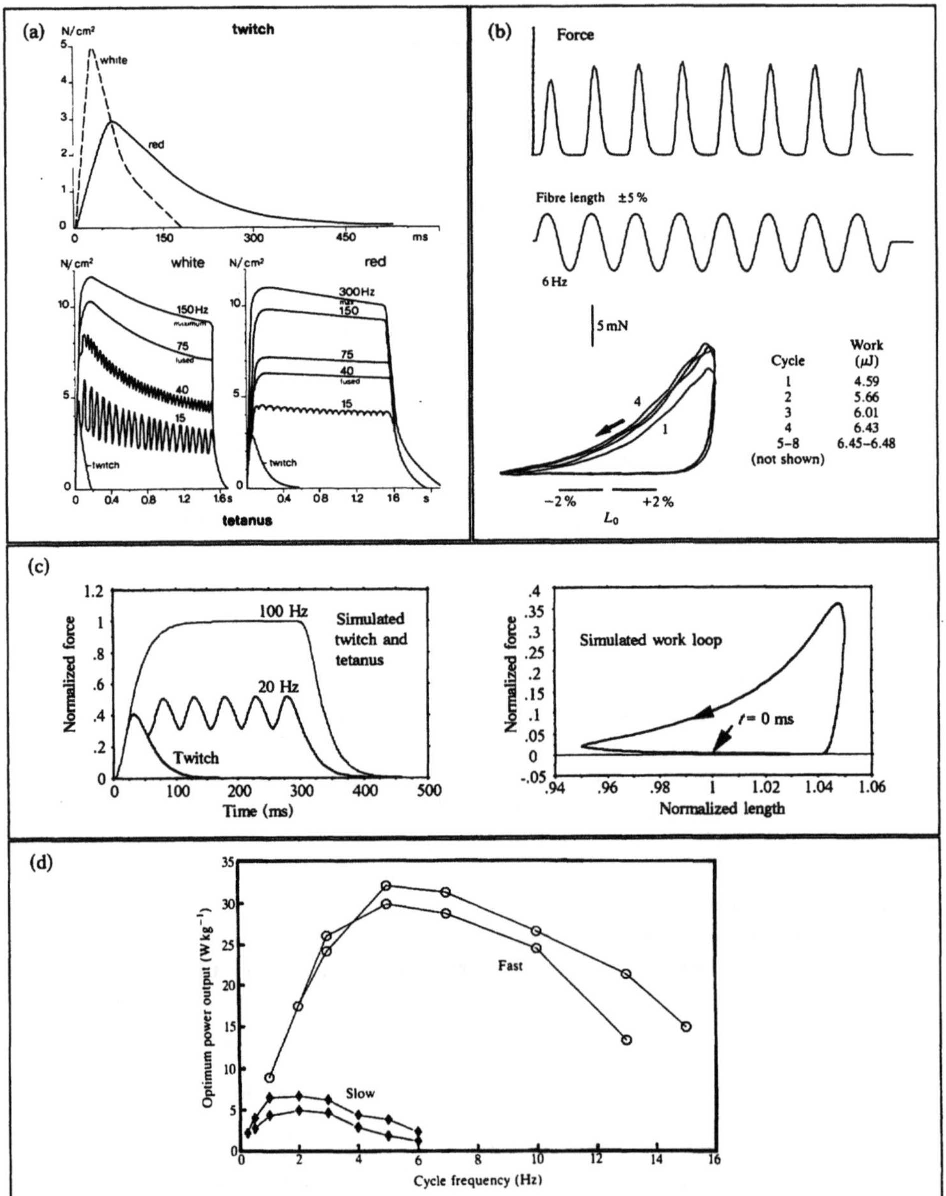

Fig. 5. **a** Schematized twitch (*top*) and tetanus (*bottom*) responses of slow (red) and fast (white) muscle fibre bundles of the carp (*Cyprinus carpio*), based on measurements from muscle-nerve preparation. (Granzier et al. 1983). **b** Measured force output of fast muscle fibres of a bullrout (*Myoxocephalus scorpius*) during a sinusoidal length change, and associated work loops (*bottom*). **c** Simulation of twitch and tetanus response for fast muscle fibre type (*left*, with stimulus frequencies as indicated), and work loop (*right side*, with a sinusoidal length change of $\pm 5\%$ about fibre optimum length, and one stimulus during stretching phase with phase angle of 54°). Parameters: $l_{act} = 1.75\,\mu m$, $l_{myo} = 1.57\,\mu m$, $l_{cbf} = 0.15\,\mu m$, $l_z = 0.06\,\mu m$, $\tau_{ca} = 6\,ms$, $\tau_1 = 15\,ms$, $\tau_{ca} = 20\,ms$, $\hat{v}_{max} = 15$ lengths/s. **d** Power output as a function of cycle frequency for slow and fast fibres of bullrout (**b** and **d** Altringham and Johnston 1990)

obtained when force is plotted against length. A clock-wise loop denotes energy absorption (negative work), whereas an anti-clock-wise loop shows that positive work in done. Figure 5b shows measured work loops with one stimulus per cycle. Figure 5c, right plot, shows a simulated work loop of a fast carp fibre (one stimulus per cycle and comparative stimulus time). In all simulations of this chapter, it is assumed that all sarcomeres of a muscle fibre have equal lengths and activations. Thus, sarcomere strain is equal to muscle fibre strain. Work loops with different shapes are obtained if the activation characteristics are changed (Sect. 4.1; Figs. 16d, e, f). One advantage of a validated model of muscle fibre work output is that predictions can be made of in vivo muscle performances, guiding experimentation. Figure 5d shows a measured plot of the maximum mean power output against frequency for slow and fast muscle fibres of a bullrout. Fast fibres have a much higher peak performance but are more quickly fatigued.

2.4 Metabolic Cost Function

Muscle energetics is discussed in Chap. 8. A metabolic cost function will be introduced here since this helps to understand why muscles are active in particular phases of their lengthening-shortening cycles. Let $\hat{P}_\sigma$ be the metabolic power per unit volume of muscle needed to produce unit mechanical stress. $\hat{P}_\sigma$ is proportional to the length of the muscle fibre. The economy of a muscle is defined as $1/\hat{P}_\sigma$. Thus, relatively short fibres produce tension economically. Heglund and Cavagna (1987) measured tension of small leg muscles of a frog (*Rana esculenta*) and the rat at a range of contraction speeds. They recorded also the O_2 consumption of these muscles, allowing to estimate metabolic power. Close to $\hat{v}_{max}$, metabolic cost per unit force is about three times the cost at zero shortening velocity. Their experiment is somewhat hampered by the problem that the ratio of muscle belly speed over muscle fibre speed could vary, owing to the presence of tendinous sheets (see Sect. 2.7). Owing to muscle pennation, muscle fibres exert force against each other. Hence, strictly speaking, the results cannot be directly applied to a single muscle fibre. This problem is avoided if the experiment could be repeated with a single fibre or a small fibre bundle. Furthermore, slow fibres produce tension more economically than fast fibres. Loosely based on the experimental data by Heglund and Cavagna (1987) and a discussion by Woledge et al. (1985), the normalized metabolic power consumption for force production as a function of the interfilamentary velocity (Fig. 3e) can be described:

$$P_\sigma^* = 1.0 - 0.7(1 - u^*)/(1 + c_1 \cdot u^*) \quad \text{for } u^* > 0 \tag{5a}$$

$$P_\sigma^* = 0.1 + 0.2(1 + u^*)/(1 - c_2 \cdot u^*) \quad \text{for } u^* < 0, \tag{5b}$$

where $c_1 = 16$, and $c_2 = 160$ are constants. The specific metabolic power usage of a muscle fibre is obtained by:

$$\hat{P}_{met} = P_\sigma^* \cdot \hat{P}_{metmax} \cdot F_a \cdot f_l / f_{lmax}, \tag{6}$$

where $\hat{P}_{\text{metmax}}$ is the maximum specific metabolic power output and f_{lmax} is the maximum filamentary overlap. Formula (6) will be useful in the discussion of the dynamic properties of the muscle–tendon complex and muscle function in terrestrial locomotion.

2.5 Connective Tissue Properties

The connective tissue of a skeletal muscle belly is traditionally divided into three structural entities. The epimysium surrounds the whole muscle, the perimysium surrounds bundles of muscle fibres, whereas the endomysium surrounds individual muscle fibres. In isolated muscle fibres, the endomysium is considered to contribute significantly to the passive properties above sarcomere optimum length. Light et al. (1985) have shown that the amount of perimysium in a muscle is generally much higher than the endomysium content (dry mass ratio varied from 2.8 to 64 in different muscles). This indicates that, in an intact muscle, the contribution of the perimysium to passive elastic forces is very significant. Purslow (1989) showed the perimysium in bovine stermomandibularis muscle to be a crossed-ply arrangement of crimped collagen fibres which reorientate and decrimp on changing sarcomere length. Mean collagen direction with respect to the muscle fibres ranged from about 80° at $l_{\text{sarc}} = 1.1\ \mu\text{m}$ to about 20° at 3.9 μm. Purslow modelled the load-sarcomere length curve of the perimysium and showed good agreement with passive elastic properties of the muscle. He concluded that the rôle of the perimysial collagen network is to prevent over-stretching of the muscle fibre bundles above $l_{\text{sarc}} \approx 3.6\ \mu\text{m}$. Most likely, the endomysium works in a similar fashion. Therefore, in an activated muscle working around optimum sarcomere length, the contribution of the endo- and perimysium to force output can generally be neglected.

2.6 Scaling Effects, Optimal Design and Muscle Fibre Types

Sarcomeres operate at different frequencies in various locomotor systems. Using asynchronous muscle fibres, a small insect may operate its wings at a frequency of several hundreds of Hz, whereas the sarcomeres vary only about 2 to 3% in length during each wing beat cycle (Pringle 1972, 1975; Ellington, 1985). Cutts (1986) estimated sarcomere length changes of about 20% in flight muscles of zebra finches and budgerigars. For slow muscle fibres, Van Leeuwen et al. (1990) estimated sarcomere length changes of about 12% in continuously swimming carp. For kick-and-glide swimming, an estimate was made of about 20%. If the operation frequency of a sarcomere decreases, more time is available for each contraction. Hence, the sarcomere length change can be expected to increase so as to maximize the mean power output in a contraction cycle. Thus, for locomotor systems of large vertebrates, relatively large changes of sarcomere length may occur, owing to the low frequencies of movement. However, experimental data are still lacking. Alternatively, more muscle fibres of a slow type can be used to increase the efficiency and endurance (Goldspink 1977). Both strategies might also be used simultaneously.

The optimal design of sarcomeres was first discussed by Otten (1987a) who examined the work output in a full contraction of sarcomeres of different geometries. A more relevant optimization criterion was formulated by Van Leeuwen (1991). He modelled the effects of different sarcomere designs and contraction ranges on the specific myofibrillar power output. The mean specific power output by a sarcomere is determined by:

(1) The efficacy by which the available myosin heads are used throughout the contraction. Cross-bridges are lost owing to thin filament overlap and collision of the thick filament with the Z-discs. Power output is further diminished since the cross-bridges do work against an internal resistance (Sect. 2.1).
(2) The myosin head density in the sarcomere. The Z-disc and the bare zone on the myosin filament consume space and therefore reduce the cross-bridge density.

Theoretically, absolute maximum power output would be delivered by an ideal (but not feasible) sarcomere with parameters $l_z = l_{bz} = 0\ \mu m$ and $l_{act} = l_{myo}$, which makes an infinitesimal contraction at the interfilamentary contraction speed for optimum cross-bridge power. Figure 6a–d shows contour plots of the mean power output as a function of l_{myo} and l_{act}, calculated with Van Leeuwen's (1991) theory, and normalized with respect to the theoretical maximum. Each point on the power surface of Fig. 6a represent a separate simulation of a sarcomere contraction, starting at a sarcomere length of $1.5l_{cmax}$ and ending at l_{cmax}, where l_{cmax} ($= l_{myo} + l_z$) is the sarcomere length with optimum cross-bridge density (Van Leeuwen 1991). The interfilamentary velocity u (set to $0.276u_{max}$ for optimum cross-bridge power) was kept constant during the contraction and was defined to be equal for all sarcomere designs. Hence, the strain rate drops with increasing filament lengths. The power surface has a skewly oriented ridge with a virtually constant l_{myo}/l_{act} ratio (just below 0.8). The power output increases slightly along the ridge in the direction of greater filament lengths. The increase gets slower with longer filament lengths. These phenomena are explained as follows. If the ratio l_{myo}/l_{act} is kept constant, the strain rate is about inversely proportional to sarcomere length, whereas sarcomere force is roughly proportional to filament length. Thus, power output is only slightly affected. Proportionality is not exact since both l_{bz} and l_z were kept constant (at 0.15 μm and 0.06 μm, respectively). Hence, the slow power rise along the ridge is caused by an increasing cross-bridge density with longer filament lengths. The shorter the filaments, the larger is the effect on the power output of a change of filament length. Figure 6b is similar to Fig. 6a, but the contraction range is now decreased to only $0.03l_{cmax}$. The l_{myo}/l_{act} ratio on the ridge is now about 1.

In locomotor activities, the dimensions of an animal and its surrounding medium (air, water or soil) determine roughly the range of operating frequencies and amplitudes of sarcomere contractions. Hence, the generation of a particular range of strain rates is required. Therefore, it is particularly interesting to prescribe the strain rate of the sarcomere contractions. The results for a large and a small contraction are shown in Fig. 6c, d. Now, a clear optimum is found, since increasing filament lengths require an increasing interfilamentary speed to keep the strain

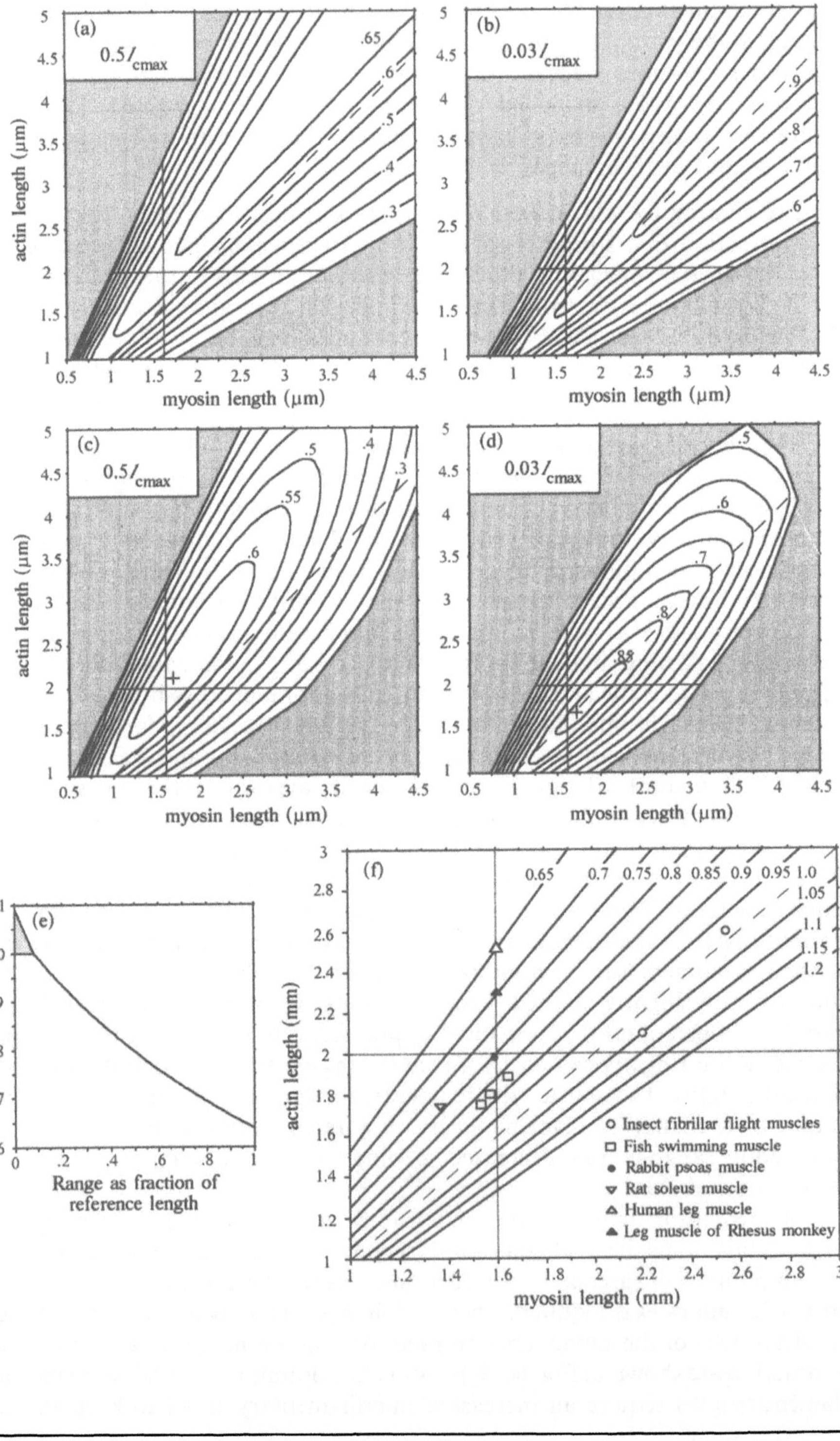
(a)
$0.5 l_{cmax}$
(b)
$0.03 l_{cmax}$
(c)
$0.5 l_{cmax}$
(d)
$0.03 l_{cmax}$
actin length (μm)
myosin length (μm)
(e)
$(l_{myo}/l_{act})_{pmax}$
Range as fraction of reference length
(f)
0.65 0.7 0.75 0.8 0.85 0.9 0.95 1.0
1.05
1.1
1.15
1.2
actin length (mm)
myosin length (mm)
Insect fibrillar flight muscles
Fish swimming muscle
Rabbit psoas muscle
Rat soleus muscle
Human leg muscle
Leg muscle of Rhesus monkey

rate constant. Quite importantly, in the neighbourhood of each of the global maxima of Fig. 6c, d, power output is quite insensitive to changes in filament lengths if the ratio l_{myo}/l_{act} is kept constant. This is understood since close to the maximum in the normalized cross-bridge power function, the power drops only slightly with a changing interfilamentary velocity (Fig. 3d). However, the power output drops very rapidly if the filaments change length so as to alter maximally the ratio l_{myo}/l_{act}. This is caused by a drop in the efficacy of the usage of myosin heads during the contraction, owing to increased effects of only partial overlap between myosin and actin filaments or overlap of opposing actin filaments.

The position of the ratio l_{myo}/l_{act} for optimum power output, $(l_{myo}/l_{act})_{pmax}$, is shifted to a higher value if the contraction amplitude decreases (Fig. 6e). For a contraction from $1.5l_{cmax}$ to l_{cmax} (Fig. 6c) (still above the estimated amplitudes for some relatively small vertebrates, cf. Alexander 1989), the $(l_{myo}/l_{act})_{pmax}$ ratio is about 0.8. This ratio is "typical" for many small vertebrates. Fish use relatively small contraction ranges in continuous swimming, so that a relatively high (l_{myo}/l_{act}) is expected. This is supported by measurements (Fig. 6f). So far, however, only relatively small specimens have been investigated. It can be expected that very large fish use larger contraction ranges. For the very short range contraction of Fig. 6d (resembling the contractions of asynchronous flight muscles), maximum power output is found for a ratio of about 1. The $(l_{myo}/l_{act})_{pmax}$ value for a high amplitude contraction from $2l_{cmax}$ to l_{cmax} of 0.639 is very close to the lower limit of this ratio found till now in vertebrate sarcomeres (Fig. 6f).

In conclusion, the complete variation in the ratio of myosin length over actin length (from about 1.05 down to 0.65, as observed in insect and vertebrate sarcomeres, Fig. 6f) can be explained as a series of adaptations for optimum power output from a small to a large contraction range, respectively. Short contraction ranges with optimal design provide excellent peak power output (for fibrillar flight muscles about 0.9 of the theoretical maximum). Power output has a less steep sarcomere length dependence, but with a much lower mean specific myofibrillar power output, in an optimum design for a larger contraction. Using the above general theory, various predictions can be made. Two examples are (requiring further research): (1) During growth of an animal, the l_{myo}/l_{act} ratio will tend to decrease in most locomotor muscles (not necessarily in all muscles owing to

◄

Fig. 6. Plot **a** shows contours of normalized mean power output $\bar{P}^*$ generated by a maximally activated sarcomere as a function of l_{myo} and l_{act}. Bare zone and Z-disc lengths were kept constant at the reference values (0.15 μm and 0.06 μm, respectively). The interfilamentary velocity was kept constant at 0.274 of the maximum. Values of $\bar{P}^*$ are given near a selection of the contours. The contraction range is $0.5l_{cmax}$. The reference sarcomere (with $l_{myo} = 1.6$ μm, and $l_{act} = 2.0$ μm) is positioned at the intersection of the *grey reference lines*. The *dashed skew line* represents $l_{myo} = l_{act}$. The *shaded areas* are of minor interest, and were not investigated. Plot **b** is similar to **a**, except that the contraction range is $0.03l_{cmax}$. Plots **c** and **d** correspond to **a** and **b**, except that now the strain rate is kept constant. Maxima are located at plus signs. Plot **e** shows the optimum l_{myo}/l_{act} as a function of contraction range. The *shaded triangle* indicates that a range of l_{myo}/l_{act} ratios gives optimum power under a particular contraction range. Plot **f** shows a scatter diagram of l_{act} against l_{myo} of measured sarcomeres from locomotor muscles (for references see Van Leeuwen 1991). **a** to **d** calculated using theory by Van Leeuwen (1991). **e** and **f** after Van Leeuwen (1991) changing functional demands). (2) The expected large contraction ranges required for large birds limit, in combination with the low operating frequencies, the specific myofibrillar power output. This may help to explain why hovering flight (requiring, in birds, larger wing stroke amplitudes and sarcomere length changes than forward flight) is only used by insects and small birds.

Muscle fibres have been classified on the basis of structural criteria (innervation, mitochondrial content, structure of the SR, filament structure, etc.) and histochemical and biochemical criteria (e.g. abundance and efficacy of various enzymes). An overview of a variety of classification attempts for mammalian muscle fibres is given by Pearson and Young (1989). Functional criteria (endurance, contraction velocity, power output, twitch characteristics, etc.) are a useful guide for classification as demonstrated in a classical paper by Burke et al. (1973). Here, only a brief discussion from a mechanical viewpoint will be given with the aid of Fig. 7. For locomotion, specific power output by the muscle fibres is a dominant characteristic. A surface is drawn of peak specific power to visualize certain trends. No quantitative information can be extracted from this picture. Power output can be delivered in bursts, during long periods, or intermediate intervals. If sustained activity is required, than the local supply of metabolic power should cover continuously the mechanical power output (Weis-Fogh and Alexander 1977). For this purpose, an extensive battery of mitochondria is required, as well as an extensive system of blood vessels and, for insects, trachea for gas exchange and in- and output of metabolites. This apparatus requires space, which reduces the specific power output. Therefore, the power surface is lower at the right side of Fig. 7 than on the left side. The gain, however, is a relatively high mean specific power output. For optimal performance, the various components of the system of power supply and consumption should be tuned relative to each other, so that none of the components forms a limiting bottle-neck due to under design, or consume extra space by over design. This state of structural design whereby the formation of structural elements is regulated to satisfy but not exceed the requirements of the functional system has been denoted as symmorphosis by Taylor

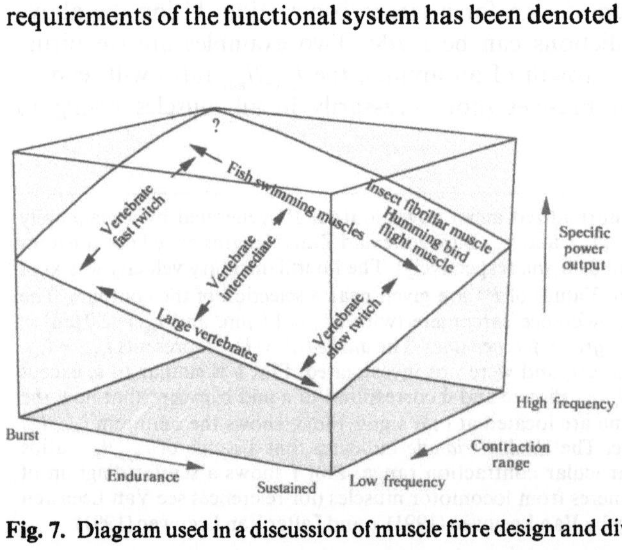

Fig. 7. Diagram used in a discussion of muscle fibre design and diversity. See text for explanation

and Weibel (1981). The specific myofibrillar power output increases if the contraction range decreases (see above). Therefore, the power surface increases from front to back. For sustained activities, the gain in specific power output from the sarcomeres by contraction range shortening is partially lost owing to increased space requirements for the energy supply apparatus. Therefore, the front-to-back slope of the power surface is steeper at the left side than on the right side. Another space-consuming structure is the SR, especially if fast response times are required as in synchronous insect flight muscles (5–20% of fibre volume, Ellington 1985). This negative effect on specific power output is largely overcome by the mechanically induced activation in asynchronous muscle fibres.

Muscle fibres designed for high power output during burst activities (fast twitch fibres, left side of power surface, Fig. 7) have a high myofibrillar content and a fast myosin-ATPase, whereas the energy supply apparatus consumes very limited space (during bursts, anaerobic metabolism prevails). Highest power output would be obtained if small contractions would be applied with appropriate sarcomere design and mechanical activation ("?" in Fig. 7). To my knowledge, this last option is not used in nature.

2.7 Dynamics of the Muscle–Tendon Complex

Muscle Architecture. The mechanical performance of a skeletal muscle is influenced by the geometrical arrangement of its muscle fibres (muscle architecture). A muscle is called parallel-fibred if its muscle fibres are parallel to the tendon(s) to which the muscle belly is attached. In pennate muscles, the fibres make particular angles with their tendon(s). Tendons do not stop abruptly at the muscle belly, but continue as tendinous sheets, to which bundles of muscle fibres (generally via a short collagenous tissue strap). The cross-sectional area of the tendinous sheet gradually decreases towards its end. This corresponds to a gradual decrease in the number of muscle fibres which can contribute to the mechanical load. Muscle fibre angulation influences the force output and length change (both vector quantities) of the muscle belly. This influence is variable, since fibre angulation increases during muscle belly shortening. Power and work output (scalar quantities) depend on muscle tissue volume, but not on muscle architecture (or only very slightly since internal energy losses may vary a little with architecture). Power output by the muscle-tendon complex (MTC) is the sum of the power output by the muscle tissue P_{mus}, the parallel elastic element in the muscle belly P_{par}, the tendinous sheets P_{ts}, and the tendons P_t. Models have been proposed to predict effects of architecture on force output and length change. Most models are planar and consider only straight muscle fibres. The considered models generally are kinematic abstractions, generally violating the force balance within the tissue. Classical papers on muscle architecture are by Stensen (Stenonis) (1667) and Benninghoff and Rollhaüser (1952) while Otten (1988) provides a recent review. All unipennate MTC models violate fundamental laws of classical mechanics. Some have an unexplained torque on the muscle belly, whereas others have an unexplained angle between tendon and tendinous sheets.

A Dynamic Muscle–Tendon Model. Here, a planar model of a MTC with a bipennate muscle belly will be considered (Fig. 8), composed of straight muscle fibres, elastic tendons and tendinous sheets, and muscle volume V_m. The symmetry of the model avoids some problems of the unipennate model, but strictly speaking, curved fibres and tendinous sheets would still be needed. The inaccuracy of the present approach is small once the pennation angle is small or when the belly is long compared to muscle fibre length. For most muscles, these conditions are fulfilled. Let all muscle fibres have an equal pennation angle α, let they be of equal length l_f, let they have the same mechanical and activation properties, and let they have equal strain ε_f. The muscle fibres incorporate all quantified properties of Sects. 2.1 and 2.2. They can, in principle, be activated asynchronously, but only synchronous activations will be discussed. The pennation angle is α_0 if the muscle is at rest (i.e. not activated and not under tension). The length of the muscle belly is denoted by l_{mb}. The physiological cross-section of the muscle fibres A_{f0} equals V_m/l_{f0}, where l_{f0} is the resting fibre length. Furthermore, let the tendinous sheets have equal lengths l_{ts}, and let l_t be the total length of the tendons with slack cross-section A_{t0}. Let $l_{tts} = l_t + l_{ts}$ be the tendon length plus one tendinous sheet length. It will be assumed that equal strains ε_t act in the tendinous sheets and tendons, so that l_{tts} can be used as the effective length of the series elastic component. This assumption is supported by measurements by Rack and Westbury (1984) for low strains and by Morgan (1977) for high strains, although some doubts remain (Proske and Morgan 1987). The total instantaneous length of the MTC is given by:

$$l_{mtc} = l_{tts} + l_f \cdot \cos\alpha = l_{tts0} \cdot (1 + \varepsilon_t) + l_{f0} \cdot (1 + \varepsilon_f) \cdot \cos\alpha, \tag{7}$$

where l_{tts0} is l_{tts} at rest. Tendons and tendinous sheets (TTS) form the dominant components of the so-called series elastic element (SEC) of a MTC. From demands of constant volume it can be recognized that:

$$\alpha = \arcsin\left\{\frac{l_{ts0} \cdot l_{f0} \cdot \sin\alpha_0}{l_{ts} \cdot l_f}\right\}, \tag{8}$$

where l_{ts0} is l_{ts} at rest. The nonlinear stress–strain relationship of tendons and tendinous sheets used in the model (Fig. 3f) is loosely based on stress–strain

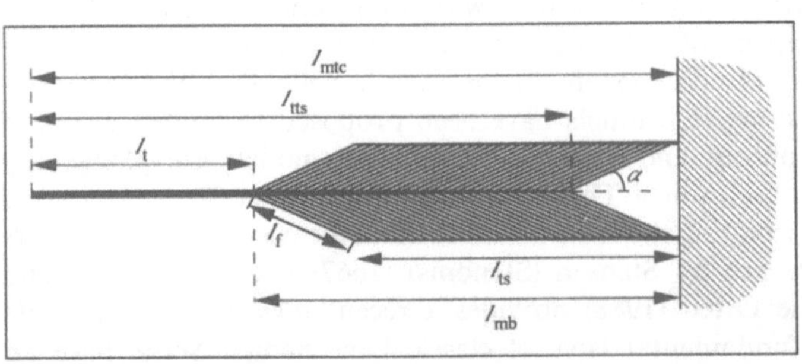

Fig. 8. Illustration of planar bipennate model of a muscle–tendon complex. Symbols are explained in Appendix A

measurements of tendons by Ker (1981) and Bennett et al (1986). Viscous effects are neglected since these are relatively small (only about 6% energy loss in cyclic movements). The stress–strain relationship (Fig. 3f) can be described by

$$\sigma_t = c_3 \cdot \varepsilon_t^{c_4} \qquad \text{for } \varepsilon_t < \varepsilon_c \tag{9a}$$

$$\sigma_t = c_5 \cdot \varepsilon_t + c_6 \quad \text{for } \varepsilon_t \geqq \varepsilon_c, \tag{9b}$$

where $c_5 = 1.5\,\text{GPa}$ (the tangent modulus), $c_6 = -22.5\,\text{MPa}$, $\varepsilon_c = 0.035$ (the critical strain above which the relationship is linear), $c_3 = c_5/(c_4 \varepsilon_c^{c_4 - 1})$, and $c_4 = c_5 \cdot \varepsilon_c/(c_5 \cdot \varepsilon_c - c_6)$. The dominant viscous effects of the muscle belly are taken account of by the velocity dependence function [Eq. (2), Fig. 3c]. Inertial effects due to accelerations of muscle and tendon mass are neglected. In computing the force of the MTC (F_{mtc}), it was used that at any time tendon force (F_t) equals muscle belly force (F_{mb}). Hence, the instantaneous lengths l_f and l_{tts} and α can be found iteratively.

The above MTC model was implemented in a computer program, allowing to simulate effects on the mechanical performance of (1) different architectures (properties of tendons and tendinous sheets and muscle fibre properties like angulation, length and cross-sectional area), (2) different types (twitch, tetanus, etc.) and timing of activation. Furthermore, the model can be used to predict optimum architecture and activation for prescribed functional demands. Experimental observations can be used to check the validity of the model. On the other hand, the model can be used to reinterpret particular experiments or to design new ones. A small number of different simulations will now be discussed. The reference muscle architecture used is loosely based on the extensor digitorum longus muscle (EDL) of the rat, as described by Ettema and Huijing (1989). Muscle fibre characteristics are from Close (1964), after reinterpretation of his data (see below). The data are listed in Table 1.

Twitch and Tetanus. Figure 9 (left side) shows a simulation of the twitch response of a fast twitch muscle fibre in four different situations. In situation 1 (s1), the fibre

Table 1. Parameter values used in a simulation of a muscle–tendon complex (Figs. 9; 10a, b; 11 series 1; 12) Data are selected so as to loosely resemble the extensor digitorum longus muscle of the rat. Ultrastructural data from Van Lookeren Campagne et al. (1988), most architectural data from Ettema and Huijing (1989), and most physiological data from Close (1964), with slight changes. A_{t0} is assumed to be $A_{f0}/50$. Fibre angle α_0 is calculated from Ettema and Huijing (1989), but is probably slightly overestimated (see Close 1964). Symbols are explained in Appendix A.

Ultrastructural data:			
$l_{act} = 2.10\,\mu m$	$l_{myo} = 1.60\,\mu m$	$l_{bz} = 0.15\,\mu m$	$l_z = 0.06\,\mu m$
Architectural data:			
$l_{f0} = 11.78\,mm$	$A_{f0} = 15.5\,mm^2$	$l_{t0} = 18.94\,mm$	$l_{ts0} = 20.95\,mm$
$A_{t0} = 0.31\,mm^2$	$\alpha_0 = 17°$		
Physiological data:			
$\sigma_{miso} = 200\,kPa$	$\tau_{ca} = 5\,ms$	$\tau_1 = 5\,ms$	$\tau_1 = 5\,ms$
$t_s = 1\,ms$	$f_s = 500\,Hz$	$\hat{P}_{metmax} = 1000\,kW/m^3$	

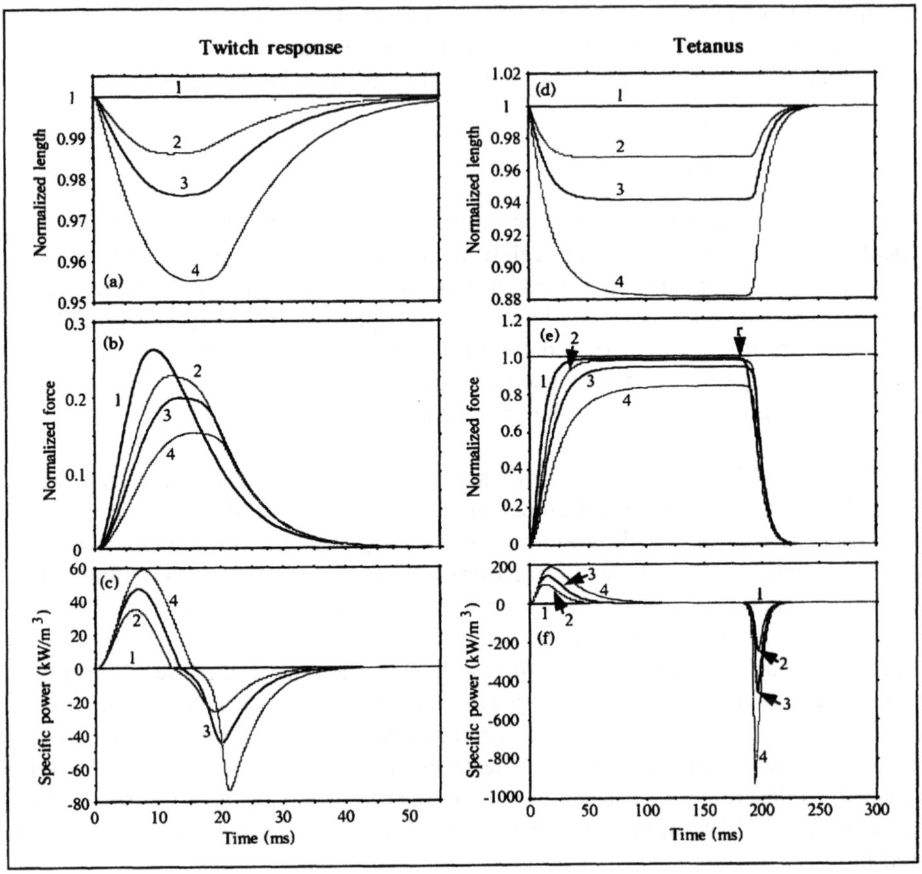

Fig. 9 a–f. Simulation of a twitch (*left side*) and tetanus response (*right side*) of *1* an isolated muscle-fibre kept at its optimum length, *2* the same fibre in a muscle belly of constant length, with tendinous sheets, but without a tendon attached, *3* the same fibre in a muscle belly with a tendon attached [reference MTC (Table 1) kept at constant length], and *4* the same fibre in a muscle belly with (compared to *3*) doubled tendinous sheet compliance, and quadrupled tendon compliance (cross-section 50%, length 200%). Muscle parameters of *1* to *3* are given in Table 1. **a** and **d** Normalized length of the muscle fibre. **b** and **e** Force production of the muscle fibre, normalized relative to maximum isometric force. **c** and **f** Specific power output of the muscle fibre. The *r* above the arrow in **e** denotes end of stimulus train

is isolated and kept at its optimum length. In situation 2 (s2), the fibre is part of a muscle belly with tendinous sheets, but without a tendon. The muscle belly is kept at the resting length, with muscle fibres at optimum length if no force is generated. Situation 3 (s3) is similar to 2, but now an external tendon is added. Situation 4 (s4) is comparable to s3, but with a doubled tendinous sheet compliance and a quadrupled tendon compliance (A_{t0} is halved and l_{t0} is doubled). The MTC is kept fixed at l_{mtc0}. As a result of the stimulus in s2 to s4, the muscle fibres shorten while producing force. The work done by the fibres is absorbed by the lengthening tendon (in s3 and s4) and tendinous sheets. Fibre length change is fastest and largest in s4, since least force is required for a particular length change. During fibre shortening, force output of the fibre is reduced compared to s1, owing to (1) the velocity dependence function and (2) (small effect) the length dependency of activation. If, in the twitch, muscle fibre activation decreases again, fibre shortening still continues for a short time, since higher forces can be produced at a lower contraction velocity. Therefore, peak force of the muscle fibre is considerably delayed and reduced relative to s1. The peak becomes also much wider. This is true, even without an external tendon. In fact, in the simulated EDL, the tendinous sheets contribute more to the SEC than the tendon does, since they are longer. The effects of TTS compliance on force output are large, because force production is very sensitive to velocity changes if the interfilamentary velocity is close to zero (Fig. 3c). Therefore, it is also important, in physiological experiments, to reduce the compliance of the measurement equipment as much as possible. For the same reason, in mechanical models, it is important to include the nonlinear elastic properties of the TTS.

Owing to tendinous sheet lengthening, the muscle belly length tends to increase. This effect is compensated by extra muscle fibre shortening and rotation (compared to the situation with a non-elastic sheet). In contrast to the time to peak tension, the apparent relaxation time of the muscle fibres in s2 to s4 seems shorter than for the isolated fibre. This might explain the relatively low value of the half relaxation time (about 8.5 ms) as measured by Close (1964) for adult rat EDL as compared to the time to peak tension (about 12.5 ms).

Figure 9c shows the specific power output of the muscle fibre. The sum of the power of all muscle fibres in the muscle belly is equal in magnitude, but opposite in sign to the power output of the TTS (contribution of parallel elastic component is negligible). During muscle fibre shortening, muscle belly power output is absorbed completely by the TTS. During muscle fibre lengthening, the absorbed energy is released from the TTS in elastic recoil and absorbed again by the muscle fibres. Highest power output of the muscle fibres is produced in s4. Most metabolic power is consumed by the muscle tissue during the highest muscle fibre shortening velocities (not shown). During lengthening, metabolic power consumption is almost negligible.

In the tetanus (other conditions equal to those of the twitch, Fig. 9d, e, f), a similar delay in force production is observed for s2 to s4, but maximum force is reduced relatively less, since the shortening velocity goes to zero while the active state is still close to one. The final reduction is mainly due to cross-bridge losses. A minor effect is due to shortening deactivation. Naturally, muscle fibre shortening, force production, and maximum strain energy in the TTS are higher in the tetanus than in the twitch. The high magnitude of the negative power peak in de deactivation phase illustrates the high capacity of muscle tissue to absorb energy. The apparent relaxation (Fig. 9e) is delayed in s2 to s4, since stretching of the muscle fibres by the TTS during deactivation has a positive effect on force output. In conclusion, owing to the complex dynamics of the MTC, muscle fibre twitch and tetanus characteristics cannot directly be derived from muscle belly (or MTC) force measurements. The above observations are also relevant for locomotion studies (see Sect. 4.3 on terrestrial locomotion).

Force and Length. Simulation of a force–length diagram of the MTC with full muscle fibre activation is shown in Fig. 10 (contraction speed → 0 m/s). Figure 10a shows F_{mtc} as a function of Δl_f, Δl_{mb}, and Δl_{mtc}, measured relative to the resting lengths. The force–length diagram is wider for Δl_{mb} than for Δl_f owing to (1) changing fibre angulation and (2) stretching of the tendinous sheets. The diagram is widest for Δl_{mtc}, since then tendon-length change contributes also. Optimum MTC force is exerted above MTC resting length, owing to (1) stretching of the TTS when force is produced, and (2) a decreasing pennation angle α with an increasing l_{mtc} (Fig. 10b), which favours muscle belly force output. Both effects are illustrated in Fig. 10c. Over the active force range, α varies from about 10° at

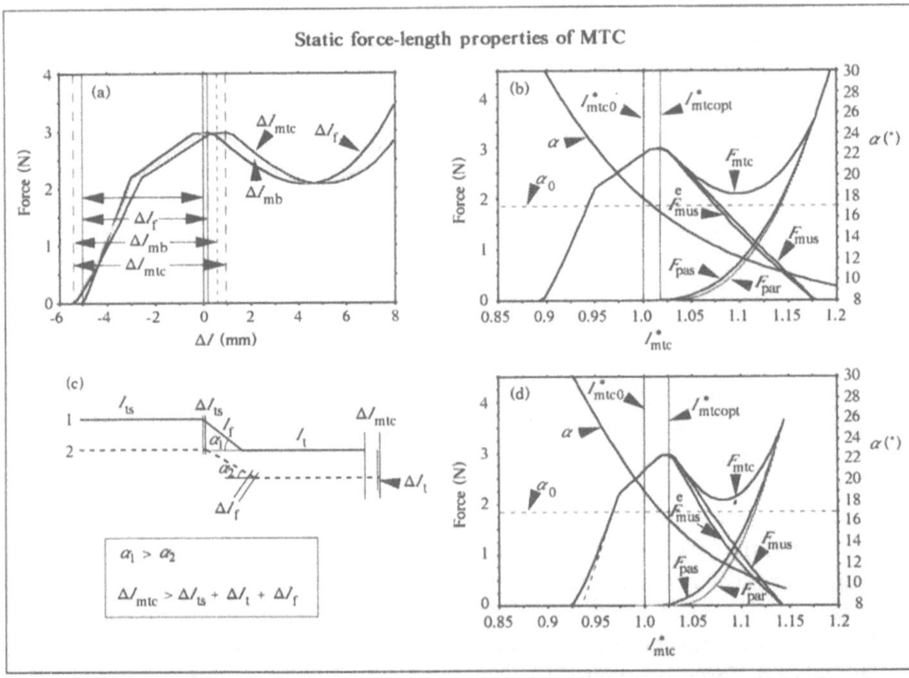

Fig. 10 a–d. Simulated static length tension characteristic of a MTC. Full activation is assumed. Muscle parameters are given in Table 1. In diagram **d**, however, l_{t0} was doubled, and A_{t0} was reduced by 50%. **a** Force output of MTC as a function of Δl_f, Δl_{mb}, and Δl_{mtc}, as denoted for each curve. *Vertical grey line* denotes position of zero length change with respect to resting value. *Other vertical lines to the right* denote excursion to optimum length, for the three length variables. Excursions are shown from minimum length till optimum length. *Unlabelled excursion* denotes value for isolated fibre. **b** Force output of the MTC F_{mtc}, of the muscle fibres F_{mus}, and of the parallel elastic elements F_{par} against normalized length of MTC. *Vertical grey line* shows position of l^*_{mtc0}. *Vertical line* shows l^*_{mtcopt}. *Horizontal dashed line* shows α_0. Passive length-tension curve of MTC is also shown: F_{pas}. The estimated muscle force (denoted with *e*) by subtracting the passive curve from the active curve deviates from the real muscle force, because F_{par} is smaller at a particular l_{mtc} when muscle fibres are active. **c** Diagram showing that in a pennate muscle a change in fibre length (from 1 to 2) is accompanied by a change in pennation angle so that $\Delta l_{mtc} > \Delta l_t + \Delta l_{ts} + \Delta l_f$. **d** Similar to **b**, but with a four times higher tendon compliance, and a doubled tendinous sheet compliance maximum length to about 30° at minimum length, reducing muscle fibre force transmission by 12% (but increasing transmission of length change) of the muscle fibres. Note that α changes fastest at shortest MTC lengths. Generally, the variation in α increases with an increasing α_0. Figure 10b shows also the separate contributions of the muscle fibres (F_{mus}) and the parallel elastic elements (F_{par}) to F_{mtc}. If the same length change is posed upon the MTC, but with inactive muscle fibres, the so called passive length–tension diagram is obtained (Fig. 10b: F_{pas}). In experimental practice, the active length–tension diagram is often obtained by subtracting F_{pas} from F_{mtc}. As discussed by Zajac (1989), this is principally incorrect, because, at the same MTC length, the parallel elastic structures are at a longer length during passive force measurements than when the muscle fibres are active (active muscle fibres cause an extra stretch of the TTS). This effect increases with the compliance of the TTS, as shown in Fig. 10d, with a doubled tendinous sheet compliance and a quadrupled tendon compliance. The situation of Fig. 10d is by no means exceptional for locomotor muscles. The rapid change in slope at the ascending limb of the force–length diagram (Fig. 10b, d) corresponds to collision of the myosin filaments on the Z-discs at short lengths. The lower part of the ascending slope is not straight (compare with dashed line in Fig. 10d) owing to variation of both α and TTS-length. A variation of muscle fibre properties may further increase this effect. The demonstrated salient features of the simulated force–length diagram are supported by measurements on cat soleus muscle by Rack and Westbury (1969) and various leg muscles of the cat by Poliacu Prosé (1985).

MTC Mechanics During Sinusoidal Length Variation. The above simulations of MTC mechanics are representative for many physiological measurements. During locomotor activities, however, muscle fibre activation and l_{mtc} change simultaneously. A convenient way of demonstrating the dynamic properties of the MTC is by prescribing a sinusoidal change of l_{mtc} around l_{mtc0} (Fig. 11b), and calculating the effects on force and power output. The MTC can only absorb power (i.e. produce negative power) when it is forcibly stretched, whereas it produces only positive power when it shortens. Muscle parameters resemble again those shown in Table 1, except that the TTS stiffness is increased (10 times) in series 2 of Fig. 11 and decreased (2 times) in series 3. Oscillation amplitude is $\pm 2.5\%$ of MTC resting length, with a frequency of 10 Hz. The contribution of the parallel elastic element was totally negligible and therefore omitted in Fig. 11. The contributions of the tendon and the tendinous sheets will be summed. In fact, the contribution to stretch and strain energy uptake is slightly larger for the tendinous sheets than for the tendon, owing to their slightly longer length. The cycle starts at minimum MTC length, so that the MTC is stretched during the first half of the cycle. Activation was by a 500 Hz train of 24 pulses, delayed by 28 ms relative to the start of the cycle. The first cycle differs slightly from subsequent cycles, owing to a small transient effect. The choice of parameters was guided by electromyographic records from the literature and the goal of demonstrating the salient dynamic features of the MTC. No special effort was made to find optimal conditions for particular functional demands.

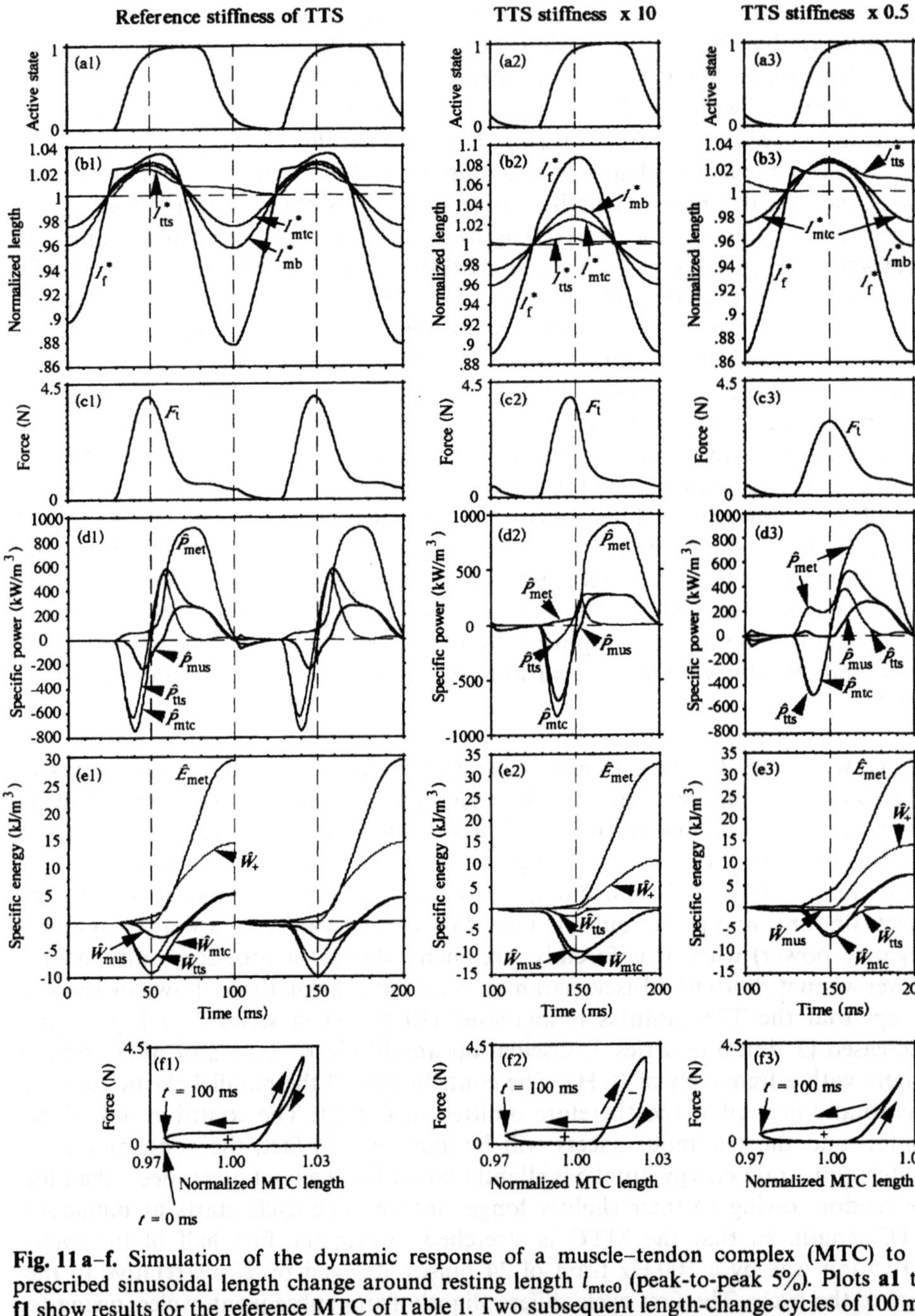

Fig. 11 a–f. Simulation of the dynamic response of a muscle–tendon complex (MTC) to a prescribed sinusoidal length change around resting length l_{mtc0} (peak-to-peak 5%). Plots **a1** to **f1** show results for the reference MTC of Table 1. Two subsequent length-change cycles of 100 ms each are examined. The length cycles start at minimum length. **a1** Active state as a result of 24 pulses at 500 Hz, started with a delay of 28 ms, **b1** Normalized length changes of MTC (l^*_{mtc}), muscle fibres (l^*_f), tendon and tendinous sheets (l^*_{tts}), and muscle belly (l^*_{mb}). **c1** Force production by the MTC. **d1** Specific mechanical power output of muscle tissue ($\hat{P}_{mus}$), TTS ($\hat{P}_{tts}$), and MTC ($\hat{P}_{mtc}$), as well as the consumed metabolic power ($\hat{P}_{met}$). **e1** Specific work and metabolic energy as

The active state rises almost to peak level during the stretching phase (Fig. 11a1); during most of the shortening phase it has a high level, but it is again close to zero at the end of the cycle. The normalized length changes of various parts of the MTC are shown in Fig. 11b1. The amplitude of normalized length change is largest in the muscle fibres. Under their optimum length, they follow a path close to a sine wave (a spectral analysis would allow a more exact statement). However, upon activation during stretching, the fibres quite abruptly reduce their speed and deviate from the sinusoidal path. Fibre length change is limited owing to force production (Fig. 11c1), so that MTC length change is largely due to stretching of the TTS (Fig. 11b1). The energetic effect is a high power absorbency by the TTS (which is stored as elastic energy) and a relatively low power absorbency by the muscle fibres (Figs. 11d1 and 11e1). Peak MTC force is produced before peak MTC length, since muscle fibre velocity drops towards the end of the stretching phase. Therefore, the TTS starts shortening before MTC shortening. Hence, muscle fibre velocity is opposite to TTS velocity. This situation continues for about 7 ms during the shortening phase. Thus, released TTS strain energy is absorbed during a short period by the muscle fibres. Although it is not shown by the simulation (it is not incorporated in the model), part of this absorbed energy could again be stored as elastic energy in the cross-bridges, since muscle fibre length change is only about 1%, which is below the limit of 3% for effective storage (Flitney and Hirst 1978). In fact, during the whole forcible stretching of the muscle fibres, the length change remains within this limit. Although the storage of strain energy in cross-bridges is a relatively small effect, it seems nevertheless useful in the transition from MTC stretching to shortening. Most TTS strain energy is released during muscle fibre and MTC shortening (first half of shortening phase). This energy shows up directly as MTC power output. The positive power output by the muscle fibres is delayed relative to this strain energy release. Thus, in the first phase, MTC power output (P_{mtc}) is dominated by strain energy fluctuations of the TTS and, finally, by muscle tissue power output. Most muscle fibre work is done at a relatively low force level. Metabolic power requirements are low during the stretching phase, and increase sharply when the muscle fibres start shortening. The MTC work output (W_{mtc}) produced during a complete length cycle can be divided by the total metabolic energy consumed per cycle to obtain an overall efficiency of 14.7% for this case. The total specific work output by the MTC during the shortening phase only ($\hat{W}_+$) is about 50% of the total metabolic energy. The work loop (Fig. 11f1) has a small negative part and a dominant positive part.

The simulation shows that the muscle fibres are mainly active at and below their optimum length for force production. This agrees with estimated sarcomere length changes in vertebrate locomotion muscles (reviewed by Alexander 1989),

---

obtained by integration of curves in **d1**. Energy levels are reset to zero at start of each cycle. The positive work done during the shortening period of the MTC is also shown ($\hat{W}_+$). **f1** Work loop of the MTC during the two cycles. Positive part is larger then negative part. Plots **a2** to **f2** are equally arranged as **a1** to **f1**, but now with a ten fold stiffer TTS, and only the second cycle is shown. Plots **a3** to **f3** show simulation results with a doubled TTS compliance (compared to series 1). Again, only the second cycle is shown and provides a high specific power output of the sarcomeres by increasing the cross-bridge density (Van Leeuwen 1991).

Now, the above simulation will be compared with the situation of a tenfold increase of the TTS stiffness (Fig. 11a2–11f2), whereas other parameters are kept constant. Only the second cycle is shown. The most conspicuous differences are:

1. Muscle fibre length follows much more the sinusoidal length variation of the MTC above resting length, leading to an increased sarcomere excursion, particularly towards higher lengths. This is caused by the more limited stretching possibilities of the TTS.
2. A faster force increase during lengthening, owing to a faster muscle fibre lengthening. Peak force falls earlier. During MTC shortening, the force decreases faster, owing to the higher muscle fibre shortening speed.
3. A much higher power uptake by the muscle fibres during MTC stretching, combined with a very limited strain energy storage by the TTS (Fig. 11d2, e2). MTC power output is now dominated by the muscle fibres.
4. MTC work output in the cycle is slightly negative, resulting in a slightly negative efficiency (– 1.7%). This is also shown by the MTC work loop with about equal negative and positive parts (Fig. 11f2).

More or less reverse effects occur when TTS compliance is doubled with respect to the reference values (Fig. 11a3–11f3):

1. With the onset of activation, muscle fibres start to shorten slightly, until a plateau is reached (Fig. 11b3). The TTS is lengthened more than the reference TTS.
2. A lower peak force is reached, since muscle fibres are not stretched. Peak force coincides with peak MTC length. Force decline is relatively slow during MTC shortening, since, for most of the time, the muscle fibres shorten slower.
3. Power absorption during MTC lengthening is almost completely due to TTC strain energy uptake. Thus, power absorption by muscle fibres is almost completely avoided by doubling the TTS compliance.
4. MTC work output in the cycle is positive, with an efficiency of 22.2% (7.5% higher than for the reference MTC). The work loop shows a positive part only.

A muscle action like that of Fig. 11 series 3 is particularly useful in saving metabolic energy when, during locomotion, an animal makes cyclic changes in kinetic and potential energy (Sect. 4.3). It demonstrates also how, by an activated prestretch, muscle performance during MTC shortening can be improved, by making effective usage of the build-in springs of the muscle. For a jump from stant still, this requires a preceding counter-movement.

In real live situations, often only a particular fraction of fibres of a muscle is activated. Compared to a full activation, these fibres "see" a relative stiff TTS. This is roughly similar to a full activation with a very stiff TTS (Fig. 11 series 2). A small fraction of active fibres, particularly if distributed over the muscle, is unlikely to be able to take much advantage of elastic effects, which will reduce the efficiency. If the fraction of active fibres is localized, a higher efficiency can be obtained, owing to a greater stretch of the associative tendinous sheet parts. The low efficiency can (partly) be compensated by usage of slow oxidative fibres during slow movements. The consequences for the co-ordination of muscle activities during locomotion will be further discussed in Sect. 4.3.

A high muscular efficiency is not always desired. If an animal wants to brake its movements suddenly, it needs a very low (i.e. high negative) efficiency. The timing of the activation has a dominant influence on the MTC power output. Figure 12a1 to 12c1 shows a simulation with an early activation (starting at 0 ms) in the stretching phase, resulting in a high power absorption by the MTC and a large negative work loop. Such activation would be useful if an animal wants to decelerate rapidly. Figure 12a2 to 12c2 shows a simulation with an activation starting at 40 ms, resulting mainly in positive power output by the MTC and a positive work loop. Such activation is useful if a high acceleration is required. The simulations demonstrate that muscles are particularly good in absorbing large amounts of work.

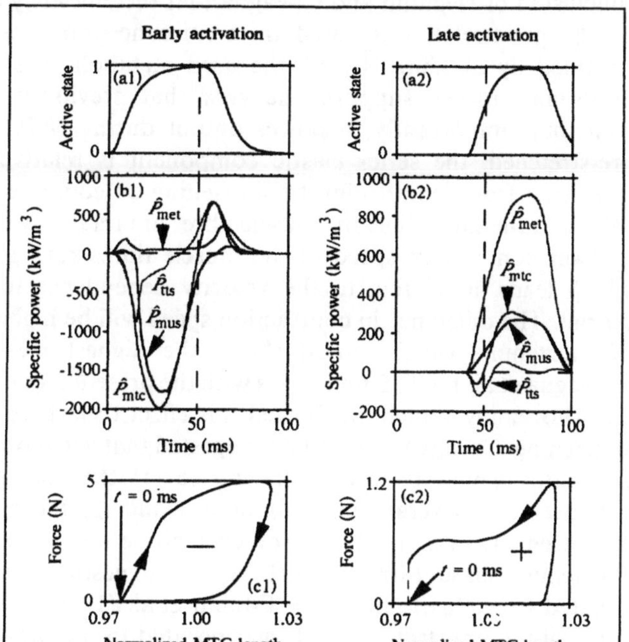

Fig. 12 a–c. Simulation of the dynamic response of a muscle–tendon complex (MTC) to a prescribed sinusoidal length change around resting length l_{mtc0} (peak-to-peak 5%). The reference MTC of Table 1 is used. Duration of cycle is 100 ms. The cycle starts at minimum MTC length. The effect of different phases of stimulation on the power output is demonstrated. Series **1** shows an early activation, resulting in a high power absorption by the MTC. Series **2** shows a late activation, which produces a high power production. **a1** Active state as a result of 24 pulses at 500 Hz, started at $t = 0$ ms. **a2** The same but activation starts at $t = 40$ ms with only 18 pulses. **b1** and **b2** Specific power output of MTC components and specific metabolic power, labelled as in Fig. 11d. Absolute peak value of peak power absorption of MTC in **b1** is much higher than peak power produced in **b2**. **c1** and **c2**. Work loops. *Arrows* on curves show direction of movement. In series **2**, the loop is not closed since the fibres are still active at the end of the cycle

In conclusion, a TTS with appropriate stiffness can increase considerably the mean MTC power output. Using the present dynamic MTC model, the required stiffness and stimulus characteristics for optimization of power output and muscle efficiency under particular functional demands can be predicted.

Measurements of MTC Mechanics. Heglund and Cavagna (1987) recorded the work output of the EDL of the rat, and estimated also metabolic power from O_2 consumption measurements. They found that $\hat{W}_+$ could be increased if a forcible pre-stretch was given to the EDL as compared to an initially isometric EDL. This can mainly be explained as follows. The work done on the preparation during stretching was partly stored as strain energy in the tendinous sheets (Heglund and Cavagna excluded the main part of the external tendon, but the tendinous sheets are of about equal importance as the external tendon in the EDL), which was again liberated during shortening. The authors measured negative work loops for the complete cycle, so that the overall efficiency of the MTC was negative. (The conclusion of Heglund and Cavagna that the efficiency of the EDL was increased by the pre-stretch was based on a misconception that the efficiency could be obtained by dividing $\hat{W}_+$ by the used metabolic energy). This experiment was important since it supports the view that previously stored strain energy can contribute importantly to power output during MTC shortening. If a MTC is pre-stretched, the series elastic component is relatively longer at the time of transition from lengthening to shortening as compared to an initially isometric MTC. At the same time, the muscle fibres are relatively shorter. Hence, on average, a lower contraction speed of the muscle fibres results during MTC shortening, which leads, according to the velocity dependence function, to a higher force output. The difference in contraction speed will be highest during the initial phase of shortening, which indeed shows the highest measured force enhancement (Cavagna et al. 1968). Simulations with the present model (using constant stretching and shortening velocities) showed this effect to last for about 30 to 40 ms during shortening. Cavagna et al. (1968) suggested that yet another effect on the contractile apparatus would be present owing to (1) the considerable duration of force enhancement (several hundreds of ms) and (2) the presence of a (slight) force enhancement, even if a 20 ms isometric phase was between the pre-stretch and the shortening phase. Firm support for this suggestion came from measurements on isolated frog muscle fibres by Edman et al. (1978). This last, yet to be defined, effect of the pre-stretch is not incorporated in the present model.

Heglund and Cavagna (1987) showed that the pre-stretch effect was much less for frog sartorius muscle than for rat EDL. A likely explanation is that the sartorius is parallel fibred muscle (one of the very few) with very short tendinous sheets, so that very little strain energy can be stored in them. In cat (Loeb et al. 1987) and man (Barrett 1962), the muscle fibres do not run continuously throughout the muscle belly, but the muscle fibres are connected to each other in series with short strips of connective tissue.

Just before this chapter was finished, an experimental study by Ettema et al. (1990) on rat medial gastrocnemius muscle was published which shows several of the phenomena discussed in this section.

3 Monitoring Muscular Performance in Vivo

This section considers methods used to record muscle performance in vivo. Direct measurements can be made of muscle performance (Sect. 3.1), or models can be used to estimate muscle performance from experimental data as external force records and cinematography (Sect. 3.2). Sometimes combinations of the two approaches are being made (Sect. 3.3).

3.1 Direct (Invasive) Experimental Techniques

Electromyography. The electrical activity of muscles can be measured by electromyography (EMG). Details about the technique can be found in textbooks on electrophysiology. For muscle function in locomotion, the differential technique gives the best results, since it allows a localized registration of muscle activity. Here, two identical electrodes are positioned closely together and the difference in potential between them is recorded. Surface electrodes have limited resolution, but are, for obvious reasons, often used in recordings from man. Needle and wire electrodes have the advantage of a higher spatial resolution. Generally, EMGs give only an impression of electrical activity, they do not show how much force is being produced. Force production lags behind electrical activity. For fast muscle fibres, the delay is about 20–30 ms, for slow fibres it is about 50–80 ms. Many attempts have been made to relate EMG to force output. For isometric cases, the rectified and low pass filtered EMG has been shown to be roughly proportional to force output. If length changes are involved, the situation is more complex. Hof and Van den Berg (1981) predicted force output of the human calf muscles, using an adaptation of the muscle model by Hill (1938). A disadvantage of their approach (and those of many others) is that (1) the muscle parameters have limited structural or physiological relevance, and (2) a calibration technique is required, since electrode placement is highly variable.

Direct Force Measurement. An important technique to measure force from a single muscle or from a muscle group with a common tendon is by the usage of a tendon buckle transducer (first proposed by Salmons 1969). This technique is limited to muscles with relatively long tendons, being exposed to surgical approach. The Achilles tendon is used most frequently. The technique has now been applied by several investigators. Early studies were by Barnes and Pinder (1974) in the horse and Walmsley et al. (1978) on cat Achilles tendon. Sherif et al. (1983) compared myoelectrical activity and muscle force during running in the cat. Biewener et al. (1988) used the tendon buckle technique to measure forces by the ankle extensors in hopping kangaroo rats. They showed that the entire range of possible muscle forces is used during normal locomotor events. At slow speeds: 20% of the maximum isometric force; at preferred speed 40% and during jumps 175% during the stretching phase of the muscles. Peak muscle stresses were found of about 350 kPa.

Direct Strain Measurement. Mercury tubes have been used to measure the strain of tendons during locomotion of the horse, using the property of a changing electrical resistance during lengthening and shortening (Van den Bogert et al. 1989). Muscle forces and elastic energy storage can be calculated if, after in vivo experimentation, the tendon properties are measured with a tensile testing machine.

3.2 Indirect Techniques

Muscle performance can be estimated with various indirect (noninvasive) techniques. Almost all methods require knowledge about muscle moment arms and muscle architecture. So, these aspects will be discussed first.

Moment arms. There are various ways to estimate muscle moment arms (An et al. (1984) provide a review). Morphometry can be applied if the position of the motion axis of the joint can be simply correlated with a curved joint surface in the specimen. If motion axis and force are at right angles, the moment arm equals the distance from axis to muscle-tendon complex (along the common perpendicular). This approach becomes problematic when the motion axis cannot be easily related to the skeleton. Axis and tendon position can be accurately determined by Röntgen photogrammetry (Selvik 1990), but this technique is laborious and often not available. Moreover, tendon constraints like retinacula and tendon sheaths can hinder the determination of the line of action of the tendon force (An et al. 1984). A muscle moment arm as a function of joint angulation $r_m(\phi)$ can be most simply and most accurately obtained from measurements of tendon displacements s as a function of ϕ in a musculoskeletal preparation. By assuming that, in an infinitesimal displacement, the work done by tendon displacement is equal to the work done by the muscle in the joint rotation (principle of virtual work), the moment arm r_m is calculated (An et al. 1984):

$$F_t \mathrm{d}s = F_t \cdot r_m \mathrm{d}\phi. \tag{10}$$

Hence:

$$r_m = \frac{\mathrm{d}s}{\mathrm{d}\phi}. \tag{11}$$

The same relationship can be derived by a kinematical analysis for planar motion (Fick and Weber 1877; Brand et al. 1975). The main advantage of this technique is that it takes implicitly account of complex joint geometries,.curved muscle paths and three-dimensional effects. If neglects, however, stretching effects of e.g. a retinaculum. Measurements of moment arms of human lower leg muscles about knee and ankle joint, using this technique, show the arms to be quite variable (Spoor et al. 1990), so that considerable errors can be made about muscle function if a constant moment arm is assumed.

Muscle architecture. Muscle architecture is still obtained best from dissections. Measurements include the following:

1. Tendon mass, slack length and cross-sectional area.
2. Slack length and mass of tendinous sheets.

3. Lengths and pennation angles of muscle fibres (or small bundles) should be measured evenly spread over the muscle belly.
4. Muscle volume and fibre cross-sectional area.
5. Sarcomere lengths of each of the dissected muscle fibres. (This can be done with a diffraction method. A cheap and convenient method is described by Dimery 1985).
6. Sarcomere structural parameters (by electron microscopy).

These structural parameters are required as input for a muscle-tendon model (Sect. 2.7). From measurements (3), (5) and (6), the muscle fibre optimum length can be calculated. Additional experiments are required to obtain physiological data. A diffraction method can also be used to estimate sarcomere lengths during a locomotor activity (Dimery 1985; Cutts 1986). To this end, killed animals are placed in positions corresponding to those from a film of the moving animal. The method is only approximate, since in vivo force patterns differ with an unknown amount from rigor forces of the muscles. If joints are well characterized, motion data can sometimes be used to calculate MTC length changes (Van den Bogert et al. 1989).

Inverse Dynamics Analysis. Muscle forces can be estimated using the so-called inverse dynamics approach (mathematical details and computer methods of this technique are described in engineering textbooks like Haug 1989). In this approach, net joint moments are calculated from

1. measured (by cinematography or accelerometry) positions, velocities and accelerations of the body segments (kinematic analysis);
2. inertial parameters of body segments.

The net joint moments are subsequently used to estimate muscle forces with the following additional information:

1. instantaneous moment arms of muscles and ligaments about the various joints;
2. muscle and ligament properties;
3. a (validated) distribution model for the forces over the concerned muscles and ligaments.

The pitfalls of this method are manifold. (1) The estimation of accelerations from noisy displacement data (owing to e.g. unwanted displacements of skin markers) easily leads to significant errors. Special computational techniques have been developed to reduce these errors as much as possible (e.g. Woltring 1986). (2) Body segments are often assumed to be rigid. (3) Moment arms are generally only poorly known. They can be quite sensitive to joint angulation (Spoor et al. 1990). (4) Presently available force distribution models are based on arbitrary principles. The best load sharing model for muscles to date seems to be the minimum fatigue criterion by Dul et al. (1984). This criterion is based on the assumption that the endurance time of muscular contraction is maximized, hence muscular fatigue is minimized. A better criterion, incorporating the dynamic properties of the MTC (including both muscle fibre types and elastic properties), will be discussed in Sect. 4.3.

Terrestrial locomotion studies often use force platforms to measure ground reaction force on the feet of the animal. These forces can be used to estimate muscle forces, if the following additional information is available:

1. Instantaneous moment arms of reaction forces, muscles and ligaments about the various joints.
2. Muscle cross-sectional areas and ligament properties.
3. A (validated) distribution model for the forces over the concerned muscles and ligaments.
4. A multi-body model of the animal to account for segmental accelerations.

In practice, ligament forces and segmental accelerations are often neglected, whereas equal stresses are assumed to act in synergistic muscle groups in combination with inactive antagonists. This last assumption is not generally acceptable (Sect. 4.3). This approach generally saves much computational effort, whereas the accuracy is generally better than with a purely inverse dynamics approach. Nevertheless, at the end, only an order of magnitude estimate can often be made for the muscle forces. Biewener et al. (1988) compared the indirect approach to estimate muscle forces with results from the tendon buckle technique. Predictions of peak forces corresponded quite well. At low force levels, relatively large differences were found.

Direct Dynamics Analysis. In a (direct) dynamics analysis (e.g. Hatze 1981), the motions of the body segments are calculated from the driving forces of the muscles, controlled by neural inputs. Basic computer methods for the engineering aspects of this analysis can be found in e.g. Haug (1989). The method is attractive, since it follows the same route as the body does (naturally, the body uses also sensory information). As a check of the model validity, calculated body segment trajectories can be compared with measured trajectories (Otten 1987b). Generally, however, the method is still hampered by the large numbers of parameters, numerical difficulties and oversimplifications owing to many unknown or ill defined factors. Modern image analysis techniques may help to reduce the number of structural unknowns (Otten 1987b).

Combined Techniques. Sometimes, it is desirable to follow a combined approach. Using the inverse dynamics analysis, net joint moments can be calculated for a moving animal. These moments can be used as input for a direct dynamics model, except for one joint which could be controlled by models of muscle-tendon complexes. Effects of parameter variation for one joint on the movements and reaction forces of the whole animal can then be studied.

4 Performance and Design of Muscles in Some Locomotor Systems

4.1 Fish Swimming Muscles

Considerable progress has been made in the understanding of fish swimming muscles in the last decade. Therefore, this topic will be discussed at some length. Fish use a great many different techniques to propel themselves. In one swimming type, the body is propelled by undulatory movements of the fins, whereas the body is kept relatively rigid. A further subdivision can be made on the basis of the relative contributions of each of the paired and unpaired fins. These swimming modes are generally used if a high efficiency is required. The dominant type of swimming is by undulation of the body, the main topic of this section. This swimming type can further be divided in three distinct modes. (1) In continuous swimming, the body repetitiously makes undulations of fixed amplitude and frequency. (2) In kick-and-glide (= intermittent) swimming, body undulations are intermitted with glides, with the body kept straight. Now, an acceleration phase can be distinguished from a deceleration phase with no power output from the trunk muscles. (3) In a fast start, the body is accelerated from rest by undulations of a relatively high amplitude. The trunk muscles deliver maximum or close to maximum mechanical power output during this swimming mode.

A major question is how the movements of the body are produced by the trunk muscles. This problem has been tackled by various methods. Blight (1976, 1977) used electromyography to study the phases of activity of the trunk muscles in newt larvae and tench at various positions along the trunk (Fig. 13a). He concluded that only a small phase shift exists in the recorded activity from anterior to posterior. Thus, at one side along the trunk, considerable overlap occurs in the activity phase of the muscle fibres. With other words, these observations suggest the caudally running wave of lateral bending of the body to be produced by an almost alternating tension development in the axial muscles from side to side. These findings got further support by work of Hess and Videler (1984). They applied the inverse dynamics approach (Sect. 3.2) to fish swimming. They recorded the movements of saithe during continuous swimming, and calculated the forces on the body using a hydrodynamic theory outlined by Lighthill (1960). From these data, they calculated the net bending moment the fish should have generated along its trunk as a function of time. Figure 13b shows the amplitude of the bending moment along the trunk. The bending moment showed hardly any phaseshift along the trunk, in contrast to the curvature of the body (Fig. 13c). These calculations cannot tell by which structures (mainly muscle fibres and connective tissue) the bending moments are produced. For this purpose, a (validated) distribution model (Sect. 3.2) is required which is not available. Nevertheless, the muscle fibres produce probably force in phase with the calculated net bending moment. Hess and Videler's analysis shows also that, anteriorly, bending moment and curvature fluctuation result in a positive mean power output. Owing to the phaseshift in curvature, the mean power output is negative in the caudal peduncle.

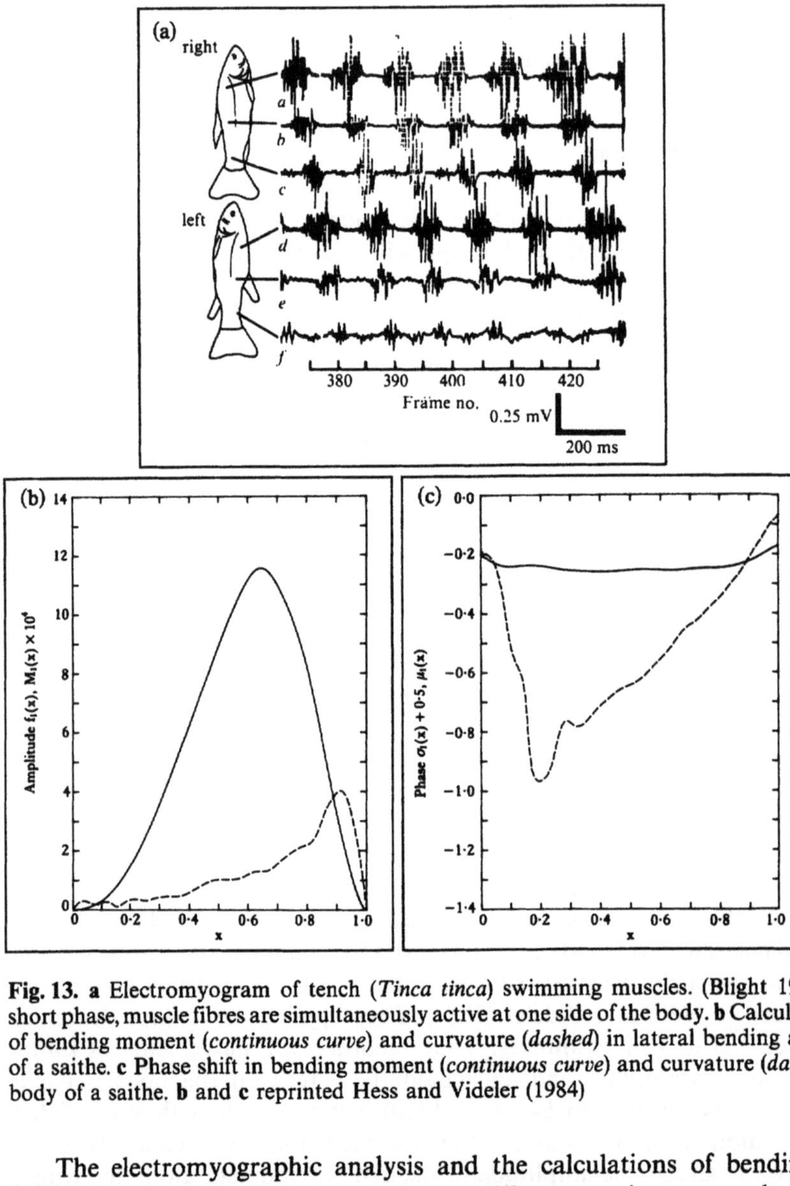

Fig. 13. **a** Electromyogram of tench (*Tinca tinca*) swimming muscles. (Blight 1976). During a short phase, muscle fibres are simultaneously active at one side of the body. **b** Calculated amplitude of bending moment (*continuous curve*) and curvature (*dashed*) in lateral bending along the body of a saithe. **c** Phase shift in bending moment (*continuous curve*) and curvature (*dashed*) along the body of a saithe. **b** and **c** reprinted Hess and Videler (1984)

The electromyographic analysis and the calculations of bending moments suggest that muscle fibres are active at different strain ranges along the body. Consequently, the muscle fibres along the body produce probably different amounts of power during the swimming cycle. Support for this suggestion came from an experimental and theoretical analysis of continuous and intermittent swimming carp by Van Leeuwen et al. (1990). They devised a method that can isolate the mechanical performance of single muscle fibres from the connective tissue. Their approach is summarized in Fig. 14. They aimed to calculate normalized forces, power and work during the swimming cycle. Only normalized quantities could be addressed since the number of involved muscle fibres could not be deduced. To this

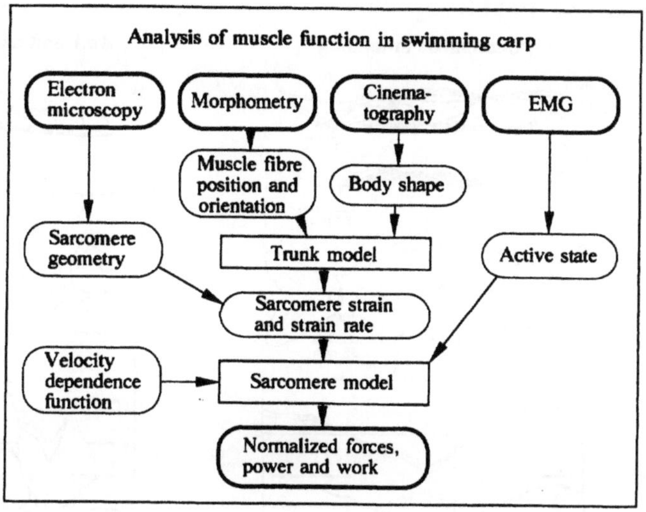

Fig. 14. Approach to study muscle fibre mechanical performance during swimming in carp as discussed in the text. (After Van Leeuwen et al. 1990)

aim a sarcomere model by Otten (1987a) was used, similar to that discussed in Sect. 2. This model requires four different types of input. First, the geometrical properties of the sarcomere (Sect. 2.1) should be known. These were obtained by electronmicroscopical observation of sarcomeres at four distributed positions along the trunk, as indicated in Fig. 15. From the measured parameters l_{myo}, l_{act}, l_{bz}, and l_z, the length dependence of force production was calculated.

Calculation of sarcomere force requires also knowledge about the velocity dependence function [Eq. (2)]. The part for positive interfilamentary sarcomere velocities was estimated from experimental data measured by Rome et al. (1988). For negative velocities (during stretching), an estimate was made on the basis of measurements by Aubert (1956). The third factor that Van Leeuwen et al. (1990) needed in their calculation was the active state. Therefore, the electrical activity of the red muscle fibres was monitored during continuous swimming at eight different positions along the trunk (Fig. 15). A differential technique was used so as to obtain records of the local activity. Measurements were made at one side of the body only, since the recruitment pattern can be assumed to be similar at both sides, but out of phase with each other. During intermittent swimming, both sides were monitored with four spots each (Fig. 15). In combination with experimental twitch data of muscle fibres of carp (time to peak tension is about 60 ms for slow fibres and 30 ms for fast fibres) by Granzier et al. (1983), the EMG records were used to estimate the active state.

Finally, data were required about the actual length fluctuations of the muscle fibres during swimming. Synchronously with the EMG records, high speed films were made of the fish swimming. A central axis of the fish was calculated from the digitized outlines of the fish. The strain and strain rate in the red (slow) muscle fibres was calculated from the local curvature of the central axis and the position and

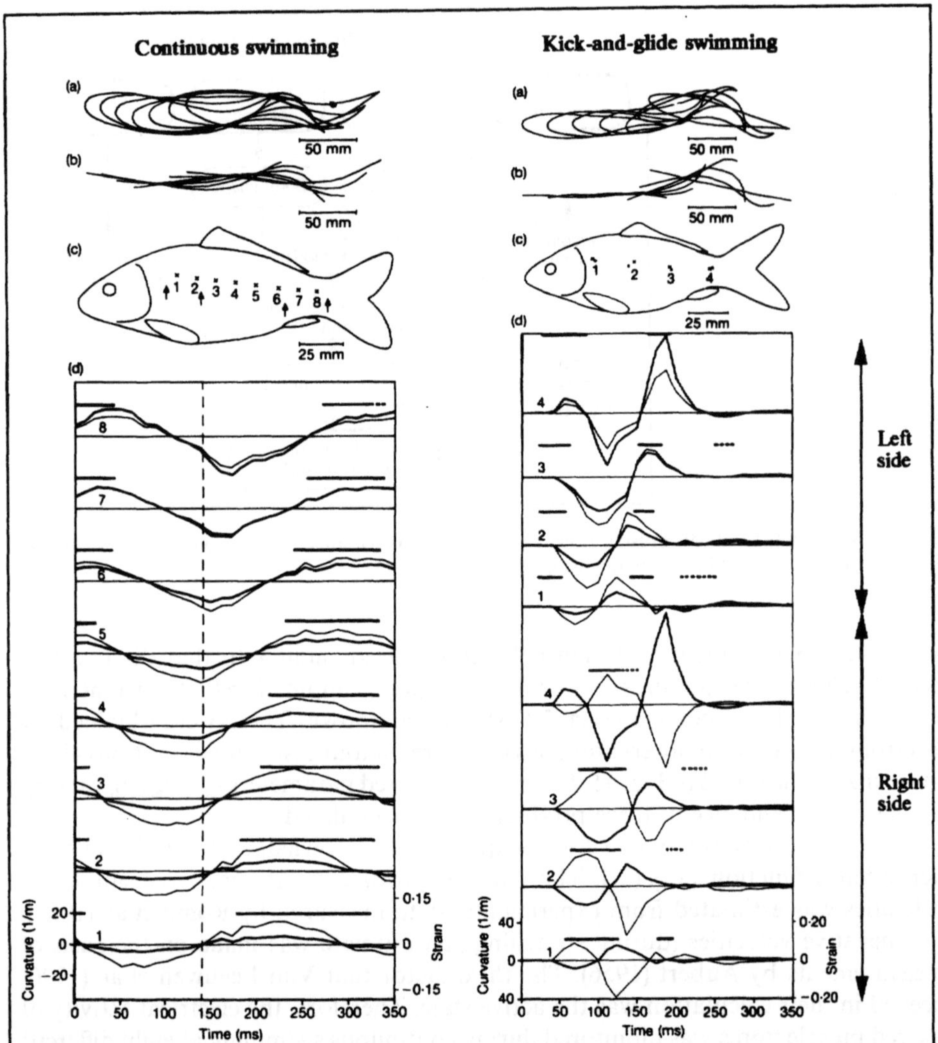

Fig. 15 a–d. *Left series of plots* Analysis of a close to continuous swimming cycle of a carp (SL = 156 mm). **a** Selection of measured outlines of the body. **b** Central axes, as calculated from the outlines. **c** Positions of electrode pairs at the *left side* along the trunk in the slow muscle fibres are numbered *1* to *8*. *Arrows* show sample positions of slow muscle fibres for ultrastructural observations. **d** Graphs of curvature of body axis (*heavy curves*, defined positive for bending to the right side), and the strain in the slow muscle fibres (*light curves*) at the electrode positions. *Numbers* correspond to those of **c**. The *horizontal bars* indicate the electrical activity of the slow muscle fibres. *Right series of plots* Analysis of a cycle of intermittent swimming of a carp (SL = 130 mm). Illustrations **a** to **d** are arranged similar to the left side. *Top four pairs* of graphs in **d** show curvature and strain on the left side of the fish (at positions corresponding to the dots in **c**). Other graphs show events from the right body side. Strain and curvature are out-of-phase at right side since curvature is defined positive for bending to the right side. *Broken horizontal bars* indicate weak electrical activity. (Van Leeuwen et al. 1990)

orientation of the muscle fibres with respect to the axis. The fibres lie as a thin (in cross-section triangular) strip directly underneath the skin. Details of the required assumptions and the calculations are discussed in Van Leeuwen et al. (1990). The calculations of the strain agree with sarcomere length measurements by Rome et al. (1988).

In continuous swimming (Fig. 15, left series of plots), the amplitude of curvature increased from anterior to posterior. However, the strain amplitude in the slow muscle fibres was constant along the trunk. Thus, in the posterior region, the higher amplitude of curvature was exactly compensated by the shorter distance between muscle fibres and body axis. During swimming without electrodes strain amplitudes were found of about ± 0.045. For swimming with electrodes this was ± 0.085, owing to an increased drag. In spite of the constant strain amplitude, the mechanical behaviour of the muscle fibres changed along the trunk since the fibres were activated at different intervals in their length cycle (Fig. 15d, left side). Anteriorly, the fibres started their electrical activity about 50 ms before peak strain, posteriorly this was about 90 ms. Anteriorly, the electrical activity lasted longest. The delay in activation over the trunk was shorter than the delay in the mechanical wave. The consequences for force and power output by the fibres will now be discussed.

Figure 16a shows the estimated active state along the trunk during the continuous swimming cycle at the 8 electrode positions, based on the data shown in Fig. 15 (simulation was started at the dashed vertical line of Fig. 15d). Activation was delayed and was shortened in a posterior direction. Active state decayed almost synchronously along the trunk. Force production (normalized to full isometric force, Fig. 16b) shows anteriorly an initial peak followed by a plateau in the shortening phase, and a final decay. Peak force (produced during forcible stretching) is below isometric force, owing to the low active state. In a posterior direction, peak force increases, whereas the plateau phase shortens. This is caused by an earlier activation in the lengthening phase, combined with a shortening in activation. Power production (normalized relative to maximum power, Fig. 16c) shows an initial negative peak over the complete trunk. This peak increases towards the tail, as expected from the length and force data. A power plateau, close to maximum power output, is present anteriorly. The plateau phase shortens towards the tail. In the caudal peduncle, only a small positive peak is left. The results may also be presented as work loops (force against length). Examples are shown for electrode position 2 (Fig. 16d, mainly positive work loop), position 6 (Fig. 16e, negative and positive part), and position 8 (Fig. 16f, negative work loop). Finally, Fig. 16g shows how the normalized work output per cycle changes along the trunk. In the anterior half of the trunk, work output is positive, it drops just in front of the anal region, whereas work output is negative at the most posterior two positions.

In intermittent swimming, the maximum curvature and muscle strain varied between successive tail strokes (Fig. 15, right series of plots). In the first tail stroke, the red fibres shortened actively at the left side of the fish (top half of Fig. 15d), while being under their resting length, except in the caudal peduncle where they were initially stretched when active. The minimum muscle strain decreased from anterior to posterior, and reached posteriorly values below -0.2. Muscle activity started almost simultaneously (within 5 ms) along the left body side. In the second stroke

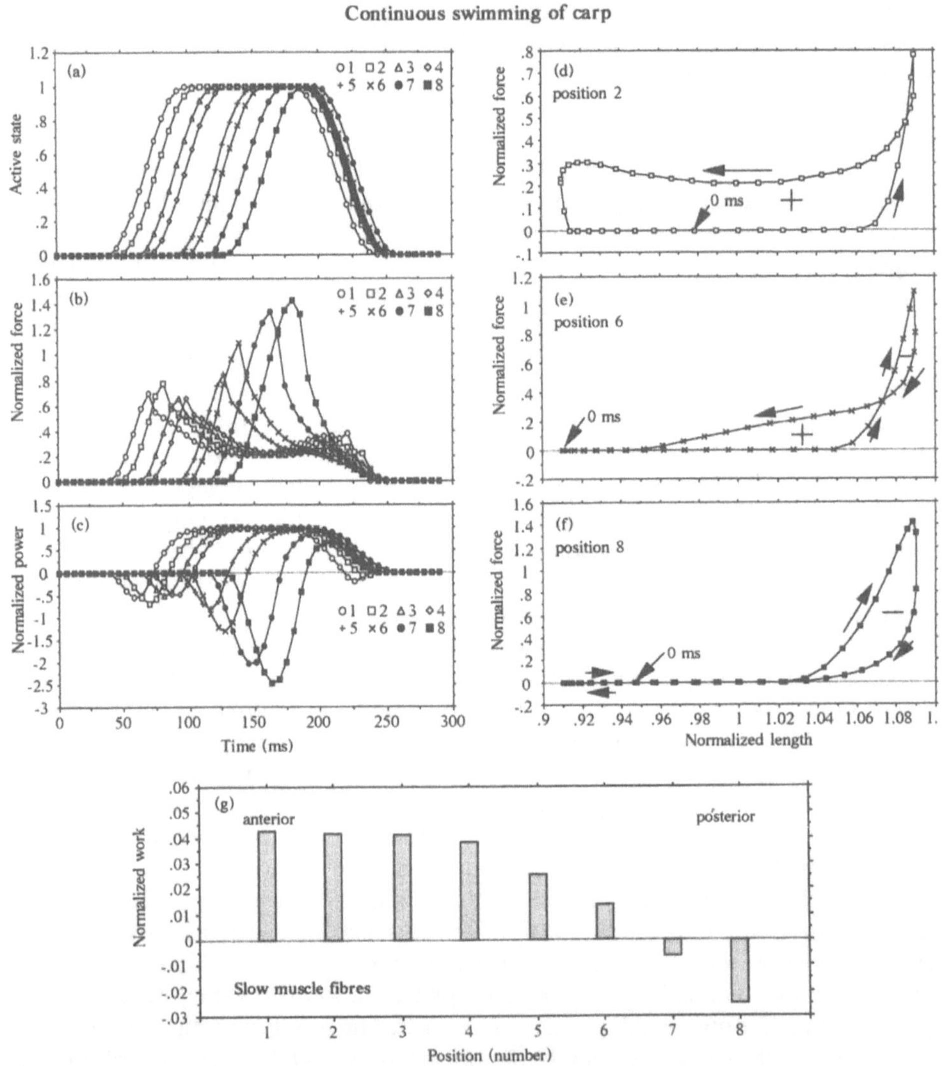

Fig. 16a–g. Results from an analysis of red muscle fibre activity along the trunk during the continuous swimming cycle of Fig. 15 (*left series* of plots). The start of the analysis ($t = 0$ ms) is denoted by the vertical dashed line in Fig. 15d. *Curve numbers* correspond to the electrode position in Fig. 15. **a** Active state. **b** Force output, normalized relative to maximum isometric force at full activation. **c** Power output, normalized with respect to maximum power output. **d** to **f** Work loops at electrode positions *2, 6, 8* respectively. **g** Normalized work output ($W^*_{sarc} = W_{sarc}/(F_0 \cdot l_{0sarc})$) at the 8 electrode positions along the trunk. **a** to **f** after Van Leeuwen et al. (1990)

already, the start of muscle activity showed a wave character (right side of the fish, see bottom half of Fig. 15d, right side). Anteriorly, the muscle fibres mainly shortened while active, from above to below their optimum length. Only initially, the fibres were stretched during the activity period. Posteriorly, they were initially

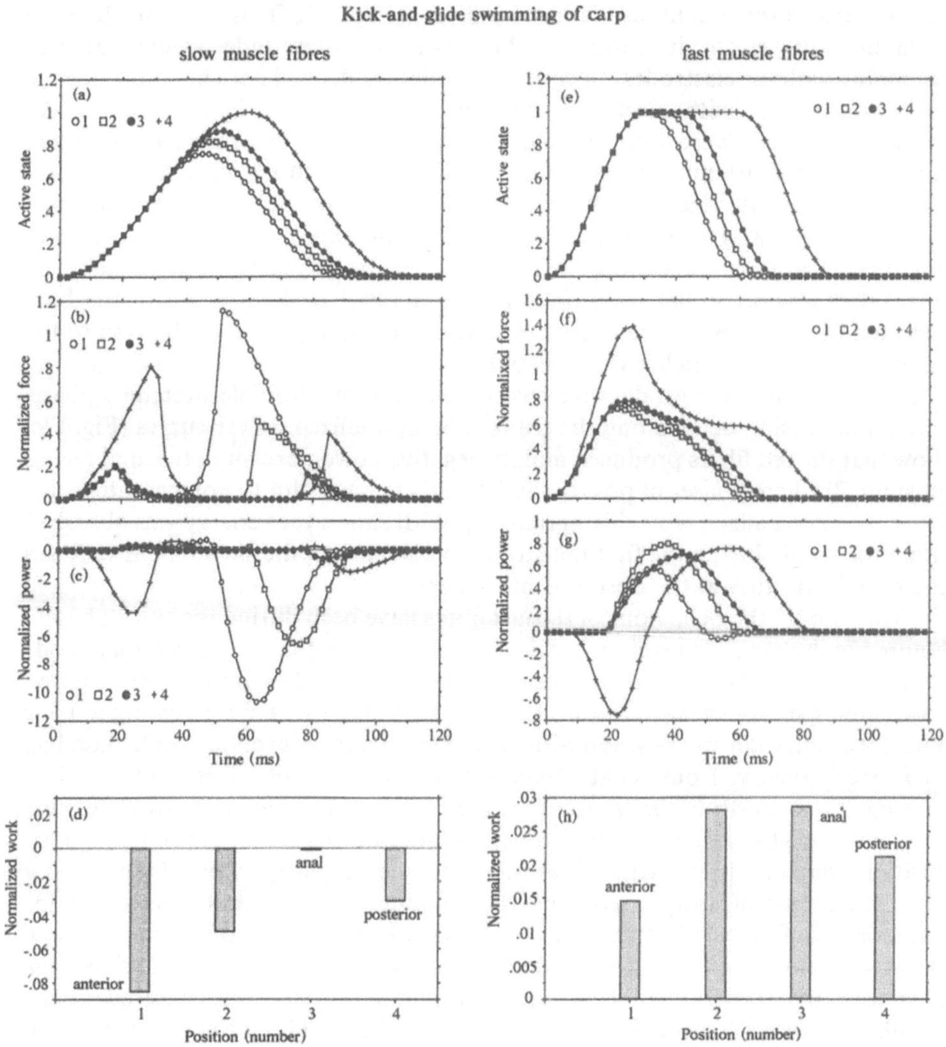

Fig. 17 a–h. Results from an analysis of muscle fibre activity along the trunk during the first tail strike of the intermittent swimming cycle of Fig. 15 (*right series* of plots). *Curve numbers* correspond to electrode positions shown in Fig. 15. **a** Active state of slow fibres. **b** Force output of slow fibres, normalized relative to maximum isometric force at full activation. **c** Power output of slow fibres, normalized with respect to maximum power output. **d** Normalized work output at the 4 electrode positions along the trunk. **e** to **h** are arranged similarly to **a** to **d**, but show predictions for white fibres. **a**, **b**, **c** based on data by Van Leeuwen et al. (1990). **e**, **f** and **g** after Van Leeuwen et al. (1990)

stretched and active above their resting length. Compared to the first stroke, this is a shift towards the pattern of continuous swimming.

The left side of Fig. 17 shows the results of a model analysis for the first tail strike after the glide phase in intermittent swimming. The swimming event shown in Fig. 15 served as a basis for this analysis. Van Leeuwen et al. (1990) reduced in their calculations all the amplitudes in the strain curves by 50%. This was done because swimming with electrodes results in about twice the amplitudes as compared to swimming without electrodes. In contrast, model results will here be shown for the actual movement with electrode positions as shown in Fig. 15. The specific contraction speeds were such that hardly any force could be produced by the red fibres in their shortening phase. Thus, to generate the movement, the fish must have used its white and pink fibres. The stimuli along the trunk were so short that only in the caudal peduncle an active state of one was reached by the red fibres (Fig. 17a). The force curves for the examined electrode positions at the left side (Fig. 17b) show an initial peak, followed by a rapid decline owing to the high contraction velocities. In the caudal peduncle, the initial peak is highest, since the fibres were forcibly stretched (top half of Fig. 15d, right series of plots and Fig. 17a). The final peak in force production was caused by a final forcible stretching phase, with an increasing delay along the trunk. The normalized power curves (Fig. 17c) show that the red fibres produced mainly negative power, except in the anal region (position 3) where phases of positive and negative power almost compensated each other. The normalized work histogram (Fig. 17d) shows that energy was absorbed along the trunk during the first tail strike. In conclusion, the slow oxidative fibres could not have driven the observed movement.

How would the behaviour of the fast fibres have been during the "kick phase"? Figure 17e–h shows the results of a calculation, assuming that the activation periods of the fast fibres were equal to those obtained for the slow fibres. The required shortening speed of the fast fibres is less than that of the slow fibres, owing to their different orientation and position with respect to the axis (Alexander 1969). Based on Alexander's theory, Rome et al. (1988) estimated that white fibres only need to shorten at one fourth of the speed of the red fibres for a particular body curvature. Sarcomere length measurements of muscle fibres from specimens, fixed in a curved position, were consistent with this estimate. In one curved specimen, I measured a contraction ratio of slow fibres over fast fibres (number of samples was to small for statistical significance) of about 3 (electrode position 2 of continuous swimming) and about 4 for two other spots (electrode positions 4 and 6 of continuous swimming). Therefore, the strain fluctuations and contraction speeds were reduced in the calculations with a factor four compared to the red fibres (This is a correction of Van Leeuwen et al. 1990, who used a reduction factor of two). For the fast fibres, $\hat{v}_{max}$ was chosen to be 15 length/s (i.e. 2.73 times $\hat{v}_{max}$ of red fibres, based on measures by Rome et al. 1988). The velocity for maximum power output, V_p, was calculated to be 4.64 lengths/s (i.e. 3.05 times V_p used for the red fibres). The delay time, τ, for full force production was set to 30 ms, i.e. half the value the used for the red fibres (see also Granzier et al. 1983). The active state reaches now a value of one over the complete trunk (Fig. 17e). Compared to the red fibres, the force curves (Fig. 17f) are single peaked, owing to the shorter response time and increased $\hat{v}_{max}$. Peak normalized force in the caudal region is now larger than one during the initial lengthening phase, owing to the shorter response time. Figure 17g shows that anterior to the caudal peduncle, only positive power is produced, except anteriorly (position 1) in the final phase. Positive net normalized work is produced over the complete length of the trunk, with a maximum in the anal region (Fig. 17h).

In both swimming modes, contraction of anterior muscle fibres results in stretching of muscle fibres in or close to the caudal peduncle. Activity of muscle fibres in the caudal region costs energy. In fact, negative work could be done more efficiently with only collagen fibres, which can release stored strain energy. However, an advantage of muscle fibres is having the possibility of regulating the stiffness so as to obtain a better hydrodynamic efficiency. For fish, using different modes of swimming, this may result in a lower overall cost of locomotion, although the negative work of the posterior fibres (with little metabolic costs) has to be compensated by anterior muscle fibres doing positive work (at a much higher metabolic energy consumption). In conclusion, to obtain a high overall muscle efficiency, it may be required to work locally with a negative efficiency.

Can fish store useful strain energy in their muscles? Measurements by Altringham and Johnston (1990) show that the parallel elastic element is too compliant for this purpose. Elastic energy can also be stored in cross-bridges (Alexander and Bennet-Clark 1977). In continuous swimming, the fibres are activated during the final stage of the stretching phase. The stretch during activation is within 3%, so that cross-bridges can remain attached and store elastic energy. This mechanism will be most prominent in the caudal peduncle, where the highest levels of activation during stretching occur. The effect is, however, small compared to strain energy storage in a MTC.

4.2 Insect Asynchronous Flight Muscles

Asynchronous flight muscles, found in flies, beetles, bees, thrips, booklice and true bugs (Ellington 1985) are designed for a high frequency, low amplitude contraction regime. These muscles are particularly interesting, because some design principles come most brilliantly into focus owing to the extreme functional demands. In Sect. 2.6, two adaptations were already mentioned that optimize specific power output:

1. the complete overlap of actin and myosin as a structural optimization for specific myofibrillar power output during a short range contraction, and
2. the mechanically induced activation, reducing the required space for the SR (activation of synchronous muscles is also length-dependent, but neural input dominates, Sect. 2.2).

With the thorax and wings, the flight muscles form a resonance system. The resonance frequency of a damped spring-mass system can generally be increased if the effective mass and damping are decreased, and the spring stiffness is increased. This is accompanied by a very low amplitude of movement. The low effective mass is assured by the small size of the locomotor system, and the low density of air. The muscle fibres have also been found to have an exceptionally high passive stiffness (Machin and Pringle 1959; Fig. 18: lower curve). A delay between stretch activation and tension development ensures that active force is predominantly produced during shortening, so that positive work is done in each wing beat cycle (see work loop of Fig. 18). The origin of the delay is not fully understood, but Wray (1979) gives

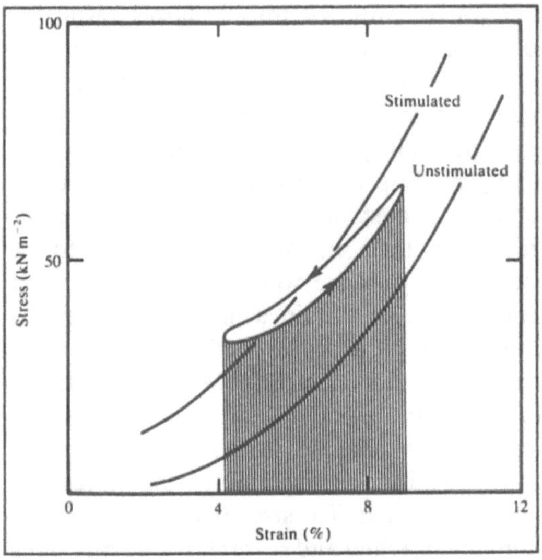

Fig. 18. Typical stress–strain curve for intact asynchronous flight muscle of *Oryctes rhinoceros* (stimulated and unstimulated) and oscillatory (work loop) conditions. *Shaded area* is proportional to absorbed work during stretch. *Loop area* represents net work done per cycle. (Ellington 1985, based on experiments by Machin and Pringle 1959)

an explanation based on filament geometry. Under natural conditions, stretching occurs dynamically by an inertial load. Machin and Pringle (1959) replaced the natural load of a flight muscle of a coconut beetle by an artificial load. Using an electronic circuit, various inertias, stiffnesses and damping influences were simulated. The self-oscillatory mechanism was disturbed if either the inertial load was very low or the damping very high. The high stiffness of the muscle, however, was kept intact. Most likely, the high stiffness originates from an exceptionally short and stiff c-filament, connecting the myosin filament to the Z-disc (White 1983). In this way, a passive spring is formed within the muscle, which can store a significant amount of elastic energy. During flight, the spring seems to be operated considerably above its slack length. This avoids the "toe" (left side in Fig. 18) of the nonlinear tension-length diagram, and assures relatively large fluctuations in the energy density function. The cross-bridges produce force in parallel to this spring and can also store elastic energy (Alexander and Bennet-Clark 1977). The strain fluctuations of the sarcomeres are so small ($\leq 3\%$, that the myosin heads can, in principle, remain attached to the actin filaments throughout the wing beat cycle. Most likely, the cross-bridges produce force continuously, thereby elevating the effective stiffness of the system. Hence, favourable conditions for the absorption of kinetic energy of the wing mass and wing virtual mass at the end of the half stroke are present. The stored elastic energy is subsequently used to accelerate the wing in the next half stroke. As mentioned, the cross-bridges produce more force during the shortening phase than during lengthening, so that overall positive work is produced.

In buff-tailed bumble bee (*Bombus terrestris*), the musculo-skeletal system realizes the wing stroke amplitudes with very small sarcomere length changes (Dudley and Ellington 1990a, b). This insect uses the same wing beat frequency (about 160 Hz) and stroke amplitude over a wide range of flight speeds, from hovering to about 4.5 m/s. Flight speed is controlled by changing other parameters, such as body angle

and stroke plane angle. Therefore, the sarcomeres of the major flight muscles seem to operate similarly at various flight speeds. Dudley and Ellington (1990b) calculated that the required power output is also constant with flight speed. They estimated that all the kinetic energy of the wings could be stored in the flight muscles. These findings are consistent with the concept of a resonance system with a relatively fixed operating frequency.

4.3 Running and Jumping in Mammals: Muscles in Concert

This section deals with muscle function in terrestrial locomotion. An animal, walking or running at (a close to) constant speed, changes cyclically the potential and kinetic energy of its body. In walking, kinetic and potential energy fluctuations are more or less out of phase, since conversions between them are made by using the legs and body as an inverted pendulum (Chap. 4). This strategy saves metabolic energy. Elastic energy storage in tendons presumably plays a minor rôle as an energy saving mechanism, owing to the low forces in the muscles.

In running and hopping, the kinetic and potential energy fluctuate much more in phase, whereas the required muscle forces are much higher. Here, the potential for the muscles to use their build-in springs as strain energy stores are much higher (for reviews see Alexander 1984, 1988). In Sect. 2.7, it is shown that TTS stiffness should be "tuned" correctly to obtain a high efficiency. At slow running speeds, body segment accelerations and peak net joint moments are still relatively low. Elastic energy storage would be far from optimal if muscle activity were distributed rather evenly among synergistic muscles, since then the population of active fibres in each muscle would be in series with a relatively stiff TTS. If, however, some muscles or just one muscle from a synergistic group were active, then overall elastic energy storage would be more favourable. To continuously optimize elastic energy storage with increasing running speed, more muscles could sequentially be activated. At low speeds it would be most efficient to use muscles with mainly slow oxidative fibres. This principle of optimum efficiency may help to understand (1) the high number of muscles present in a leg, (2) distribution patterns of muscle fibre types over various muscles and (3) to predict activation patterns of synergistic muscle groups. It may also help to understand structural subdivisions within muscles. Similar to the swimming muscles of the carp, it is important to consider the overall efficiency of the muscular system.

The optimum efficiency hypothesis is supported by measurements of electrical activity, forces (using the tendon buckle method), and length changes of soleus (SOL) and medial gastrocnemius (MG) muscles in walking and trotting cat by Walmsley et al. (1978). They showed that the SOL is active at all speeds. Peak force production varies relatively little with locomotion speed. The MG, however, increases its peak force with running speed (Fig. 19d). The mechanism of strain energy storage during activated pre-stretch is used during running and jumping. Even during a fast trot, peak force in the MG (≈ 20 N) is relatively low compared to that in maximal jumping (Fig. 19c). Peak force is generated before peak MG length.

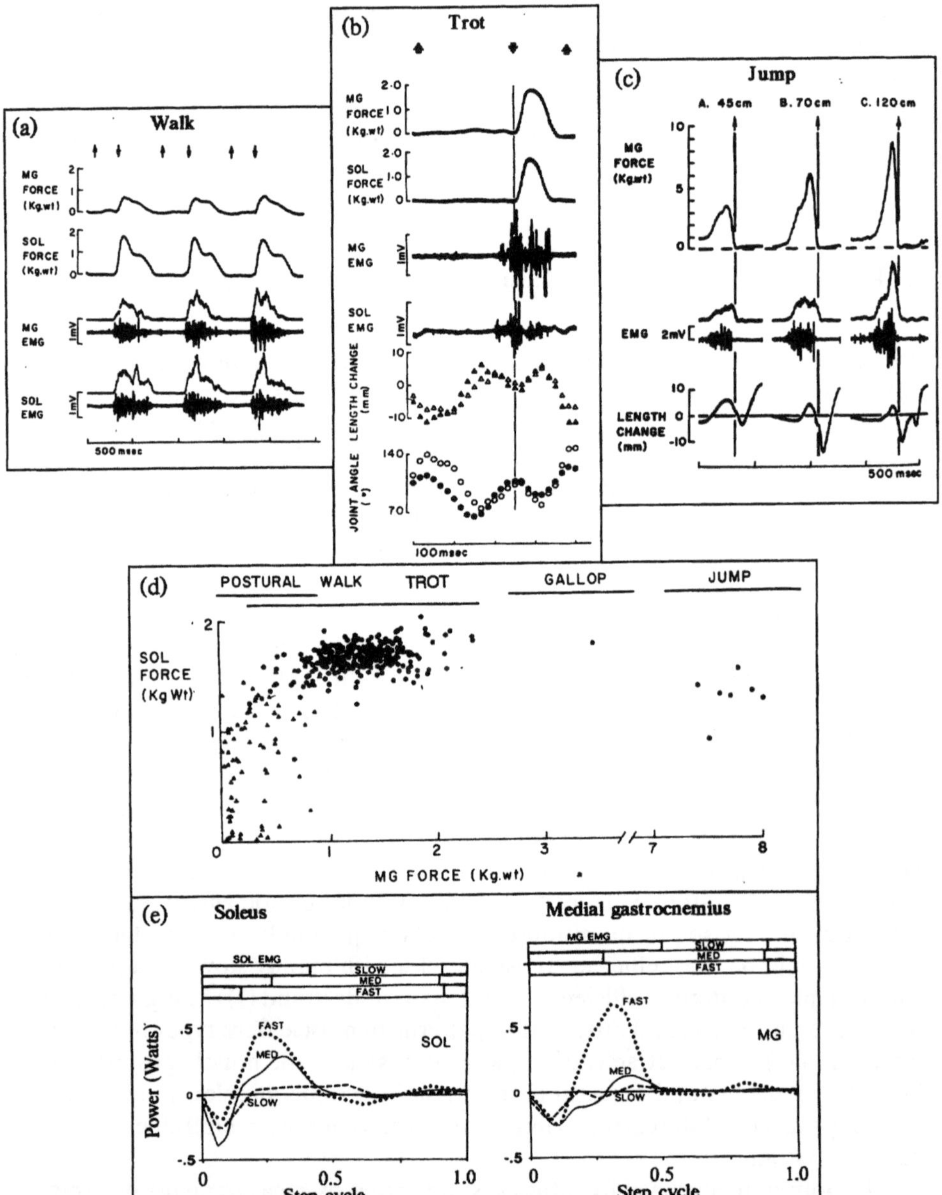

Fig. 19 a–e. Force and EMG (*upper EMG traces* in **a** and **c** are rectified and smoothed) records of cat soleus (*SOL*) and medial head of m. gastrocnemius (*MG*) during walking (**a**), trotting (**b**), and jumping (**c**). Plot **d** shows peak force of MG against that of SOL during different activities. Read for *gallop* transition from trot to gallop. Force units are incorrect and should be multiplied by about 9.8 for conversion to N. (Walmsley et al. 1978). **e** Power output of SOL and MG during running at three different running speeds (Whiting et al. 1984)

During a transition from trotting to galloping, peak force was higher (about 35 N), which was accompanied with a shorter delay of peak length. During a high jump, peak force was highest (≈ 85 N), whereas peak length occurred simultaneously, or slightly later. These events are predicted by the MTC model of Sect. 2.7, as shown in the simulations of Fig. 11. With submaximal MG activation, the TTS is relatively stiff, compared to muscle fibre force. This would roughly resemble the simulation of Fig. 11 series 2. (In the simulations compared in Fig. 11, TTS stiffness is varied, in stead of activation levels. The relative effects, however, are similar). The relative TTS compliance is lowest in the jump, resembling roughly the situation of Fig. 11 series 3, with either coinciding force and length peaks of the MTC, or peak force slightly delayed to peak length. It is likely that the elastic properties of the TTS are best used during the jump, so as to obtain a (close to) optimum efficiency. Peak force of the SOL during the jump is lower than during running. This is caused by the high contraction velocities (Remember that the SOL contains mainly slow fibres). The cat would be able to jump higher if it had a faster SOL, but it would walk and trot less efficiently.

Further evidence for the present predictions comes from the work by Whiting et al. (1984), who studied the same muscles of the cat. They used also the tendon buckle technique to measure forces, but improved the measurements of length changes of both MTCs, so that estimates could be made of instantaneous power output (Fig. 19e). For the SOL, peak force and peak negative power decrease with an increasing running speed, whereas the positive power peak increases. This can be explained with the present MTC model. If only a fraction of the muscle fibres is activated, a relatively stiff TTS is experienced by them, resulting in a high muscle fibre stretching, more initial negative work of these muscle fibres, less elastic energy storage and presumably a lower MTC efficiency than with full activation. For the MG, it is mainly the positive power peak that increases with speed. More firm conclusions could be drawn if model simulations were made with the actual muscle parameters, in combination with various activation patterns.

In conclusion, activation patterns of muscles during locomotion are likely to be optimized so as to obtain optimum overall muscular efficiency during various levels of mechanical loading. Load sharing between synergistic muscles can so be predicted.

In terrestrial locomotion, most locomotor muscles work (in co-ordination with other muscles) in kinematic linkages such as legs. In principle, muscles might be constrained in their performance by the nature of the kinematic chain. The following discussion is mainly based on the work done on human jumping by Van Ingen Schenau and colleagues (reviewed by Van Ingen Schenau 1989), measurements of muscle moment arms by Spoor et al. (1990), and personal view points. Figure 20a illustrates some of the principle human leg muscles involved in a jump. The gluteus group (GLG) tends to extend the hip. The gluteus medius (GLME) and minimus (GLMI) are monoarticular. The gluteus maximus (GLMAX) inserts (with the biarticular m. tensor fasciae latae) for more than 50% on the tractus ilio-tibialis, a long tendinous part which passes the knee joint. Preliminary observations indicate that both muscles have an extending moment arm about the knee. (The treatment of the GLMAX as a strictly monoarticular muscle by Van Ingen Schenau c.s. and

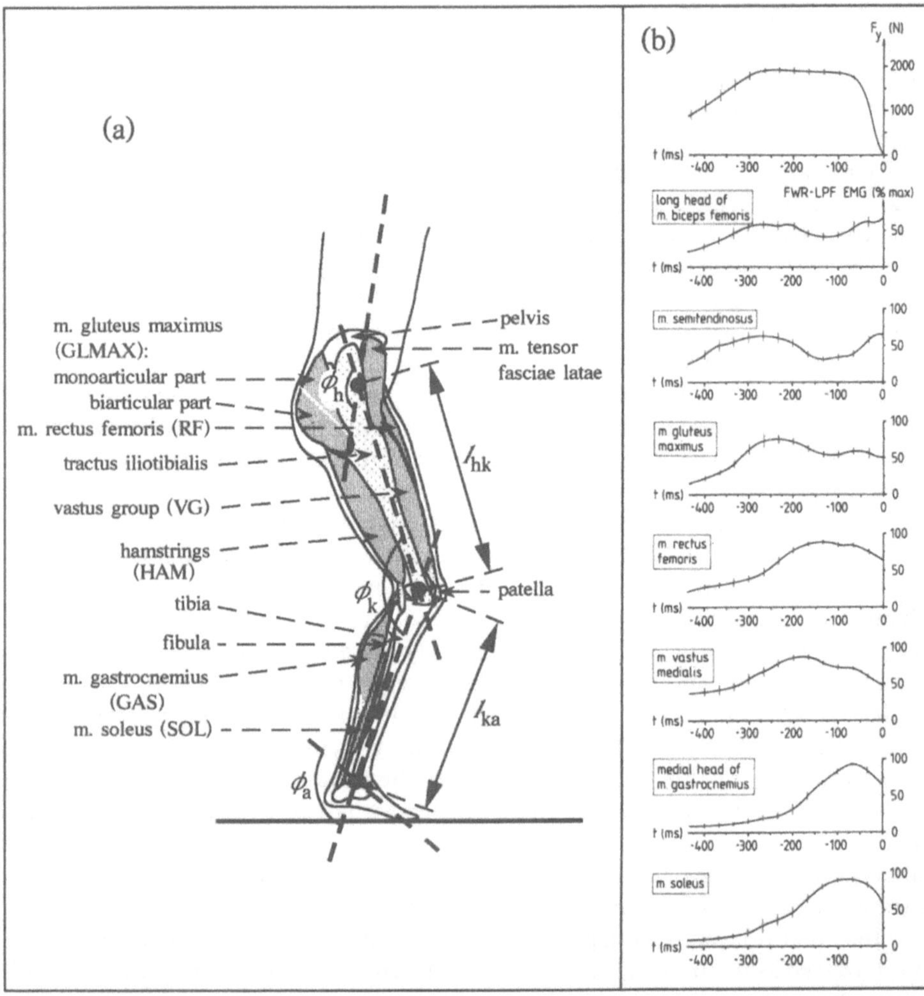

Fig. 20. **a** Diagram of important muscle groups involved in human jumping. Symbols ϕ_a, ϕ_k, and ϕ_h denote angles of ankle, knee and hip. *Dashed lines of heavy type* run through instantaneous centres of rotation of joints. **b** Ensemble averages ($n = 10$) of (1) vertical force on feet F_y and (2) rectified and low pass filtered electromyograms of some major leg muscles during the jump. Long head of m. biceps femoris and m. semitendinosus are parts of the hamstrings. *Vertical bars* show standard deviations. **b** Bobbert and Van Ingen Schenau (1988)

Pandy and Zajac (1990) is erroneous). The lesser part of the GLMAX is monoarticular. The vastus group (VG) tends to extend the knee. The SOL tends to extend the ankle (this usually is called plantar flexion; extension is used here since the rotation lengthens the leg). Several other muscles pass two joints. The hamstrings (HAM) tend to extend the hip and to bend the knee. The rectus femoris (RF) tends to flex the hip and extend the knee. The m. gastrocnemius (GAS) tends to flex the knee and to extend the ankle. The phrase 'tend' is used because quite often joint rotations are opposite in direction to a particular muscle moment about the joint. For one-joint muscles, this always involves forcible stretching. For a biarticular muscle this may be the case.

Generally, the joints allow predominantly rotations to occur. Simple rules for the transformation from joint rotations to trunk rise can be derived if, for a moment, the joints are considered as planar hinges (Van Ingen Schenau 1989). First, we will consider a leg with only two segments. The hip height h in jumping can now be expressed as (using the law of cosines):

$$h = \sqrt{l_{\mathrm{hk}}^2 + l_{\mathrm{ka}}^2 - 2l_{\mathrm{hk}} \cdot l_{\mathrm{ka}} \cdot \cos\phi_{\mathrm{k}}}, \tag{12}$$

where ϕ_{k} is knee angle, l_{hk} is thigh length (distance from hip joint to knee) and l_{ka} is crus length (distance from knee to ankle). The velocity of the hip is obtained by differentiation of formula (12):

$$v_{\mathrm{h}} = \frac{\mathrm{d}h}{\mathrm{d}t} = [l_{\mathrm{hk}} \cdot l_{\mathrm{ka}} \cdot \sin\phi_{\mathrm{k}} / \sqrt{l_{\mathrm{hk}}^2 + l_{\mathrm{ka}}^2 - 2l_{\mathrm{hk}} \cdot l_{\mathrm{ka}} \cdot \cos\phi_{\mathrm{k}}}] \cdot \frac{\mathrm{d}\phi_{\mathrm{k}}}{\mathrm{d}t}. \tag{13}$$

The expression between square brackets is the transfer function H from angular velocity to translational velocity v_{h}. Function H decreases to zero at $\phi_{\mathrm{k}} = 180°$. Therefore, a jumper with a two-segment leg will be airborne before the knees are fully extended. For two reasons, this is unfavourable:

1. Towards full knee extension, a useful contribution to jumping could only be made with a sharply rising angular velocity. The muscles involved would not be able to keep up with this demand. Thus, the period when the most voluminous leg muscles would be able to contribute useful power to the jump is shortened.
2. At take-off, the thigh and crus have considerable angular velocities, representing an important source of kinetic energy which is not made available in a useful way to the jumping height. In fact, the segments would have to be decelerated to avoid knee damage. Thus, the efficiency would be low.

The restrictions to muscular performance by the above geometrical constraint are largely avoided by (1) the presence of a third segment (the foot), (2) the presence of biarticular muscles, and (3) appropriate changes in muscle moment arms with joint angulation.

Figure 21a shows the measured joint angles, joint angular velocities, net joint moments and net joint powers during a two-legged jump with a preparatory counter-movement (Bobbert and Van Ingen Schenau 1988). The subjects were encouraged to jump as high as possible. The jumping performance can be expressed as the sum of the potential and kinetic energy of the centre of mass of the body (CM) at the time of toe-off ($t_{\mathrm{to}} = 0$ ms). The push-off phase, defined by an upward CM acceleration, started about 300 ms before t_{to} (-300 ms). During early push-off, hip extension is most prominent, knee extension starts at about $t = -200$ ms, whereas ankle extension starts at about -100 ms. The time sequence of the onset of joint extensions is associated with increasing activation levels of monoarticular muscles crossing the corresponding joints (Fig. 20b; remember that the mechanical response is delayed relative to the electrical activity, Sect. 3.1). All joint angular velocities diminish

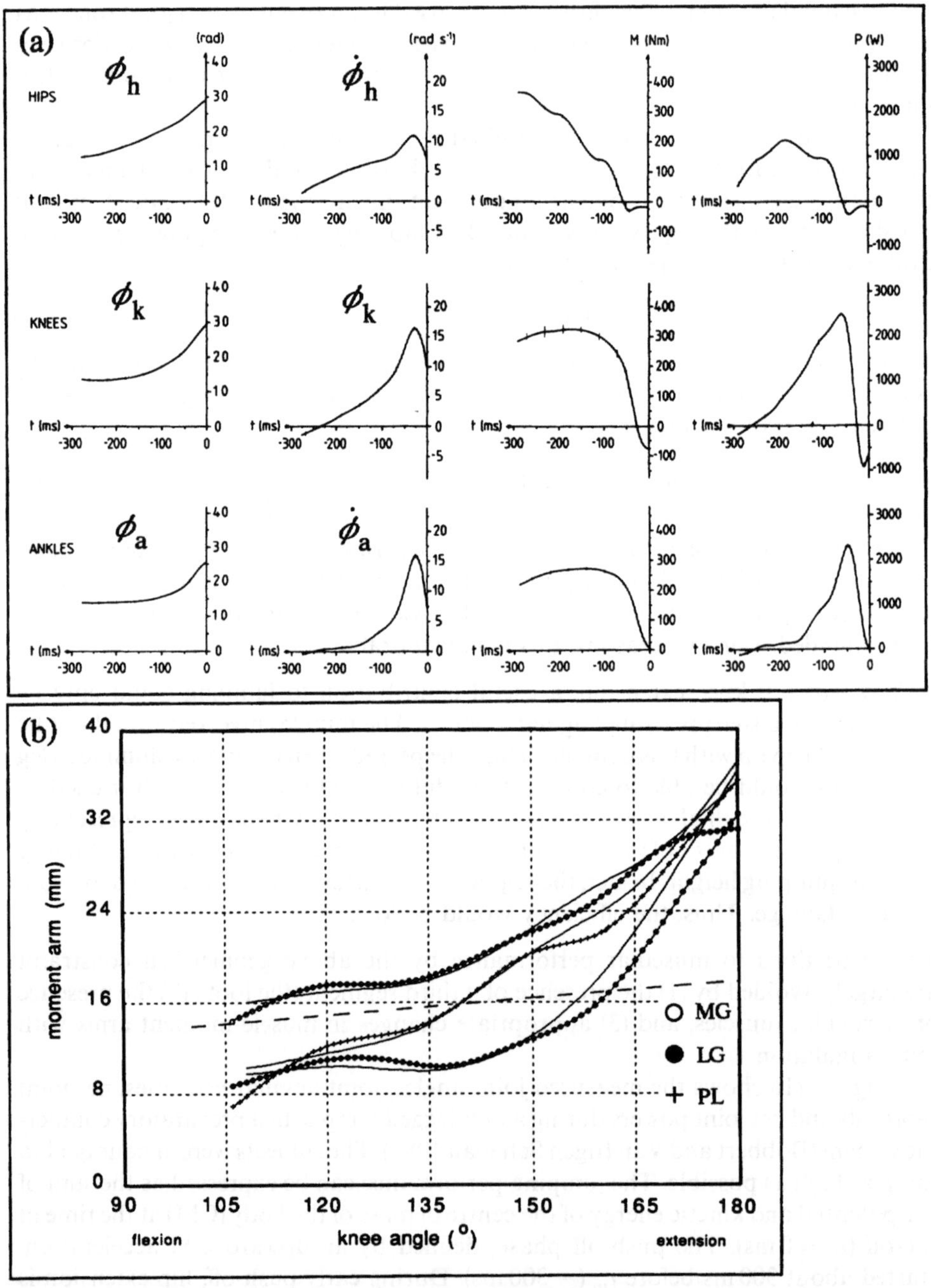

Fig. 21. **a** Mean time histories ($n = 10$) of hip, knee and ankle angles (for definition see Fig. 20), angular velocities, net moments of force (M) and net power output (P) at the joints. Time t is expressed relative to the instant of toe-off ($t = 0$ ms). Bobbert and Van Ingen Schenau (1988). **b** Measured moment arms of medial and lateral heads of gastrocnemius (*MG* and *LG*), and m. plantaris (*PL*) as a function of knee angle. *Unlabelled curves* show results of a repetition of the measurements. *Dashed line* is based on an (inaccurate) estimate by Grieve et al. (1978). (Spoor et al. 1990)

about 30 ms before take-off. In hip and knee, this is accompanied with negative moments, so that negative net power is produced in these joints. Peak net joint power is highest in the knee (≈ 2500 W), it is only slightly lower in the ankle and considerably lower in the hip (≈ 1500 W). This is in contrast to the volumes of the muscle crossing the joints (muscle fibre power output is proportional to muscle fibre mass). A consideration of muscle volume shows that the muscle fibres of the ankle extensors would not be able to generate the high peak net power output. The discrepancy is explained by (1) the actions of the biarticular muscles, evoking a mechanical coupling of joints and (2) rapid release of stored elastic energy in Achilles tendon and tendinous sheets of GAS and soleus.

The coupling action of a biarticular muscle is complicated owing to its complex dynamic properties (Sect. 2.7). The coupling is influenced for instance by the activation of the muscle fibres, instantaneous MTC length and velocity. Thus, coupling is by no means as strict as it would be with an inextensible string. The GLMAX and the HAM belong to the first muscles to become active during push-off (Fig. 20b). This helps to extend the hip. Initially, the knee angle varies very little during push-off (Fig. 21). Thus, upward acceleration of the CM is initiated by extension of the trunk. Knee extension is negligible in this phase, since a large external moment acts on the knee (the ground force has a much larger moment arm about the knee than about the hip in this phase). This effect is even enlarged by the active HAM (opposing knee extension), preventing an inertial force on the trunk which would hamper a high angular acceleration. The transfer from hip rotation to vertical translation of the trunk diminishes towards the vertical position. Knee extension comes into play when this transfer becomes less effective. Knee extension is supported by an increasing activation of the VG and RF and is favoured by a decreasing activity of the HAM. The contribution of the RF needs further consideration. The RF is a mechanical coupler of hip and knee. Hip extension during early push-off results in stretching of the RF, since the knee angle varies vary little up to -200 ms. During stretching, the activation level of the RF rises. Thus, strain energy will be stored in the RF. This energy originates (partly) from the GLG and HAM. Although the RF is continuously active, it is uncertain (although it seems likely) whether this strain energy is usefully released (catapult effect) in a subsequent shortening phase before t_{to}, since the net length change during the simultaneous extension of hip and knee is not known. A fine tuning of the activation is required to avoid energy losses owing to stretching of muscle fibres in the RF. Nevertheless, it can act as a force transmitter for the hip extensors, so that power generated by the hip extensors can show up in the net knee joint power. This helps to explain the simultaneous activation of both antagonistic groups. During knee extension, the transfer from rotation to translation becomes gradually less favourable [Eq. (13)]. An increasing activation of the GAS accomplishes that (1) kinetic energy due to the angular velocities of thigh and crus can be used to power ankle extension with still a favourable transfer from rotation to translation, and (2) muscle power from proximal muscles (mainly VG, but also GLG and others) can be used for ankle extension. This construction functions at the same time as an elegant shock absorber for the knee joint. Recent measurements of moment arms of the MG and LG by Spoor et al. (1990) support this view. Towards knee extension, the moment arms of

GAS about the knee almost double, owing to the build in shift of the axis of rotation of the knee joint. For the same reason, the moment arm of the quadriceps is likely to decrease towards full knee extension. This secures, with changing muscle activations, a rapid increase of the flexing moment about the knee. Again, part of the energy can be temporarily stored in the TTS of the triceps surae (Bobbert and Van Ingen Schenau 1988). This helps to explain also the overlap in activity between VG and GAS. A disadvantage of the usage of biarticular muscles as force transmitters are energy losses owing to viscous properties in the tendons and the muscle belly. It is of crucial importance that substantial stretching of muscle fibres is avoided (cf. Figs. 11 and 12). Finally, the foot has a low mass and therefore does not gain a high amount of kinetic energy.

In conclusion, it is likely that:

1. Biarticular muscles can help to improve the effective power output of monoarticular muscles, once muscle fibre stretching is avoided by a fine tuning of activation. Nevertheless, it remains to be proven that the energetic advantages dominate the energy losses in biarticular muscles.
2. Muscle moment arms about the knee vary so as to decelerate the knee angle towards full extension, and to make use of segmental kinetic energy to further heighten the CM through enhanced net ankle power output. Via the force transmitting limb bones, this power is transported to proximal segments. As explained by Pandy and Zajac (1991), the massive trunk segment gains most energy during the jumps, so that one can speak of an energy flow from the leg muscles in a proximal direction.
3. Kinetic and potential energy can be converted to strain energy in the various MTCs in a preceding counter-movement, and can therefore increase jumping performance.

A more complete understanding of the co-ordination of muscle activity during jumping can be obtained by the application of a direct dynamics analysis with appropriate MTC models for all leg muscles and including instantaneous variations of muscle moment arms. This would show the complete flow of mechanical power in the jumping apparatus. Search techniques could be employed to find optimum muscle activation patterns. A valuable step towards this goal was made by Pandy et al. (1990) and Pandy and Zajac (1991). They formulated a four-segment model, jointed together with frictionless revolutes, and driven by eight musculo–tendon actuators. The authors challenged the idea that jumping performance could be increased by the unique biarticular action of the GAS, by comparing simulated jumps (1) with all muscles, (2) with all biarticular muscles removed and (3) with all biarticular muscles removed, but with addition of a monoarticular GAS. Jumping performance was found highest in case (3) and lowest in case (2). For the following reasons, I doubt whether these findings allow the statement about the GAS:

a) The moment arms of the simulated muscles are very inaccurate, resulting for instance in a vanishing flexor knee moment of GAS towards knee extension (Pandy et al. 1990; Fig. 4b). This is a contrast to the increasing moment arms with knee extension as measured by Spoor et al. (1990). Since GAS is mainly active just before toe-off, I would indeed expect only a small calculated difference between a monoarticular and a biarticular GAS. In case (1), RF was calculated to have a significant negative work contribution. Therefore, the absence of RF in case (2) gives the highest jumping performance.

b) The linear stress–strain relationship of the tendons used by Pandy and Zajac (1991) might have led to a more than realistic muscle fibre stretching and, therefore, to an increased energy loss in biarticular muscles. Furthermore, contraction speed and activation dynamics were the same for all muscle–tendon actuators, whereas the well-known fast properties of the GAS may be essential to avoid an initial forcible muscle fibre stretching.

In conclusion, the direct dynamics approach is very promising, but it requires implementation of more accurate model description of muscles and joints before the issue of the rôle of biarticular muscles can be settled.

The above analysis considered jumping in a fairly large cursorial mammal. Owing to the required preceding counter-movement and the relatively compliant tendons, the jump is rather a time-consuming process. Small mammals (< 1 kg) apply several principles that allow a much quicker jump (i.e. with a higher, but shorter acceleration). First, they operate their legs in a much more flexed range of positions (Biewener 1989), with immediately a favourable transfer from rotation to translation [not requiring a substantial counter-movement, Eq. (13)]. Second, the tendons and aponeuroses of the muscles seem to be relatively stiff (Biewener et al. 1981), so that less time is consumed by an initial stretching of TTSs (similar to Fig. 9). Hence, muscle fibre power output appears with less delay as MTC power output, allowing a faster rise of CM kinetic and potential energy. Third, small mammals have a relatively higher number of fast muscle fibres (Goldspink 1977). Thus, small mammals, seem to be build for high accelerations (which might help to avoid predation by an escape response) as advocated by Biewener and Blickhan (1988) for the kangaroo rat. The disadvantage is a relatively high metabolic power consumption (Alexander and Ker 1990; Kram and Taylor 1990) since (1) elastic energy storage can only play a small rôle in locomotion (Biewener et al. 1981), (2) the noncursorial posture is inefficient and (3) fast muscle fibres are less economic than slow fibres (Goldspink 1977).

Even higher jumping accelerations can be obtained with yet another strategy, which uses the property that natural springs can release previously stored strain energy much quicker than a muscle can produce power. This strategy involves a relatively long process of energy storage in a natural spring (like an apodeme) by a muscle (as in Fig. 9 d–f). The jump is started with a click mechanism which triggers the quick release of the elastic energy. This strategy is especially important for small animals like insects (which have, owing to the short legs, very short acceleration times). A classical study of such a mechanism in the jumping locust is by Bennet-Clark (1975).

5 Conclusions and Perspectives

This chapter has shown how several well-known structural and functional properties of sarcomeres, muscles fibres and muscle–tendon complexes are effectively used in various locomotor systems. The diversity of structural design of sarcomeres was interpreted as a series of adaptations for optimum power output. In the last decade, much improvement has been made in the understanding of the timing of muscle activation of various locomotor systems. When actively stretched, muscle fibres produce much higher forces at very low metabolic cost than during shortening. This property allows the stiffness of the caudal peduncle of a swimming fish to be controlled at relatively low cost. It allows also an effective conversion of potential (associated with gravity) and kinetic energy of the body in elastic energy of stretched tendons and tendinous sheets, which can be released during subsequent shortening of the muscle–tendon complex (MTC). This requires an appropriate timing of activation as demonstrated with a dynamic MTC model, incorporating nonlinear properties of the contractile and tendinous system. Based on this model, an optimum efficiency principle was formulated for load sharing of synergistic muscles. The discussion of human jumping has indicated that biarticular muscles can distribute power from monoarticular muscles over various joints, and can help to convert segmental rotational energy to centre of mass energy.

The understanding of muscle function in locomotor systems is still largely based on EMG, force and strain measurements, muscle models and inverse dynamics analyses. Substantial further improvement may be obtained by a wider application of direct dynamics methods.

Acknowledgments. I thank Dr H.A. Akster, Professor R. McNeill Alexander, Drs. B. Gerritsen, Dr. M. Muller and Dr. C.W. Spoor for their helpful comments and interest. Professor M.G. Pandy and Professor F.E. Zajac are thanked for sending me papers that were still in press.

References

Aidley DJ (1975) Excitation–contraction coupling and mechanical properties. In: Usherwood PNR (eds) Insect muscle. Academic Press, London, pp 337–356

Akster HA (1981) Ultrastructure of muscle fibres in the head and axial muscles of the perch (*Perca fluviatilis* L.). A quantitative study. Cell Tissue Res 219: 111–131

Alexander R McN (1969) The orientation of muscle fibres in the myomeres of fishes. J Mar Biol Ass 49: 263–289

Alexander R McN (1984) Elastic energy stores in running vertebrates. Am Zool 24: 85–94

Alexander R McN (1988) Elastic mechanisms in animal movement. Cambridge Univ Press, Cambridge

Alexander R McN (1989) Muscles for the job. New Sci 1660: 50–54

Alexander R McN, Bennet-Clark HC (1977) Storage of elastic strain energy in muscle and other tissues. Nature (Lond) 265: 114–117

Alexander R McN, Ker RF (1990) Running is priced by the step. Nature (Lond) 346: 220–221

Altringham JD, Johnston IA (1990) Modelling power output by a swimming fish. J Exp Biol 148: 395–402
An KN, Takahashi K, Harrigan TP, Chao EY (1984) Determination of muscle orientation and moment arms. J Biomech Eng 106: 280–282
Aubert X (1956) Le couplage énergetique de la contraction musculaire. Thesis, Arscia, Brussels
Barnes GRG, Pinder DN (1974) In vivo tension and bone strain measurement and correlation. J Biomech 7: 35–42
Barrett B (1962) The length and mode of termination of individual muscle fibres in the human sartorius and posterior femoral muscles. Acta Anat 48: 242–257
Bennet-Clark HC (1975) The energetics of the jump of the locust, *Schistocerca gregaria*. J Exp Biol 63: 53–83
Bennett MB, Ker RF, Dimery NJ, Alexander R McN (1986) Mechanical properties of various mammalian tendons. J Zool Lond A 209: 537–548
Benninghoff A, Rollhaüser H (1952) Zur inneren Mechanik des gefiederten Muskels. Pflügers Archiv 254: 527–548
Biewener AA (1989) Scaling body support in mammals: limb posture and muscle mechanics. Science 245: 45–48
Biewener AA, Blickhan R (1988) Kangaroo rat locomotion: design for elastic energy storage or acceleration? J Exp Biol 140: 243–255
Biewener AA, Alexander R McN, Heglund NC (1981) Elastic energy storage in the hopping of kangaroo rats (*Dipodomys spectabilis*). J Zool Lond 195: 369–383
Biewener AA, Blickhan R, Perry AK, Heglund NC, Taylor CR (1988) Muscle force during locomotion in kangaroo rats: force platform and tendon buckle measurements compared. J Exp Biol 137: 191–205
Blight AR (1976) Undulatory swimming with and without waves of contraction. Nature (Lond) 264: 352–354
Blight AR (1977) The muscular control of vertebrate swimming movements. Biol Rev 52: 181–218
Bobbert MF, Van Ingen Schenau GJ (1988) Coordination in vertical jumping. J Biomech 21: 249–262
Bone Q (1978) Locomotor muscle. In: Hoar WS, Randall DJ (eds) Fish physiology, vol VII. Academic Press, New York, pp 361–424
Brand PW, Cranor KC, Ellis JC (1975) Tendon and pulleys at the metacarpophalangeal joint of a finger. J Bone Jt Surg 57 (A): 779–784
Brown LM, Hill L (1982) Mercuric chloride in alcohol and chloroform used as a rapidly acting fixative for contracting muscle fibres. J Microsc 125: 319–336
Burke RE, Levine DN, Tsaeris P, Zajack FE (1973) Physiological types and histochemical profiles in motorunits of the cat gastrocnemius. J Physiol Lond 234: 723–748
Cavagna GA, Dusman B, Margaria R (1968) Positive work done by a previously stretched muscle. J Appl Physiol 24: 21–32
Close R (1964) Dynamic properties of fast and slow skeletal muscles of the rat during development. J Physiol 173: 74–95
Cutts A (1986) Sarcomere length changes in the wing muscles during the wing beat cycle of two bird species. J Zool Lond A 209: 183–185
Dimery NJ (1985) Muscle and sarcomere lengths in the hind limb of the rabbit (*Oryctolagus cuniculus*) during a galloping stride. J Zool Lond A 205: 373–383
Dudley R, Ellington CP (1990a) Mechanics of forward flight in bumblebees I. Kinematics and morphology. J Exp Biol 148: 19–52
Dudley R, Ellington CP (1990b) Mechanics of forward flight in bumblebees II. Quasi-steady lift and power requirements. J Exp Biol 148: 53–88
Dul J, Johnston GE, Shiavi R, Townsend MA (1984) Muscular synergism-II. A minimum-fatigue criterion for load sharing between synergistic muscles. J Biomech 17: 675–684
Ebashi S, Endo M (1968) Calcium ion and muscular contraction. Prog Biophys Mol Biol 18: 125–183
Edman KAP (1979) The velocity of unloaded shortening and its relation to sarcomere length and isometric force in vertebrate muscle fibres. J Physiol 291: 143–159
Edman KAP, Elzinga G, Noble MIM (1978) Enhancement of mechanical performance by stretch during tetanic contractions of vertebrate skeletal muscle fibres. J Physiol 281: 139–155
Ellington CP (1985) Power and efficiency of insect flight muscle. J Exp Biol 115: 293–304

Ettema GJC, Huijing PA (1989) Properties of the tendinous structures and series elastic component of EDL muscle–tendon complex of the rat. J Biomech 22: 1209–1215
Ettema GJC, Huijing PA, Van Ingen Schenau GJ, De Haan A (1990) Effects of prestretch at the onset of stimulation on mechanical work output of rat medial gastrocnemius muscle–tendon complex. J Exp Biol 152: 333–351
Fick AE, Weber E (1887) Anatomisch-mechanische Studie über die Schultermuskeln. Verh Phys Med Ges Würzb 11: 123–152
Flitney FW, Hirst DG (1978) Cross-bridge attachment and sarcomere "give" during stretch of active frog's muscle. J Physiol 276: 449–465
Goldspink G (1977) Muscle energetics and animal locomotion. In: Alexander R McN, Goldspink G (eds) Mechanics and energetics of animal locomotion Chap. 3, Chapman and Hall, London
Gordon AM, Huxley AF, Julian FJ (1966) The variation in isometric tension with sarcomere length in vertebrate muscle fibres. J Physiol 184: 170–192
Granzier HLM, Wiersma J, Akster HA, Osse JWM (1983) Contractile properties of a white- and red-fibre type of the m. hyohyoideus of the carp (*Cyprinus carpio* L.). J Comp Physiol 149: 441–449
Granzier HLM, Burns DH, Pollack GH (1989) Sarcomere length dependence of the force-velocity relation in single frog muscle fibers. Biophys J 55: 499–507
Grieve DW, Pheasant S, Cavanagh PR (1978) Prediction of gastrocnemius length from knee and ankle joint posture. In: Asmussen E, Jorgensen K (eds) Biomechanics-VI-A, International Series on Biomechanics, Vol. 2A, Univ Park Press, Baltimore, pp 405–412
Hatze H (1981) Myocybernetic control models of skeletal muscle. Characteristics and applications. Univ South Africa, Pretoria
Haug EJ (1989) Computer aided kinematics and dynamics of mechanical systems. Volume 1. basic methods. Allyn and Bacon, Boston
Heglund NC, Cavagna GA (1987) Mechanical work, oxygen consumption, and efficiency in isolated frog and rat muscle. Am J Physiol 253: C22–C29
Heilmann C, Pette D (1979) Molecular transformations in sarcoplasmic reticulum of fast-twitch muscle by electro-stimulation. Eur J Biochem 93: 437–446
Hess F, Videler JJ (1984) Fast continuous swimming of saithe (*Pollachius virens*): a dynamic analysis of bending moments and muscle power. J Exp Biol 109: 229–251
Hill AV (1938) The heat of shortening and the dynamic constants of muscle. Proc R Soc B 136: 399–420
Hof AL, Van den Berg JW (1981) EMG to force processing I: an electrical analogue of the Hill muscle model. J Biomech 14: 747–758
Huxley AF, Niedergerke R (1954) Structural changes in muscle during contraction. Interference microscopy of living muscle fibres. Nature (Lond) 173: 971–973
Huxley HE, Hanson J (1954) Changes in cross-striations of muscle during contraction and stretch and their structural interpretation. Nature (Lond) 173: 973–976
Huxley AF, Simmons RM (1971) Proposed mechanism of force generation in striated muscle. Nature (Lond) 233: 533–538
Josephson RK (1985) Mechanical power output from striated muscle during cyclic contraction. J Exp Biol 114: 493–512
Julian FJ, Rome LC, Stephenson DG, Striz S (1986) The influence of free calcium on the maximum speed of shortening in skinned frog muscle fibres. J Physiol 380: 257–273
Ker RF (1981) Dynamic tensile properties of the plantaris tendon of sheep (*Ovis aries*). J Exp Biol 93: 283–302
Kram R, Taylor CR (1990) Energetics of running: a new perspective. Nature (Lond) 346: 265–267
Light N, Champion AE, Voyle C, Bailey AJ (1985) The role of the epimysial, perimysial and endomysial collagen in determining texture in six bovine muscles. Meat Sci 13: 137–149
Lighthill MJ (1960) Note on the swimming of slender fish. J Fluid Mech 9: 305–317
Loeb GE, Pratt CA, Chanaud CM, Richmond FJR (1987) Distribution and innervation of short, interdigitated muscle fibres in parallel fibered muscles of the cat hindlimb. J. Morphol 191: 1–17
Machin KE, Pringle JWS (1959) The physiology of insect fibrillar flight muscle II. Mechanical properties of a beetle flight muscle. Proc R Soc B 151: 204–225
Machin KE, Pringle JWS (1960) The physiology of insect fibrillar muscle. III. The effect of sinusoidal changes of length on a beetle flight muscle. Proc R Soc B 152: 311–330

Morgan DL (1977) Separation of active and passive components short-range stiffness of muscle. Am J Physiol 232: C45–C49
Offer G (1987) Myosin filaments. In: Squire JM, Vibert PJ (eds) Fibrous protein structure. Academic Press, London
Otten E (1987a) Optimal design of vertebrate and insect sarcomeres. J Morphol 191: 49–62
Otten E (1987b) A myocybernetic model of the jaw system of the rat. J Neurosci Methods 21: 287–302
Otten E (1988) Concepts and models of functional architecture in skeletal muscle. In: Pandolf KB (ed) Exercise Sport Sci Rev, vol 16. MacMillan, New York, pp 89–139
Page SG, Huxley HE (1963) Filament lengths in striated muscle. J Cell Biol 19: 369–390
Pandy MG, Zajac FE (1991) Optimal muscular coordination strategies for jumping. J Biomech 24: 1–10
Pandy MG, Zajac FE, Sim E, Levine WS (1990) An optimal control model for maximum-height human jumping. J Biomech 23: 1185–1198
Pearson AM, Young RB (1989) Muscle and meat biochemistry. Food science and technology. A series of monographs. Academic Press, San Diego, pp 235–265
Podolsky RJ (1964) The maximum sarcomere length for contraction of isolated myofibrils. J Physiol 170: 110–123
Poliacu Prosé L (1985) De functionele stabiliteit van de kat. Een experimentele en biomechanische analyse van de bijdrage van ligamenten en spieren. Thesis, Free Univ Amsterdam and Free Univ Brussels (in Dutch)
Pringle JWS (1972) Arthropod muscle. In: Bourne GH (ed) The structure and function of muscle, vol 1, 2nd edn. Academic Press, New York, pp 491–541
Pringle JWS (1975) Insect fibrillar muscle and the problem of contractility. In: Bolis L, Maddrell HP, Schmidt-Nielsen K (eds) Comparative physiology, functional aspects of structural materials. North-Holland, Amsterdam, pp 139–152
Proske U, Morgan DL (1987) Tendon stiffness: methods of measurement and significance for the control of movement. A review. J Biomech 20: 75–82
Purslow PP (1989) Strain-induced reorientation of an intramuscular connective tissue network: implications for passive muscle elasticity. J Biomech 22: 21–31
Rack PMH, Westbury DR (1969) The effects of length and stimulus rate on tension in the isometric cat soleus muscle. J Physiol 204: 443–460
Rack PMH, Westbury DR (1984) Elastic properties of the cat soleus tendon and their functional importance. J Physiol 347: 479–495
Rapoport I (1972) Mechanical properties of the sarcolemma and myoplasm in frog muscle as a function of sarcomere length. J Gen Physiol 59: 559–585
Rome LC, Funke RP, Alexander R McN, Lutz G, Aldridge H, Scott F, Freadman M (1988) Why animals have different muscle fibre types. Nature (Lond) 335: 824–827
Salmons S (1969) Report on the 8th International Conference on Medical and Biological Engineering. Biomed Eng 4: 467–474
Selvik G (1990) Roentgen stereophotogrammetric analysis – a review article. Acta Radiol 31: 113–126
Sherif MA, Gregor RJ, Liu LM, Roy RR, Hager CL (1983) Correlation of myoelectric activity and muscle force during selected cat treadmill locomotion. J Biomech 16: 691–701
Spoor CW, Van Leeuwen JL, Meskers CGM, Titulaer AF, Huson A (1990) Estimation of instantaneous moment arms of lower-leg muscles. J Biomech 23: 1247–1259
Stenonis N (Stensen N) (1667) Elementorum Myologiæ Specimen, seu Musculi Descriptio Geometrica, vol 2. Stellae, Florence, pp 61–111
Stephenson DG, Williams DA (1982) Effects of sarcomere length on the force-pCa relation in fast and slow-twitch skinned muscle fibres from the rat. J Physiol 333: 637–653
Stevens ED (1988) Effect of pH and stimulus phase on work done by isolated frog sartorius muscle during cyclical contraction. J Muscle Res Cell Motil 9: 329–333
Stevenson RD, Josephson RK (1990) Effects of operating frequency and temperature on mechanical power output from moth flight muscle. J Exp Biol 149: 61–78
Street SF (1983) Lateral transmission of tension in frog myofibrils: a myofibrillar network and transverse cytoskeletal connections are possible transmitters. J Cell Physiol 114: 346–364
Taylor CR, Weibel ER (1981) Design of the mammalian respiratory system. I. Problem and strategy. Respir Physiol 44: 1–10

Usherwood PNR (1967) Insect neuromuscular mechanisms. Am Zool 7: 553–582
Van den Bogert AJ, Hartman W, Schamhardt HC, Sauren AAHJ (1989) In vivo relationship between force, EMG and length change in the deep digital flexor muscle of the horse. In: Biomechanics XI-A, Free Univ Press, Amsterdam, pp 68–74
Van Ingen Schenau GJ (1989) From rotation to translation: constraints on multiple joint movements and the unique action of biarticular muscles. Human Movement Sci 8: 301–337
Van Leeuwen JL (1991) Optimum power output and structural design of sarcomeres. J Theor Biol 149: 229–256
Van Leeuwen JL, Lankheet MJM, Akster HA, Osse JWM (1990) Function of red axial muscles of carp (*Cyprinus carpio* L.): Recruitment and normalized power output during swimming in different modes. J Zool Lond 220: 123–145
Van Lookeren Campagne AAH, Te Kronnie G, Huijing PA (1988) Muscle and fibre length force characteristics and filament length of rat semimembranosus and gastrocnemius muscle. In: Carraro U (ed) Sarcomeric and nonsarcomeric muscles: basic and applied research prospects for the 90's. Unipress Padova, Padova, pp 525–530
Vrbová G, Gordon T, Jones R (1978) Nerve-muscle interaction. Chapman and Hall, London
Wallinga-DeJonge W (1980) Force development in rat skeletal muscle: measurements and modeling. Thesis, Enschede
Walmsley B, Hodgson JA, Burke RE (1978) Forces produced by medial gastrocnemius and soleus muscles during locomotion in freely moving cats. J Neurophysiol 41: 1203–1216
Wang K (1985) Sarcomere-associated cytoskeletal lattices in striated muscle. In: Shay JW (ed) Cell and muscle motility, vol 6. Plenum. New York, pp 315–369
Weis-Fogh T, Alexander R McN (1977) The sustained power output from striated muscle. In: Pedley TJ (ed) Scale effects in animal locomotion. Academic Press, London
White DCS (1983) The elasticity of relaxed insect fibrillar flight muscles. J Physiol 343: 31–57
Whiting WC, Gregor RJ, Roy RR, Edgerton VR (1984) A technique for estimating mechanical work of individual muscles in the cat during treadmill locomotion. J Biomech 17: 685–694
Woledge RC, Curtin NA, Homsher E (1985) Energetics aspects of muscle contraction. Monographs of the Physiological Society 41. Academic Press, London
Woltring HJ (1986) A Fortran package for generalized, cross-validatory spline smoothing and differentiation. Adv Eng Softw 8: 104–113
Wray JS (1979) Filament geometry and the activation of insect flight muscles. Nature (Lond) 280: 325–326
Zajac FE (1989) Muscle and tendon: properties, models, scaling, and application to biomechanics and motor control. In: Bourne JRV (ed) CRC critical reviews in biomedical engineering, vol 17. CRC Press, Baco Raton, pp 359–411

Appendix A
Symbols and Abbreviations

Symbols

Symbols labelled with * represent normalized quantities. Symbols labelled with represent specific quantities. "At rest" denotes not activated and not under tension

A_{f0}	physiological cross-section of the muscle fibres of a MTC: $A_{f0} = V_m/l_{f0}$
A_{t0}	slack cross-section of tendon(s) and tendinous sheets of a MTC
c	constant in formula (16)
c_1, c_2	constants in formula (5)

c_3, c_4, c_5, c_6	constants in formula (9)
d	$[Ca^{2+}]_{0.5}$ at optimum sarcomere length [formula (16)]
$\hat{E}_{met}$	specific metabolic energy consumption of a muscle belly during a lengthening-shortening cycle of a MTC (in J per unit muscle volume)
F_0	maximum active force generated by a sarcomere, under isometric conditions and full activation
$F_a; F_{a0}$	active state, $F_{a0} \leqq F_a \leqq 1$; active state of a resting muscle fibre
F_{cb0}	average active force generated by the available cross-bridges in a sarcomere, under zero shortening speed and full activation
$F_{cu0}; F^*_{cu0}$	active force of a sarcomere, under conditions of zero shortening speed and full activation; $F^*_{cu0} = F_{cu0}/F_0$
$f_l; f_{lmax}$	filamentary overlap function; maximum filamentary overlap
F_{mb}	force exerted by a muscle belly (sum of the forces of parallel elastic element and muscle tissue)
F_{mtc}	force exerted by a MTC
F_{mus}	force exerted by the muscle fibres of a muscle belly (not including force of parallel elastic element).
F_{pas}	passive length-tension diagram of a MTC
$F_{pas1}; F^*_{pas1}$	passive force of a sarcomere at short lengths, resisting the cross-bridge forces and assumed to be independent of contraction velocity; $F^*_{pas1} = F_{pas1}/F_0$
$F_{pas2}; F^*_{pas2}$	passive force of a sarcomere above optimum length, acting in parallel to the cross-bridge forces; idem, but normalized
F_q	force of muscle fibre in relation to $[Ca^{2+}]$ under steady-state conditions
f_s	stimulus frequency
F_{sarc}	"dynamic" force generated by a sarcomere [formula (3)]
F_t	force exerted on a tendon
F_u	interfilamentary velocity dependence function
h	hip height
H	transfer function from angular velocity to translational velocity
k	constant in formula (2)
l_{act}	length of actin filament in two adjacent sarcomere halves, minus l_z
l_{bz}	length of bare zone (devoid of myosin heads) on myosin filament
l_c	length of connecting filament
l_{cmax}	sarcomere length with maximal effective cross-bridge density; $l_{cmax} = l_{myo} + l_z$
$l_f; l_{f0}; l^*_f$	muscle fibre length; idem but at rest; $l^*_f = l_f/l_{f0}$
l_{hk}	thigh length
l_{ka}	crus length
$l_{mb}; l_{mb0}; l^*_{mb}$	length of muscle belly; idem but at rest; $l^*_{mb} = l_{mb}/l_{mb0}$
l_{min}	minimum sarcomere length in active force-length diagram (Fig. 3a, b). Note that l_{min} is a function of u
$l_{mtc}; l_{mtc0}; l^*_{mtc}$	length of MTC; idem but at rest; $l^*_{mtc} = l_{mtc}/l_{mtc0}$
$l_{mtcopt}; l^*_{mtcopt}$	optimum length of MTC; $l^*_{mtcopt} = l_{mtcopt}/l_{mtc0}$

l_{myo}	length of myosin filament
$(l_{myo}/l_{act})_{pmax}$	filament ratio for maximum power output
l_{0sarc}	sarcomere optimum length: length at which the maximum active force can be generated under isometric conditions and full activation
$l_{sarc}; l^*_{sarc}$	length of sarcomere including the thickness of one Z-disc; $l^*_{sarc} = l_{sarc}/l_{0sarc}$
$l_t; l_{t0}; l^*_t$	length of tendon(s) of a MTC; idem but at rest; $l^*_t = l_t/l_{t0}$
$l_{ts}; l_{ts0}; l^*_{ts}$	tendinous sheet length of a MTC; idem but at rest; $l^*_{ts} = l_{ts}/l_{ts0}$
$l_{tts}; l_{tt\delta 0}; l^*_{tts}$	$l_{tts} = l_t + l_{ts}$; idem but at rest; $l^*_{tts} = l_{tts}/l_{tts0}$
l_z	thickness of the Z-disc of a sarcomere
n_{acb}	number of available cross-bridges in sarcomere subunit
$\hat{P}_{met}$	specific metabolic power consumption of a muscle fibre (in W per unit muscle volume)
$\hat{P}_{metmax}$	maximum value of $\hat{P}_{met}$
$\hat{P}_{mtc}$	specific power output by a MTC (in W per unit muscle volume)
$P_{mus}; \hat{P}_{mus}$	power output by the muscle tissue of a muscle belly; idem but specific (in W per unit muscle volume)
$P_{par}; \hat{P}_{par}$	power output by the parallel elastic element of a muscle belly; idem, but specific (in W per unit muscle volume)
P_{sarc}	sarcomere power output
$P_t, P_{ts}; \hat{P}_t, \hat{P}_{ts}$	power output of the tendon(s) and tendinous sheets of a MTC, respectively; idem, but specific (in W per unit muscle volume)
$\hat{P}_{tts}$	$\hat{P}_{tts} = \hat{P}_t + \hat{P}_{ts}$
$\hat{P}_\sigma; P^*_\sigma$	metabolic power consumption per unit volume of muscle needed to produce unit mechanical stress; idem but normalized
R	normalized stimulus of muscle fibre
r_m	muscle moment arm
s	tendon displacement
t	time
t_s	duration of stimulus
t_{to}	moment of toe-off in a jump
$u; u^*$	within one sarcomere, the velocity of the myosin filament relative to the actin filament; $u^* = u/u_{max}$
u_{max}	the maximum value of u
u_{ref}	interfilamentary velocity for reference (i.e. frog) sarcomere
u_{sarc}	absolute contraction velocity of sarcomere; $u_{sarc} = 2u$
v_h	translational velocity of the hip
$\hat{v}_{max}$	maximum specific shortening velocity of a muscle fibre in optimum fibre lengths/s
V_m	muscle volume
$W_{sarc}; W^*_{sarc}$	mechanical work output by sarcomere from $t = t_a$ to t_b; $W_{sarc} = \int_{l_{sarc}(t_a)}^{l_{sarc}(t_b)} F_{sarc} dl_{sarc}$; idem, but normalized: $W^*_{sarc} = W_{sarc}/(F_0 \cdot l_{0sarc})$

$\hat{W}_{mtc}$	specific work output by a MTC during a lengthening-shortening cycle (in J per unit muscle volume)
$\hat{W}_{mus}$	specific work output by the muscle tissue of a muscle belly during a lengthening-shortening cycle of a MTC (in J per unit muscle volume)
$\hat{W}_{par}$	specific work output by the parallel elastic element of a muscle belly during a lengthening-shortening cycle of a MTC (in J per unit muscle volume)
$\hat{W}_t, \hat{W}_{ts}$	specific work output of the tendon(s) and tendinous sheets of a MTC during a lengthening shortening cycle of a MTC, respectively (in J per unit muscle volume)
$\hat{W}_{tts}$	$\hat{W}_{tts} = \hat{W}_t + \hat{W}_{ts}$
$\hat{W}_+$	specific work output by a MTC during the shortening phase of a lengthening-shortening cycle (in J per unit muscle volume)
$\alpha; \alpha_0$	pennation angle of the muscle fibres in a muscle belly; idem but at rest
Δt	duration of a sarcomere contraction
ε	sarcomere strain [Eq. (16)]
ε_c	critical strain of tendon and tendinous sheets above which a linear stress-strain relationship is assumed
ε_f	muscle fibre strain
ε_t	strain of tendon(s) and tendinous sheets of a MTC
$\phi; \phi_a; \phi_h; \phi_k$	joint angle; ankle angle; hip angle; knee angle
λ	number of myosin heads per unit length on the myosin filament
σ_{miso}	maximum isometric stress of muscle fibres at optimum length
σ_t	mechanical stress of tendon(s) and tendinous sheets of a MTC
τ_{ca}	time constant in process of Ca^{2+} release from the SR (Appendix B)
τ_1, τ_2	time constants in formula (17) (Appendix B)

Abbreviations

CM	centre of mass
EDL	m. extensor digitorum longus
EMG	electromyography
LG	lateral head of m. gastrocnemius
GAS	m. gastrocnemius
GLG	gluteus group of muscles
GLMAX	m. gluteus maximus
GLME	m. gluteus medius
GLMI	m. gluteus minimus
HAM	hamstrings muscles
MG	medial head of m. gastrocnemius
MTC	muscle–tendon complex
PL	m. plantaris
RF	m. rectus femoris
SEC	series elastic component
SOL	m. soleus
SR	sarcoplasmic reticulum
TTS	tendon(s) and tendinous sheets of a MTC
VG	vastus group of muscles

Appendix B
Excitation–Contraction Coupling

The Ca^{2+} release from the SR can be described with a first order differential equation (Otten 1987b):

$$d[Ca^{2+}]/dt = \{[Ca^{2+}]_{max} \cdot R(t) - [Ca^{2+}]\}/\tau_{ca}, \tag{14}$$

where $[Ca^{2+}]$ is the instantaneous Ca^{2+} concentration in the sarcoplasm, $[Ca^{2+}]_{max}$ is the maximum concentration, τ_{ca} is the time constant of the process, and $R(t)$ is the normalized stimulus as a function of time ($0 \leqq R \leqq 1$). For fast muscle fibres, τ_{ca} is shorter than for slow muscle fibres. This corresponds to a greater SR-surface (e.g. Akster 1981) and faster ATPase in the SR-membrane (Heilmann and Pette 1979) of fast fibres, enabling a faster control of $[Ca^{2+}]$. Under steady state conditions, the relationship between normalized force of a muscle fibre and pCa ($= -\log_{10}[Ca^{2+}]$) is a sigmoid curve (Stephenson and Williams 1982; Julian et al. 1986), which can be described by (Otten 1987b):

$$F_q = [Ca^{2+}]^b / \{[Ca^{2+}]^b_{0.5} + [Ca^{2+}]^b\}, \tag{15}$$

where $[Ca^{2+}]_{0.5}$ is the Ca^{2+} concentration for a 50% activation. For rat muscle fibres, Stephenson and Williams (1982) showed $[Ca^{2+}]_{0.5}$ to vary with sarcomere length. Their experimental data can be described by:

$$[Ca^{2+}]_{0.5} = 10^{-(c \cdot \varepsilon + d)} M, \tag{16}$$

where ε is the sarcomere strain, d is $[Ca^{2+}]_{0.5}$ at optimum sarcomere length, and c is a constant which depends on the fibre type. For slow rat muscle fibres at 20 °C: $c = 0.65$. Fast fibres are less sensitive to sarcomere length changes. A good fit of the experimental data is produced with $c = 2.6$. Different values are needed with other temperatures.

F_q can be viewed as a "drive" for force development under dynamic conditions. Myofibrillar activation under these conditions is adequately described by inserting F_q into a second order linear differential equation, describing an overcritically damped system:

$$\tau_1 \cdot \tau_2 \cdot d^2 F_a/dt^2 + (\tau_1 + \tau_2) \cdot dF_a/dt + F_a = F_q, \tag{17}$$

where F_a is the active state ($F_{a0} \leqq F_a \leqq 1$), and τ_1 and τ_2 are the time constants which depend on the muscle fibre type. This equation can describe the experimental observation that, after stimulation, the rise time of force production is shorter than the relaxation time. The above simple phenomenological description of the activation process is useful in the study of power and work output of muscle fibres during locomotor activities (Sect. 4.1). An exception has to be made for asynchronous flight muscles of insects which are relatively insensitive to the free $[Ca^{2+}]$ (e.g. Aidley 1975), and which rely for a great deal on mechanical activation (Sect. 4.2). More detailed discussions on the excitation–contraction coupling can be found in Ebashi and Endo (1968), Hatze (1981), and Wallinga-DeJonge (1980).

Chapter 8

Energetics of Locomotion

T.M. Casey

Contents

1 Introduction

The capacity for sustained activity varies widely in animals. The metabolic organization of endotherms, ectotherms, and insects varies dramatically. In ectothermic vertebrates the maximal aerobic energy expenditure is about ten times the resting metabolic rate and just about equivalent to the resting metabolic rate of endothermic vertebrates (Bennett and Ruben 1979). For endotherms the maximal

Department of Entomology, Cook College, Rutgers University, New Brunswick, New Jersey 08903, USA energy expenditure is an order of magnitude greater than that of terrestrial ectotherms although some "athletic endotherms" are capable of nearly double the normal value (Taylor et al. 1987). Insects combine resting metabolic rates in the range typical of resting ectotherms while their energy expenditure during continuous flight exceeds the maximum rate of oxygen uptake (VO_2 max) of endothermic vertebrates (Bartholomew and Casey 1978). Despite these differences, insects, ectothermic and endothermic vertebrates must operate at similar performance levels under widely different physical constraints depending on the locomotor mode.

The energetics of locomotion is an exciting, interdisciplinary area of research. In the past decade measurements of energy expenditure have been obtained along with simultaneous kinematic information to allow estimates of mechanical power output. Several reviews of energetics of locomotion of vertebrates are available (Goldspink 1977; Bennett 1985). This chapter will examine the energy cost of locomotion both in terms of performance by the animal and in comparison to the functioning of the muscle. I want to specifically address the problem of understanding metabolic cost in relation to mechanical performance with particular reference to the effects of size and speed on efficiency of locomotion. I will also be concerned with comparing estimates of mechanical power output (obtained either from biomechanical studies or muscle preparations) to metabolic rates to examine overall aerobic efficiency of animals and mechanical efficiency of muscle. While physical and mechanical constraints on organisms vary widely for different locomotor modes, some generalizations about energetic performance are possible across taxa, regardless of locomotor mode.

2 Energetics in Relation to Mechanics

To examine energetics in relation to mechanics one must quantify the fate of energy during locomotion. Total power input which is available to the muscle represents the energy metabolism minus the amount needed to power the respiratory and circulatory system and cover maintenance costs (Tucker 1973). In insects, this component is assumed to be negligible but it can represent a significant fraction of the total energy expenditure in vertebrates, particularly endotherms. The portion of the power input remaining which actually gets to the working muscle of the animal is subdivided into two major components, mechanical power output and heat, which is a byproduct of muscle contraction. The mechanical efficiency of the muscle represents the ratio of the total mechanical power output to the power input to the muscle. Measurements of mechanical efficiency of isolated muscles during isotonic or isometric contraction are usually assumed to be about 25% (Hill 1950). The coincidence between muscle efficiency studies of humans which also are about 25% efficient and Hill's classical muscle studies on frog muscle are probably the source of this assumption. Mechanical efficiency of muscle varies from about 10% in rat soleus muscle to about 37% in tortoise iliofibrularis muscle.

It is significant that muscles of smaller animals which operate at higher contraction frequencies are less efficient (Woledge 1989).

Efficiency is a contentious subject in biology and recently a symposium has been convened to examine its definition, utility and assumptions (Blake 1991). For the purpose of this chapter whole animal efficiency is defined as the body mass-specific mechanical power requirements divided by the body mass-specific energy expenditure. Assumptions involved for different locomotor modes will be pointed out during the discussion.

The mechanical power requirements are usually subdivided into two major categories – the power required to move the body through the medium (i.e. do work on the medium) and the power required to move the propulsion producing appendages. The magnitude of this requirement can often exceed the requirements for useful power output (Sotavalta 1954; Weis-Fogh 1973). Moreover, it is difficult to assess the magnitude of this cost because it can be modified directly depending on the degree of storage and release of elastic energy. Although difficult to quantify, elasticity is clearly important in runners (Alexander 1984), fliers and probably swimmers (Alexander 1988). Uncertainties concerning the magnitude of elastic storage and recovery reduce the precision of estimates of total mechanical power requirements and mechanical efficiency.

The central position of mechanical power output between the energy expended (above) and the power required (below) makes this variable amenable to analysis by widely different means. Physiological studies often utilize the "top down approach" – by making assumptions about efficiency of the working muscle (Webb 1971a; Casey 1981a) estimates of the mechanical power output can be made from measurements of oxygen consumption. The "bottom up" method used in biomechanical analysis determines the power requirements to make a direct estimate of the mechanical power output. Each method relies on a variety of assumptions and although the former is more straightforward, uncertainty concerning the assumption of a particular mechanical efficiency reduces confidence in this approach. Perhaps the best way to examine this phenomenon is to use both approaches. By estimating mechanical power requirements and measuring metabolic rates of the same animals and same experiments the mechanical efficiency of the muscle may be estimated directly. Finally, the "side view" represents a way of examining the muscle properties directly, which provides an independent estimate of the mechanical power output and energy expenditure of muscle. Using the work loop technique developed by Machin and Pringle (1959) for asynchronous muscle and modified for synchronous muscle by Josephson (1985a), several laboratories have examined maximal mechanical power production of muscle at frequencies, phase, strain, and temperatures approximating actual conditions of working muscles during locomotion (Josephson 1985a; Stevens 1988; Stokes and Josephson 1988; Altringham and Johnston 1990b). These data are useful because they yield an independent estimate of the mechanical power output of muscle which may be used to examined biomechanical estimates of mechanical power requirements.

3 Metabolism Versus Speed: Mechanics and Energetics

3.1 Flight

Power requirements during forward flight have been predicted to be a U-shaped function of forward speed (Pennycuick 1969; Rayner 1979). At any given speed the aerodynamic power requirements represent the sum of three components. Induced power is the power required to overcome gravity. In general, this component decreases with increasing forward speed and represents a theoretical maximum at zero speed (hovering) because under these conditions all momentum imparted to the air must be supplied by the beating wings. Much experimental work on has been performed on quantifying induced power in insects and vertebrates using several different theories [blade element (Weis-Fogh 1973), momentum jet (Pennycuick 1969; Tucker 1973), vortex theory (Ellington 1977, 1984; Rayner 1979)]. Parasite power represents the cost of body drag and is an increasing function of forward speed (Pennycuick, 1969). Profile power represents the power required to overcome drag on the beating wings. This is the most difficult power requirement to estimate (see Spedding, Chap. 3, this Vol.).

Several studies have measured oxygen consumption in relation to flight speed in birds and bats. Metabolic rates are U-shaped in several species (Tucker 1968; Thomas 1975; Carpenter 1985) but in others the rates are either much flattened or show no significant change with flight speed (Tucker 1972; Bernstein et al. 1973; Torre–Bueno and LaRochelle 1978; Hudson and Bernstein 1983; Rothe et al. 1987). Although the theoretical estimates often predict an accurate mean power over a limited speed range, the shapes of the theoretical and experimental curves do not agree closely in a single case. These discrepencies have not yet been resolved, but they may occur either as a result of changes in morphology and or kinematics as a function of flight speed, or changes in the efficiency of the muscles. Moreover, some studies have examined flight metabolism over a limited range of speeds and relatively constant metabolic data may result because they occur at speeds in the bottom of the U-shaped curve. A major problem in interpretation of these results in that most birds and bats cannot fly at very slow speeds, where changes in flight metabolism are predicted to be dramatic. However, hummingbirds routinely hover and fly slowly as well as rapidly. Hummingbirds show no change in flight metabolism over a speed range of 0–8 $m\,s^{-1}$ and show an increase at higher speed, which presumably reflects the importance of parasite drag at higher speeds (Berger 1985).

Insects also routinely hover and fly slowly as well as rapidly but until recently, technical difficulties prevented measurement of oxygen consumption at any speed except hovering. Ellington et al. (1990) measured oxygen consumption of bumblebees in free flight at speeds ranging from 0–4 $m\,s^{-1}$ in a wind tunnel. Like the hummingbirds, metabolism showed no change with forward speed. Wing stroke frequencies and amplitudes were likewise independent of speed. Change in body angle (becoming less with increased speed) and stroke plane angle (increasing with increased speed) represent the only significant kinematic changes during these flights (Casey and Ellington 1989). Kinematic and aerodynamic analysis of bumblebees in the wind tunnel (Dudley and Ellington 1990a, b) allowed power requirements to be calculated. Unlike previous estimates, those for bumblebees are essentially flat over the range of speeds examined, due primarily to an increase in profile power with increasing speed. These data suggest that efficiency does not change with flight speed. The constancy of frequency and amplitude of the wings suggests similar performance by the flight muscle. A similar analysis combining aerodynamic and energetic measurements for flying vertebrates is sorely needed and until such data become available, extrapolation of energetic values from existing theories should be done with caution.

3.2 Swimming

Mechanical power requirements during swimming are a curvilinear function of speed due to increased drag. In theory, cost to overcome body drag can be calculated from knowledge of the dimensions of the animal. Measurements of towed animals routinely exceed the estimates based on calculations of equivalent flat plates (Williams and Kooyman 1985). The actual thrust required to overcome body drag will be greater than drag of a towed object since the transverse components of the fish's movements thins the boundary layer (Lighthill 1971).

Like flight, swimming animals are immersed in a fluid and the properties of the fluid and the size and shapes of their bodies have important effects on their power requirements and energy expenditure (see Daniel 1991; and Daniel et al. Chap. 2, this Vol.). The energy cost of swimming has been examined in a variety of adult teleost fishes (reviewed by Bennett 1985) as well as in several aquatic and semi-aquatic animals. In general, oxygen consumption increases exponentially with speed up to the critical swimming speed, which represents the highest speed at which the animal can sustain performance aerobically (Webb 1971a). Because of the shape of the curve, U_{crit} speed is well above the speed at which the animal can operate most economically (obtained by drawing a line tangent to the metabolism speed curve which intersects the origin of the plot).

Swimming on top of the water causes additional power requirements associated with the production of a surface wake. Surface swimming has been examined in humans, turtles (Prange 1976), ducks (Prange and Schmidt-Nielsen 1970), marine iguanas (Gleeson 1979; Vleck et al. 1981), minks (Williams 1983a), muskrats (Fish 1982), seals, (Williams and Kooyman 1985) and penguins (Hui 1988b). In all cases metabolic requirements are much greater than for submerged swimming. Streamlining, as well as position of the topswimmer in the water, affects the energy cost.

Although oxygen consumption has been examined for a variety of aquatic and semi-aquatic forms, far fewer studies are currently available which examine the mechanics and energetics of the same animals. Perhaps the most complete data for swimming energetics and mechanics are those for trout (Webb 1971a, b). Oxygen consumption and mechanical power requirements were determined for the same fish. Assuming a constant muscle efficiency, mechanical power output was calculated from metabolic data and compared with biomechanical estimates.

By attaching plates to the trout which affected the drag, Webb measured different rates of oxygen consumption at a given speed. He also measured changes in the energy metabolism which closely correlated with predicted change in mechanical power requirements (Webb 1971b).

In a recent study, Webb et al. (1984) reexamined the kinematics of swimming to examine the effects of size and swimming speed on mechanical performance and energetics. Their findings indicate that tail beat frequency is the major determinant of power and speed and that tail beat amplitude is relatively constant except at very low speeds. Thus, power output and input increase with increased frequency of muscle contraction and frequency is linearly related to speed. Moreover, they also report that whole animal efficiency increases with increasing speed and that the change in efficiency at the low end of the speed range is quite steep. Efficiency also increases as speed increases during swimming in seals (Williams and Kooyman 1985) and in penguins (Hui 1988b).

Fish utilize different muscle fiber types to vary their performance (Goldspink 1977). During swimming at slow speeds small power requirements are met by slow muscle fibers which appear to be entirely aerobic (see van Leeuwen, Chap. 7, this Vol.). As speed increases and mechanical power requirements rise, the fast muscles are activated (Johnston et al. 1977). It is of interest that slow muscles produce their maximum power output at low frequencies (about 2 Hz). In Bullrot *Myoxocephalus scorpius* L. of 23–29 cm length, fast muscles have optimal frequency of 5–7 Hz and at these frequencies produce about five times the power of the slow muscle and constitute about 90% of the myotomal mass (Altringham and Johnston 1990b).

3.3 Running

When an animal is running at constant speed it is said to do no work, since positive and negative changes in energy fluctuation during a stride tend to cancel. However, both the body and limbs as well as the center of mass undergo cyclic changes during each stride. The kinetic energy applied relative to the center of mass is the internal work. Other potential and kinetic energy changes during the stride represent external work (see Bennett, Chap. 5, this Vol.). Mechanical power requirements during running are composed of the sum of these two variables. As speed increases the total mechanical power requirements increase in an approximately linear fashion (Cavagna et al. 1977; Heglung et al. 1982).

In most terrestrial animals energy metabolism is linear function of speed. The total energetic cost is a combination of a resting metabolic rate, an offset cost (usually asssociated with a cost for maintaining posture) and an incremental cost, which represents the change in energy metabolism with changes in speed. The incremental cost increases linearly with speed, leading to the improbable conclusion that the minimum cost of transport is the same for a given animal, regardless of running speed (Taylor et al. 1970, 1982). Within vertebrates, the incremental cost of transport is strongly size-dependent and similar for ectotherms (amphibians [Full 1986; Full et al. 1988] and reptiles [Gleeson 1979; John-Alder and Bennett

1981; John-Alder et al. 1982]) and endotherms (Taylor et al. 1982) of equivalent mass. A quantitatively similar pattern occurs in insects (Herreid et al. 1981; Lighton 1985; Full et al. 1990) and crustaceans (Full and Herreid 1983, 1984; Full 1987). No obvious correlations exist between energetics of running and limb configuration (Taylor et al. 1974), phylogenetic sequence, or leg number. Ants, crabs, cockroaches, beetles and even centipedes show metabolic rates generally similar to those predicted by extrapolating the equations based on mammalian data to smaller size (Full 1989).

There are several conspicuous exceptions to the pattern of metabolism vs. speed described above. In ponies trained to walk, trot, and gallop over a range of speeds, oxygen consumption is a curvilinear function of speed within any gait pattern. This results in an optimal speed in relation to gait where the minimum cost of transport is lowest, and above and below that speed the cost of transport rises sharply (Hoyt and Taylor 1981). This pattern is consistent with actual speeds and gait patterns used by animals in the field (Pennycuick 1975). This study provides an explanation for gait patterns used by animals as a set of tuned springs with the optimal speed indicating the point at which elastic storage and release is maximized (Taylor 1985).

Kangaroos also depart markedly from the normal pattern of energy metabolism in relation to forward speed. In their two-legged "hopping gait", metabolism actually decreases slightly with increased running speed. This appears to be the result of elastic storage and release of energy at each stride. Stride frequency remains constant while stride length increases with forward speed (Dawson and Taylor 1973). As the speed increases, the effectiveness of elastic storage/release increases to match increased kinetic energy such that the cost per contraction is essentially constant (Alexander and Vernon 1975; Cavagna et al. 1977). Biewener et al. (1981) estimate elastic energy saving of kangaroos is about 54%! Similar elastic energy saving has been measured in human runners (Cavagna et al. 1964). Due to the scaling of leg structures, small hopping animals cannot save as much energy elastically as large hoppers (Biewener et al. 1981).

The pattern of oxygen consumption of hopping is also closely coupled to gait in Australian hopping mice (Dawson 1976) as well as in minks, which use a bounding gait (Williams 1983b). Hoyt and Kenagy (1988) have demonstrated different relationships between metabolism and speed of walking and running (trotting and galloping combined) ground squirrels, and suggest that this pattern may be more widespread than previously appreciated. This is intuitively more appealing than VO_2 being a linear function regardless of gait pattern, since it relates to different muscle operation during different gait patterns (Taylor 1985).

As in swimming (see above), mechanical efficiency increases with increasing speed (Heglung et al. 1982). For larger vertebrates, the whole body efficiency (total power output/power input) can exceed 50%! Since mechanical efficiency of the muscle probably does not exceed 25%, these data suggest that elasticity becomes increasingly important as speed increases for larger animals.

4 Size Effects on Locomotor Energetics

4.1 Hovering Flight

In a broad survey of hovering animals, Weis-Fogh (1973) calculated mechanical power requirements for insects, birds and bats. His analysis indicated that there was no systematic size effect on mechanical power requirements. Like Sotavalta (1954), his analysis indicated that inertial power requirements (to accelerate and decelerate the wing mass) were very high. Comparing calculated mechanical power (equals aerodynamic requirements plus inertial requirements) with available metabolic data from tethered animals, he concluded that elastic storage must be an important feature for flying insects because otherwise muscle efficiency would be unreasonably high.

Casey (1981a) compared estimates of mechanical power requirements based on morphology and flight mechanics to those obtained by equating flight metabolism to mechanical power output (assuming 20% efficiency of flight muscle). These results suggested that mechanical power output was similar to estimates based on energy metabolism of sphinx moths over a wide size range. Inertial power was greater in magnitude than aerodynamic power and when the two components were added together, they accounted for the inverse relationship between mass-specific energy metabolism and body mass. The assumption of constant mechanical efficiency is required to support these conclusions, while no assumption was made about an upper limit for maximum mechanical power output.

Ellington (1984) produced a quantitative interpretation of the power budget from a different perspective. Instead of assuming that insect flight muscle had a constant mechanical efficiency, he assumed that the maximum mechanical power output was fixed at about 250 W Kg^{-1} of muscle (Weis-Fogh and Alexander 1977). However, since the fixed maximum mechanical power output was well below the values of 20% of the metabolic rate during hovering, his ratio of total aerodynamic power requirements to total mechanical power output was much greater than that of Casey (1981a). His analysis suggested that inertial power requirements were substantially reduced due to elastic storage and release of energy. However, since he postulated a fixed maximum mass-specific mechanical power output while mass-specific metabolism is inversely related to body mass, a progressive reduction in mechanical efficiency of the muscle as size decreases must occur. Insects either produce unreasonably high mechanical power output or they rely on elastic storage to reduce inertial power requirements but their muscles operate at very low mechanical efficiency. Thus, while the assumptions of the two investigators are each reasonable in themselves, they are mutually exclusive. Clearly, the assumptions need to be evaluated before the conclusions of these studies are addressed.

An evaluation of these assumptions has recently been obtained by in vivo measurements of muscle function in the sphingid moth, *Manduca sexta* (Stevenson and Josephson 1990). Attaching flight muscle of an intact moth to an ergometer, they measured maximal power output during cyclic contraction. Their data indicate that the maximum mechanical power output of the flight muscle of *M. sexta* operating at optimal temperature, contraction frequency and strain was about $100\,W\,kg^{-1}$ (muscle). Aerodynamic power requirements (assuming perfect elastic storage) amounted to about $90\,W\,kg^{-1}$ (muscle), indicating that almost all of the inertial power must be reduced due to elastic storage. Comparison of these data with metabolic rates of *Manduca sexta* during free hovering flight yield mechanical efficiency of about 8–11%, thereby verifying the predictions of Weis-Fogh (1973) and Ellington (1984) about the importance of elastic storage. Josephson and Stevenson (1991) report direct measurements of both the mechanical power output and oxygen consumption of locust flight muscle during cyclic contraction from the same experiments and indicate mechanical efficiencies are low – on the order of about 4–8%. Similar low efficiencies have been reported for crab muscle (Stokes and Josephson 1988). These studies conclude that when muscles function under conditions equivalent to locomotion, they are appreciably below the expected 20% value.

The importance of elasticity is clear for asynchronous muscle (Machin and Pringle 1959; Alexander and Bennett–Clark 1977; Ellington 1984; Alexander 1988). In Euglossine bees, mass specific mechanical power requirements were size independent while mass specific metabolic rates were strongly inversely correlated with size. Assuming perfect elastic storage, efficiency varied from 4–16% and the mechanical power output was about $180\,W\,kg^{-1}$ (muscle) (Casey and Ellington 1989).

4.2 Swimming

Although swimming animals, like flying animals, are immersed in a fluid whose properties strongly affect power requirements, the properties of air and water are sufficiently different that there are major differences in the constraints imposed by the two. For flying animals, density and viscosity are sufficiently low that only at the very extremes of the size range do these properties impose important effects (Weis-Fogh 1973). For swimmers, however, particularly those at medium ranges of Reynolds numbers, the hydrodynamic forces have been difficult to model (see Daniel and Webb 1987 and Daniel et al. Chap. 2. this Vol.).

Aerobic metabolism in fishes is clearly related to size with small animals expending more energy per unit body mass than large ones at any given speed (Videler 1989). Kaufmann's (1990) work on larviform fishes extends these findings and leads to similar conclusions. In contrast, mechanical power requirements of trout ranging in size from 5.5 to 56 cm at the critical swimming speed show little size dependence (Webb et al. 1984). Aerobic efficiency of power production in swimming trout is strongly size-dependent, with larger fish showing apparent whole body efficiency well in excess of the 25% predicted from studies of isolated muscle. They suggest that muscle efficiency must be size-dependent, since Froude efficiency at V_{crit} is essentially size-independent. Since tail beat frequency decreases with increasing size in trout (Bainbridge 1958), they suggest that muscle efficiency is associated with frequency of tail beat.

4.3 Terrestrial Locomotion

In a large series of running birds and mammals the mass-specific mechanical power requirements (excluding elastic energy) at a physiologically equivalent speed (i.e. trot gallop interface) were found to be independent of body size (Heglund et al. 1982). In contrast, the mass specific metabolic rates increase as body size decreases (Taylor et al. 1982). Thus, whole animal efficiency is a direct function of body mass. Recent experiments on arthropods have extended these findings (Blickham and Full 1987; Full and Tu 1990). The pattern of movement of arthropods is similar to that of vertebrates and the mass-specific cost to move the center of mass is size-independent and similar in magnitude for the largest vertebrates to the smallest insects (Full 1989). The degree of energy stored and recovered elastically in the arthropods has not been determined. However, given the prevalence of resilin in the arthropod cuticle (Weis-Fogh 1973), it is reasonable to assume that there can be significant energy saving due to elasticity in pedestrian arthropods.

As size decreases, smaller animals expend more energy to produce the same mechanical power (assuming theories for calculating power output are correct). Smaller animals are more inefficient than large animals. Mechanical efficiency is size and/or frequency-dependent. Possible explanations for these data are the cost of calcium cycling at the sarcoplasmic reticulum (calcium activation is undoubtedly an important component, it may amount to up to 30% of the energy cost of an isometric twitch, but this magnitude is not sufficient to explain the large changes in efficiency). In addition, the cost of cycling cross-bridges between actin and myosin accouts for the additional energy expenditure (Rall 1986).

Mass-specific metabolic rates during locomotion are strongly inversely related to size for swimming (Videler 1989), running (Taylor et al. 1982), and hovering flight (Casey 1989). Mechanical power output, on the other hand, is size-independent at least in particular groups, including running vertebrates (Heglund et al. 1982), running invertebrates (Full 1989), swimming fish (Webb et al. 1984) and flying insects (Weis-Fogh 1973; Casey and Ellington 1989). As a result, whole body efficiency is directly correlated with size in these groups. These striking parallels between swimming, running and flight lead to the following conclusions. (1) Mechanical efficiency is not constant across the size range for any locomotor mode but rather, decreases progressively with decreasing size and/or frequency of muscle contraction. (2) Values of efficiency obtained from isolated muscle studies (isotonic or isometric contraction) may not accurately reflect the situation for steady-state locomotion.

5 Energetics in Relation to Muscle Performance

5.1 Effects of Temperature on Locomotory Energetics

The reviews of Josephson (1981) for insects, and Bennett (1984) for ectothermic vertebrates generally provide similar conclusions on the effects of temperature on muscles performance. Rate function terms are strongly thermally sensitive while maximal force production are less thermally sensitive or completely insensitive over considerable ranges of temperature. Maximal power output (a combination of the two) tends to be temperature-dependent. Power shows particularly high Q_{10}'s at low temperatures because tension development as well as frequency is more thermally sensitive (Bennett 1984).

The incremental cost of locomotion is comparable regardless of temperature in ectothermic insects and vertebrates (Bennett 1985; Full 1991). How does an animal operating over a range of temperatures handle performance when muscle function is thermally dependent? In swimming fish, different muscle fibers operate depending on the power requirements. At low performance they rely primarily on red muscle fibers. At higher levels red plus white fibers are employed. As the temperature is lowered there is a compression in the recruitment order of the muscle. At lower temperature, white muscle which would normally kick in at high swimming speeds operates at lower speed to compensate for the reduced performance of red muscle (Rome et al. 1984). A qualitatively similar pattern occurs in running lizards. Jayne et al. (1990) examined kinematics and electromyography of running varanid lizards. They report that kinematics (and therefore the mechanical power requirements) are temperature independent. At lower temperatures greater number of fibers are recruited to compensate for decreased temperature. EMGs in red fibers at 25 °C had a higher spike amplitude than those at 35 °C. The data suggest that swimming fishes and running lizards use similar mechanisms to produce equivalent mechanical and energetic performance to compensate for the reduced muscle function as temperature decreases.

Under most circumstances animals do not operate at VO_2 max and if they do, they operate at preferred body temperature. A conspicuous exception to this situation is the case of endothermic insects which must warm their flight muscle prior to take off [a temperature change of as much as 20 °C in a few minutes (Heinrich and Bartholomew 1981)]. During preflight warm-up, cycle frequency is maximal and closely coupled to thoracic temperature. As temperature increases, the frequency for maximum mechanical power output also increases (Stevenson and Josephson 1990) and is closely correlated to the actual in vivo frequency exhibited by *M. sexta* during warm-up (Heinrich and Bartholomew 1971). Over a wide range of temperatures power is determined primarily by frequency. In vivo measurements of oxygen consmption and frequency during preflight warm-up are consistent with this interpretation and indicate only at relatively low temperatures (15–20 °C) does the energy cost per cycle decline (Casey and Hegel-Little 1987). This appear to be associated with lower amplitude muscle potentials (Goller and Esche 1990) and presumably decreased work output.

5.2 Energy Cost per Locomotor Cycle

Metabolism or power input has often been divided by contraction frequency to determine the energy expended per locomotor cycle. Indeed, Gold (1973) suggested that this parameter was a constant among animals and that energetic data could be determined by measurement of locomotor cycle frequency. Although this is an oversimplification, there are strong parallels between energy expenditure and frequency in all locomotor modes. Variously reported as stroke work (Casey 1980, 1981b), stroke work input (Casey and Hegel-Little 1987), cost per stride in runners (Heglund and Taylor 1988) and cost per tailflip in swimmers (Davis et al. 1985), this parameter is rarely compared across locomotor modes or taxa. Sotavalta and Laulajainen (1961) examined oxygen consumption and wing stroke frequency of synchronous and asynchronous fliers together and determined that energy per cycle scaled with the 1.4 power of body mass. This exponent was considered unlikely and the conclusion was rejected (Weis-Fogh 1973). Data for sphingid moths (synchronous fliers) indicated that energy per cycle was size-independent (Bartholomew and Casey 1978). Several other moth species with widely differing size, wing stroke frequencies and energy metabolism during flight exhibited energy per cycle which is equivalent to the sphingid moths (Casey 1980, 1981b). If Sotavalta and Laulajainen's (1961) data for synchronous fliers only are included they sustain the size independence of energy cost per cycle (Casey 1989).

Asynchronous fliers also show scaling of energy per cycle which is size-independent (Casey et al. 1985). However, the magnitude for asynchronous insect is well below that of synchronous fliers. For any given body mass, a sphinx moth (synchronous) operates at a much lower stroke frequency than a bee (asynchronous) but at comparable energy metabolism (Casey et al. 1985). This is undoubtedly related to the difference in muscle strain (length change as a percentage of resting length) that the respective muscles produce during cyclic contraction. Strain of synchronous muscle is on the order of 6% of the resting length (Stevenson and Josephson 1990), while that of asynchronous muscles is about 1–2% of its resting length (Ellington 1985; Casey and Ellington 1989). The reduced cost per cycle of asynchronous muscles refects the tradeoff in developing muscles which are capable of operating at high frequency for the asynchronous fliers. Body mass-specific energy per cycle for synchronous and asynchronous muscle during flight are about 11 and 4 $J\,kg^{-1}$ $cycle^{-1}$ respectively (Casey 1989), paralleling the difference in muscle strain in the two groups.

In terrestrial locomotion, the similarity of scaling of cycle frequency and oxygen consumption with size is well known (Heglund et al. 1974). Recently, Heglund and Taylor (1988) examined this relationship in greater detail. At physiologically equivalent levels of effort (such as the trot-gallop interface) the cost per cycle is remarkably size-independent. The size-independent relationship holds true at five other physiologically equivalent levels [i.e. preferred trotting speed, preferred galloping speed, maximum sustained aerobic speed etc., see Heglund and Taylor (1988) for further discussion]. The implication is that for a wide variety of conditions, cost is determined primarily by the frequency of muscle contraction. However, the metabolic rate is a linear function of speed, while the cycle frequency shows different relationships to speed in different gaits (it invariably shows lower slopes of frequency vs. speed at higher speeds). Given the mismatch between frequency and power input, the cost per contraction does increase as the speed increases (Heglund and Taylor 1988). The increase cost per cycle is modest, varying from 5.0 $J\,kg^{-1}$ (body mass) $cycle^{-1}$ at the preferred trotting speed to about 9.4 $J\,kg^{-1}$ (body mass) cycle^{-1} at the maximum galloping speed. In running insects, the cost per contraction appears to be speed-independent and comparable in magnitude to the vertebrate values. Unlike the vertebrates, arthropods exhibit no different gait patterns and stride frequency and VO_2 vary in a similar manner over the size range examined (Full et al. 1990).

Kram and Taylor (1990) recently reported that if the metabolic rate is divided by the ground contact time (i.e. the actual time when the leg muscles are generating most of the forces and are supporting the weight of the animal) rather than by cycle frequency, the cost coefficient (about 0.2 JN^{-1}) is comparable across a wide size range (30 g rats to 140 kg ponies) and is independent of speed. Their data indicate that the rate of energy utilization is inversely related to the time the leg applies forces to the ground. Smaller animals operating at higher cycle frequencies must utilize faster, shorter muscles which cycle cross-bridges more rapidly. The metabolic expenditure should be associated with the rate of cross-bridge cycling rather than to the mechanical power required. Their results suggest that the cost of running is determined primarily by the cost of supporting the animal's weight and the time course of generating that force (Kram and Taylor 1990).

Fewer comparable data are available for swimming animals but they do support the generalization cited above. In swimmers, increases in speed are associated with increases in tail beat frequency (in teleosts, Webb et al. 1984; sharks, Graham et al. 1990; seals, Davis et al. 1985) and for a given speed, tail beat frequency is size-dependent in teleosts (Bainbridge 1958) as in terrestrial runners and flying insects. Smaller fishes have higher tail beat frequency than larger fish at any given absolute speed (Graham et al. 1990) and the metabolic cost for smaller fish is also correspondingly higher. While it is possible that the cost per contraction (equals tail beat) is generally similar in fish regardless of size, further data are needed to test this suggestion.

6 Energetic Correlates with Muscle Ultrastructure

As size changes, there is a problem with muscle packing involving a trade-off between elements in the muscle which produce the force (myofibrils) and those which generate ATP (mitochondria). Additional problems arise with high frequency, since the muscle needs more sarcoplasmic reticulum for faster cycling of calcium, thereby leaving less space available for other components. While this may have little consequence for a fish or a reptile whose muscle aerobic performance is moderate and mitochondrial component is relatively small (Johnston 1985), it may be critical and limiting to highly aerobic performers. Among the vertebrates, hummingbirds have the highest mitochondrial density for any vertebrate muscle (about 50% – Lasiewski et al. 1965). Packing muscle with any more mitochondria appears to be self-defeating, since increases in mitochondria require decreases in force-generating elements (Pennycuick and Rezende 1984).

6.1 Muscle Power Output

Two models are available which use force velocity properties and assumed muscle strain to predict the maximum mechanical power output for muscle (Weis-Fogh and Alexander 1977; Pennycuick and Rezende 1984). These models assume that muscle is composed primarily of myofibrils and mitochondria and each model predicts that the power should increase with increased cycle frequency. The maximum values predicted by these two theories (250 W kg^{-1} muscle and 430 W kg^{-1} muscle respectively) are much higher than values obtained in mammal, fish, frog, and insect preparations (Woledge 1989). Ellington (1985) revised these estimates downward to about 100 W kg^{-1} (muscle), pointing out several faulty assumptions of the models, and argued that muscle efficiency of insect flight muscle is low (less than about 10%). This has been experimentally verified in katydid singing muscle by comparison of metabolic rates during singing (Stevens and Josephson 1977) with measurements of maximum mechanical power output (Josephson 1985b) and in sphingid moth flight muscle using a similar approach (see above), and by direct, simultaneous measurements of mechanical power and oxygen consumption on locust muscle (Josephson and Stevenson 1991). In all cases the mechanical power output was somewhat below Ellington's maximum estimate and efficiency was appreciably below 10%. Thus, it appears that small animals operating at high frequency are not capable of producing as much power as was previously estimated. This is consistent with biomechanical estimates of similar mass-specific mechanical power output over a wide size range despite very large changes in cycle frequency (Heglund et al. 1982; Casey and Ellington 1989). Maximum mechanical power output of frog sartorius, rat diaphragm, bumblebee flight muscle, sphingid flight muscle, tettigoniid singing muscle, crab respiratory muscle and locust flight muscle are comparable when temperature of the muscle is taken into account (Stevenson and Josephson 1990). Relatively high power output has been reported for fish white muscle (Altringham and Johnston 1990a, b). This is perhaps a consequence of higher percentage of force-generating myofibrils than for more aerobic muscles and hence involving an ultrastructural tradeoff in the muscle between myofibrils, SR, and mitochondrial volume fractions (Johnston 1985; Altringham and Johnston 1990b). If the force production were normalized per unit of myofibril these values would be more comparable.

Why is there not a strong dependence of maximal mechanical power output on cycle frequency as predicted by the two models? Josephson (1985b) suggests two reasons. The previous models do not account for the volume of sarcoplasmic reticulum and T-tubules which must necessarily increase for muscles operating at higher frequencies. This decreases the volume of myofibrils in the muscles and therefore the work output per unit mass of muscle. Secondly, he suggests that the assumption of both models that activation and inactivation is instantaneous is unlikely. Power losses when turning the muscle on and off are liable to be more important as frequency increases. The latter explanation is particularly attractive for asynchronous insect flight muscle, which like vertebrate skeletal muscle shows strong size-related muscle efficiency changes but has reduced SR (and presumably minimal cost involving cycling calcium) and operates at very high frequencies (Josephson and Young 1981; Ellington 1985).

6.2 Muscle Power Input

The general concensus among exercise physiologists is that the central circulation represents the limit to aerobic performance (Astrand and Rodahl 1977). The comparative approach by C.R. Taylor, E.R. Weibel, H. Hoppler and associates (Weibel and Taylor 1981; Taylor et al. 1987) indicate that the mitochondria are the ultimate determinate of aerobic performance. Over several orders of magnitude in size, the VO_2 max and total mitochondrial volume exhibit similar scaling such that at maximal levels of oxygen uptake the oxygen consumption per unit mitochondrion ranges from 3–5 ml O_2 ml^{-1} mito min^{-1} and is surprisingly size-independent for equivalent muscles over a very wide size range for mammals running at VO_2 max on the treadmill (Hoppeller et al. 1987). For animals capable of different performance levels, the whole body mitochondrial content varies in such a way as to keep the mitochondrial oxygen uptake essentially constant. Thus, the oxidative capacity of the muscle is matched to the aerobic capacity of the individual. Comparisons of animals of similar size but different performance capacity support this contention. The VO_2 max of horses is about 2.7 times greater than that of steers. This difference is matched by a 2.4-fold difference in volume of total skeletal muscle mitochondria (Kayar et al. 1989).

At VO_2 max, the rate of oxygen uptake per unit mitochondrion is essentially constant across the size range (Hoppeler et al. 1987; Schwerzmann et al. 1989). Frequency and oxygen consumption scale similarly across the size range, yielding a size-independent cost per contraction (Heglund and Taylor 1988). The rate of oxygen uptake of the mitochondria is the product of the frequency of muscle contraction and the uptake per contraction. Consequently, the oxygen uptake per contraction by the mitochondria must decrease as size decreases. Therefore, in order for mitochondrial oxygen consumption to be essentially constant over a wide size range, there must be a relationship between mitochondrial volume and muscle contraction frequency, as predicted by Pennycuick and Rezende (1984)

The relatively constant uptake of oxygen per unit of mitochondria also occurs in lower vertebrates. Data obtained from calling muscles suggest a lower oxygen uptake per unit mitochondrion (Marsh and Taigen 1987). However, with appropriate assumptions about Q_{10}'s of various mitochondrial components, they calculate that frog muscle yields mitochondrial performances equivalent to mammalian muscle under appropriate thermal conditions. Comparing reptiles to mammals over a size range, Else and Hulbert (1985) report that allometric exponents of mitochondrial parameters are comparable in the two groups and that mitochondrial volume density represents the major factor involved in changing mass-specific metabolism of tissues both as a function of size and in the evolution of endothermy in mammals from reptiles. Increases in internal mitochondrial surface density were also important but to a lesser extent.

The relative simplicity of the insect flight motor provides an attractive alternative to the vertebrate model for understanding muscle function. In vertebrate locomotion such a variety of different muscle groups is involved to varying degrees that it is difficult to determine oxygen uptake of the muscle much less to a unit of mitochondria within it. During hovering flight, virtually all the oxygen is utilized by the flight muscles, two groups of antagonistic muscles. Oxygen transport is also greatly simplified in the insects. Oxygen delivery in vertebrates requires two diffusive step and two convective steps (Weibel and Taylor 1981). In the tracheal system only a single diffusive step and convective step occur, eliminating the entire cardiovascular sequence (gas exchanger to respiratory pigments to pump to tissue). Fibrillar muscle is homogeneous, eliminating consideration of muscle fiber types with different physiological characteristics (Johnston 1985). The muscle is composed of only two basic elements, mitochondria and myofibrils. Moreover, since the sarcoplasmic reticulum is reduced (Ellington 1985), the cost of cycling calcium (Rall 1986) should be reduced in these muscles.

Data for Euglossine bees suggest that asynchronous muscle not only differs in magnitude of oxygen consumption from the vertebrates but also in scaling (Casey and Ellington 1989). Mitochondrial oxygen consumption varies from 6 to 16 ml O_2 ml^{-1} mito min^{-1}. The problem for comparison with vertebrate muscle is exacerbated because it is likely that these levels are not maximal rates (unlike mammalian estimates). Insects operate at much higher metabolic rates (i.e. as much as 150 ml g^{-1} body mass h^{-1}). Insects may have different capacities for mitochondrial oxygen uptake. In the bees, inner mitochondrial volume density of cristae is size-independent and therefore does not explain the differences in oxygen uptake per mitochondria with size. Inner mitochondrial membrane density is very high (density of 50 m^2 cm^{-3}) (J. Gabriel, pers. commun.) compared with about 35 for mammals (Hoppeler et al. 1987). Thus, some of the differences in oxygen uptake may be directly related to structure of the mitochondria.

In vertebrates, ultrafine sectioning and three-dimensional reconstruction indicate that in highly aerobic muscles the mitochondria are not the simple sausage-shaped structures portrayed in textbooks but rather, that they are much more elaborate (Kayar et al. 1988). Given the obvious aerobic capacity of fibrillar flight muscle, it is reasonable to hypothesize that their mitochondria are rather more elaborate than in most tissues. More detailed electron microscopy studies of insect flight muscle would be useful to evaluate this suggestion.

7 Metabolism and Force Production

Muscle activity in terrrestrial locomotion is terribly complicated. While some muscles contract and do useful work to move the animal, others may act as springs to store and release elastic energy with their tendons, some may be contracting isometrically to stabilize joints thereby producing no work, while still others are being stretched at the same time (performing negative work). Thus, even though all muscles are utilizing ATP, they do not necessarily produce equivalent quantities of work (Biewener 1990).

Taylor et al. (1980) examined size effects and load carried by training animals to run on treadmills with loaded backpacks. Their data indicate that there is a 1 to 1 ratio between the change in oxygen consumed to the change in load carried. The cost to generate a gram of force varies with size of the animal. Since the kinematics did not change, they hypothesize that the cost of generating force rather than the mechanical power output is the major determinant of power input (see also Taylor 1985). Full et al. (1990) examined metabolism and gait characteristics of three insects (beetle, cricket and cockroach). They report large variation in the cost of transport (beetles use only about half the energy used by crickets and roaches). After normalizing for the rate of force production, the cost per stride of these species was comparable to mammalian values, though interspecific variation remained. Further, normalizing for ground contact time also failed to account for the interspecific variation.

Variation in the total amount of muscular force generated for a given ground reaction force may be important determinent of energy expenditure. In mammals, there is about a tenfold decrease in mechanical advantage of the limbs (Biewener 1983, 1989) from horses (1000 kg) to mice (0.05 kg). The smaller mammals run with crouched postures while larger mammals run more upright. The upright postures allows large mammals to align their limbs more closely with the ground reaction force, thereby reducing the force required of the muscles (Biewener 1989). This is perhaps consistent with the ecology of the smaller mammals. Small mammals often do not run for long periods and it may be important for them to accelerate rapidly or jump. Their posture, while more expensive, allows them to do these behaviors instantly (Alexander and Ker 1990).

Few data are available on the cost of carrying loads in flight. Heinrich (1975) reported that bumblebees could carry substantial loads (as much as 100% of their empty mass) and that the increased cost increased linearly with the load carried. Similarly Wolf et al. (1989) report that flight metabolism of honeybees also increases linearly with load carried. Unfortunately, neither study reported any kinematic data, so the comparison with other data as to mechanism is not possible. Female mosquitos substantially increase metabolism during flight after blood feeding to accommodate the additional load. Like the running vertebrates, cycle frequency did not change and the mass-specific cost per cycle changed to accommodate the additional load (Casey and Lance, in prep.). The ratio of delta VO_2 to delta load (loaded – unloaded) was 1 to 1, indicating that the muscle increased work output per cycle while maintaining the "appropriate frequency" to which the muscles were tuned to operate at for a given performance level (Taylor 1985).

8 Morphological Adaptation for Locomotion

It is obvious from looking at the size and shape of animals that they are adapted for locomotor performance. Moreover, for any given size, differences in shape can have important effects on locomotor performance and cost. This is perhaps most clearly demonstrated for flying animals. Sphingid moths have much smaller wings than saturniid moths and they operate these at higher frequencies and higher energy metabolism (Bartholomew and Casey 1978). Comparing gypsy moths with tent caterpillar moths, for animals of a given mass their metabolic rates during hovering differ by a factor of 2.6 ($45\,ml\,O_2\,g^{-1}\,h^{-1}$ in the former and $125\,ml\,O_2\,g^{-1}\,h^{-1}$ in the latter). Morphology is the major determinant of wing-stroke frequency, which to a large extent determines aerodynamic and inertial power requirements and ultimately the energetic cost. This comparison is only a small portion of a much wider continuum and could be extended further to include bumblebees on the one hand, which operate at cycle frequencies of as much as 200 Hz, and the Rhopalocera, which sport huge wings and flap them only a few times per second.

It is perhaps intuitively obvious that this should also be the case in aquatic locomotion. The importance of streamlining in aquatic locomotion is well documented and aquatic animals exhibit well-known special adaptations for fast swimming (Lighthill 1971; see also discussion by Webb 1984). Morphological and physiological adaptations are apparent for fast swimming, but the energetic significance of these adaptations has not been systematically examined. It seems reasonable to suppose that adaptations for speed have general energetic significance. By reducing body drag and increasing speed the cost of transport should be reduced. Trade-offs in morphological adaptation yield less economical performance (Webb 1984). Fishes have very reduced costs of transport compared with animals swimming on top of the water. Semi-aquatic birds and mammals (ducks, minks, muskrats and humans) have costs of transport 10 to 20 times that of a fish of equivalent body mass. Sea otters are more economical, and penguins and seals with their obvious streamlining are considerably lower than other semi-aquatic forms but still with two or three times the costs for swimming fish (Davis et al. 1985).

In terrestrial locomotion this idea should also hold but it has not always been easy to demonstrate. For animals of a given size (Taylor et al. 1982) regardless of the size and limb configuration, the metabolism versus speed curves are quite comparable [but see the recent study by Steudel (1990)]. Full and associates have extended these findings to include invertebrate (insects, crabs, centipedes) and demonstrate essentially the same size dependence of oxygen consumption versus speed as demonstrated for mammals by Taylor et al. (1982). Nevertheless, different animals of the same size which vary in limb configuration also vary in muscle properties, VO_2 max and the range of speed they can manage. There can be wide divergence in the cost of locomotion for a given body size (as much as fourfold), but the biomechanical significance of these energetic differences has not yet been clearly established (Full 1989). Size is clearly a major determinant for the cost of terrestrial locomotion. For pedestrians, the predicted minimum cost of transport is strongly size-dependent, and predictions for a given body size are surprisingly insensitive to the taxon of animals (mammals, reptiles, birds, amphibians, insects and other arthropods).

Kram and Taylor's (1990) findings provide an energetic link for those animals which are obviously built for speed (e.g. antelopes with long legs and cheetahs with flexible spines). These morphological adaptations are thought to increase the stride length for a given frequency. This has previously been associated with an increased work at higher stride lengths. If the cost of running is "priced by the step", the longer legs and flexible backs may reduce the metabolic cost of running at a given speed (Alexander and Ker 1990).

9 Cost of Transport

The cost of transport is the energy cost to an animal to move a given mass over a given distance, obtained by dividing the metabolic cost by the speed of movement. Adapted to animals by Tucker (1970) and Schmidt-Nielsen (1972), this parameter is useful for comparing distance-specific economy across taxa, modes of locomotion, and wide size ranges. In general, larger animals operate at higher speeds than small ones and the cost of transport decreases as size increases for all modes of locomotion. The costs for different modes of locomotion are not equivalent. For any given body size, swimming is cheapest, followed by flight, followed by terrestrial locomotion (Tucker 1970). This has significant implications for migration and competition (Schmidt-Nielsen 1972). Differences in the shape of the curves of metabolism vs. speed yield differences in the minimum cost of transport as well as the predicted cruising speeds. Swimming and flight yield a curvilinear relationship, suggesting a cruising speed at which energetic cost will be minimized (economy is greatest). At lower and higher and higher speeds the cost of transport will increase – thereby reducing economy.

For flying animals there is very little energetic data available for any species over a wide enough speed range of confidently predict cost of transport. Original estimates by Tucker (1970) and Schmidt-Nielsen (1972) had very few species in free flight. Although calculated power requirements of fliers show U-shaped curves with pronounced minima, the energetic costs are often speed-independent. Thus, the cost of transport will decrease with increasing speed up to the point at which parasite power (and possibly profile power – see Dudley and Ellington 1990b) requirements begin to increase sharply with speed. Many of the insects they used in the original analysis were tethered or hovering, further reducing the confidence in allometry of cost of transport in flight. We also need to learn more about the speeds which animals normally use in nature and duplicate those speeds in empirical studies of energetics and biomechanics. Although the cost of transport for animals was formulated two decades ago and represents an important ecological and evolutionary evolution of animal energetics, insufficient work has been done to refine the estimates and evaluate the assumptions used in its derivation.

In most running animals total cost of transport is the sum of three components, the resting or maintenance cost, an offset cost and an incremental cost. A second value (net cost of transport) represents the Active VO_2-Rest VO_2/speed and may represent a more realistic ecological evaluation of the energetics of terrestrial locomotion (Bennett 1985). For understanding biomechanics and energetics (as well as muscle performance), the incremental cost of transport (total cost minus resting and offset cost) would seem to represent a more realistic estimate of the cost of movement. However, the validity of baseline subtraction for all speeds has been questioned (Stainsby et al. 1980). Cost of transport for terrestrial locomotion is analyzed in detail by Full (1991).

10 Conclusions and Future Directions

Although mechanical power requirements and rates of energy expenditure of animals in locomotion are correlated, they vary systematically with size such that the whole animal efficiency increases progressively with size. The energetic cost of locomotion in animals, regardless of the medium, is tied up in the frequency of muscle contraction. Frequency in turn is determined to a great extent by the morphology and size of the animal. Limitations placed by medium results in significantly different power requirements and ranges of speeds possible. Energetic estimates are of interest in themselves in examining ecological and evolutionary adaptations (e.g. Bennett 1985); they are also useful to place biomechanical estimates of cost into biological perspective.

Over the size range muscles are adapted to operate at appropriate frequencies. Differences in temperature and size largely explain differences in mechanical power generation and energy expenditure. As muscles operate at higher frequencies, the energy metabolism increases but the maximum mechanical power output does not. Instead, it appears that there is a trade-off between the frequency of muscle contraction and the work which can be done per contraction.

Why should work output go down with size? Muscle strain appears to be somewhat size-dependent. Across the size range muscle strain may be as great as 25% in large mammals and probably about 15% for flight muscle of cruising birds (Pennycuick and Rezende 1984; Cutts 1986), about 5–6% in small fish (Altringham and Johnston 1990a, b), 6% for sphingid moths and locusts (Stevenson and Josephson 1990; Josephson and Stevenson 1991), and only 1–2% for insect fibrillar muscle (Casey and Ellington 1989). Future studies are needed to systematically examine muscle strain across the size range since it must be a major determinant of work output of the muscle.

We must also address the quantitative significance of elastic storage and recovery of energy as it appears to be important as size changes in terrestrial locomotion and also in insect flight. Without an accurate accounting of the contribution of elastic storage a complete description of the mechanical power requirements and whole animal efficiency cannot be made.

In such a brief survey it is impossible to cover all the energetic studies from the wide variety of taxa which are currently available. Earthworms, octopi, snakes, caterpillars, snails, shrimp and water beetles (to name just a few) exhibit very different patterns of movement and energetics. I have tried to draw parallels between taxa, locomotor modes and metabolic organization but it is obvious that there are a wide range of patterns of locomotion which I could not address.

Acknowledgments. Grant support from NSF # DCB-8802443 and by the New Jersey Agricultural Experiment Station (Project 08337) are gratefully acknowledged. I thank Drs. T. Daniel, C.P. Ellington, R.J. Full, C. Hui, H.B. John-Alder, R.K. Josephson, S. Kayar, C.J. Pennycuick, G. Spedding, R.D. Stevenson, T. Taigen, P. Webb, and T.M. Williams for useful discussions and Drs. T.L. Daniel and R.J. Full, J.M. Gabriel, R.K. Josephson and R.D. Stevenson for access to manuscripts currently in press or unpublished observations. I also thank Professor R. McNeill Alexander for giving me the opportunity to prepare this paper and for his infinite patience.

References

Alexander R McN (1982) Size, shape, and structure for running and flight. In: Taylor CR, Johansen K, A companion to animal physiology. Cambridge Univ Press, pp 309–324

Alexander R McN (1984) Elastic energy stores in running vertebrates. Am Zool. 24: 85–94

Alexander R McN (1988) Elastic mechanisms in animal movement. Cambridge Univ Press, 144 pp

Alexander R McN, Bennet-Clark HC (1977) Storage of elastic strain energy in muscle and other tissues. Nature (Lond) 265: 114–117

Alexander R McN, Ker RF (1990) Running is priced by the step. Nature (Lond) 346: 220–221

Alexander R McN, Veron A (1975) Mechanics of hopping by kangaroos (Macropodidae). J Zool Lond 177: 265–303

Altringham JD, Johnston IA (1990a) Scaling effects on muscle function: power output of isolated fish muscle fibres performing work. J Exp Biol 151: 453–467

Altringham JD, Johnston IA (1990b) Modelling muscle power output in a swimming fish. J Exp Biol 148: 395–402

Astrand PO, Rodahl K (1977) Textbook of work physiology: physiological bases of exercise, 2nd edn. McGraw-Hill, New York

Bainbridge R (1958) The speed of swimming of fish as related to size and to the frequency and amplitude of the tail beat. J Exp Biol 35: 109–133

Bartholomew GA, Casey TM (1978) Oxygen consumption of moths during rest, preflight warm-up, and flight in relation to body size and wing morphology. J Exp Biol 76: 11–25

Bennett AF (1984) Thermal dependence of muscle function. Am J Physiol (Regulatory Integrative Comp Physiol) 16: R217–R229

Bennett AF (1985) Energetics and locomotion. In: Hildebrand U, Bramble DM, Liem KF, Wake DB (eds). Functional vertebrate morphology. Belknap, Harvard Univ Press, Cambridge, pp 173–184

Bennett AF, Ruben JA (1979) Endothermy and activity in vertebrates. Science 206: 649–654

Bernstein MH, Thomas SP, Schmidt-Nielsen (1973) Power input during flight in the fish crow, *Corvus ossifragus*. J Exp Biol 58: 401–410

Berger M (1985) Sauerstoffsverbrauch von Kolibris (*Colibri coruscans* and *C. thalassinus*) beim Horizontalflug. In: Nachtigall W (ed) BIONA Report 3. Acad Wiss, Mainz, Fisher, pp 307–314

Biewener AA (1983) Allometry of quadrupedal locomotion: the scaling of duty factor, bone curvature and limb orientation to body size. J Exp Biol 105: 147–171

Biewener AA (1989) Scaling body support in mammals: limb posture and muscle mechanics. Science 245: 45–48

Biewener AA (1990) Biomechanics of mammalian terrestrial locomotion. Science 250: 1097–1103

Biewener JD, Alexander R McN, Heglund NC (1981) Elastic storage in the hopping kangaroo rat (*Dipodomys spectabilis*). J Zool Lond 195: 369–383

Blake RW (1991) Efficiency, economy and related concepts in comparative animal physiology. Cambridge Univ Press (in press)

Blickhan R, Full RJ (1987) Locomotion energetics of the ghost crab: II. Mechanics of the center of mass during walking and running. J Exp Biol 130: 155–174

Carpenter R (1975) Flight metabolism of flying foxes. In: Wu TY, Brokaw CJ, Brennen C (eds) Swimming and flying in nature, vol 2. Plenum Press, New York, pp 883–889

Casey TM (1980) Flight energetics and heat exchange of gypsy moths in relation to air temperature. J Exp Biol 88: 133–145

Casey TM (1981a) Comparison of mechanical and energetic estimates of flight cost for hovering sphinx moths. J Exp Biol 91: 117–129

Casey TM (1981b) Energetics and thermoregulation of *Melacosoma americanum* (Lepidoptera: Lasiocampidae) during hovering flight. Physiol Zool 54: 362–371

Casey TM (1989) Oxygen consumption during flight. In: Goldsworthy G, Wheeler C (eds) CRC Uniscience series: Insect flight. CRC Press, Boca Raton, pp 257–272

Casey TM, Ellington CP (1989) Energetics of insect flight. In: Wieser W, Gnaiger E (eds) Energy transformation in cells and animals. Thieme, Stuttgart, pp 200–210

Casey TM, Hegel-Little JR (1987) Instantaneous oxygen consumption and muscle energetics in *Malacosoma americanum* during pre-flight warm-up. J Exp Biol 127: 389–400

Casey TM, May ML, Morgan KR (1985) Flight energetics of Euglossine bees in relation to morphology and wing stroke frequency. J Exp Biol 116: 271–289

Cavagna GA, Saibene FP, Margaria R (1964) Mechanical work in running. J Appl Physiol 19(2): 249–256

Cavagna GA, Heglund NC, Taylor CR (1977) Mechanical work in terrestrial locomotion: two basic mechanisms for minimizing energy expenditure. Am J Physiol 233(5): R243–R261

Cutts A (1986) Sarcomere length changes in the wing muscles during the wing beat cycle of two bird species. J Zool Lond 209: 183–185

Daniel TL (1991) Efficiency in aquatic locomotion: limitations from single cells to animals. In: Blake RW (ed) Efficiency, economy and related concepts in comparative animal physiology. Cambridge Univ Press (in press)

Daniel TL, Webb PW (1987) Physical determinants of locomotion. In: Dejours P, Bolis L, Taylor CR, Weibel ER (eds) Comparative physiology: life in water and on land. Liviana Press, Padova, pp 343–369

Davis RW, Williams TM, Kooyman GL (1985) Swimming metabolism of yearling and adult harbor seals. Physiol Zool 58(5):: 590–596

Dawson TJ (1976) Energetic cost of locomotion in hopping mice. Nature (Lond) 259: 305–306

Dawson TJ, Taylor CR (1973) Energy cost of locomotion in kangaroos. Nature (Lond) 246: 313–314

Dudley R, Ellington CP (1990a) Mechanics of forward flight in bumblebees I. Kinematics and morphology. J Exp Biol 148: 19–52

Dudley R, Ellington CP (1990a) Mechanics of forward flight in bumblebees II. Quasi-steady lift and power requirements. J Exp Biol 148: 53–88

Elllington CP (1977) The aerodynamics of normal hovering flight: three approaches. In: Schmidt-Neilsen K, Bolis L, Maddrell SHP (eds) Comparative physiology – water, ions and fluid mechanics. Cambridge Univ Press, pp 327–345

Ellington CP (1984) The aerodynamics of hovering insect flight. VI. Lift and power requirements. Phil Trans R Soc Lond B 305: 145–181

Ellington CP (1985) Power and efficiency of insect flight muscle. J Exp Biol 115: 293–304

Ellington CP, Machin KE, Casey TM (1990) Oxygen consumption of bumblebees in forward flight. Nature (Lond) 347: 472–473

Else PL, Hulbert AJ (1985) An allometric comparison of the mitochondria of mammalian and reptilian tissues: the implications for the evolution of endothermy. J Comp Physiol B 156: 3–11

Fish FE (1982) Aerobic energetics of surface swimming in the muskrat *Ondatra zibethicus*. Physiol Zool 55: 180–189

Full RJ (1986) Locomotion without lungs: energetics and performance of a lungless salamander, *Plethodon jordani*. Am J Physiol 251: R775–R780

Full RJ (1987) Locomotion energetics of the ghost crab: 1. Metabolic cost and endurance. J Exp Biol 130: 137–154

Full RJ (1989) Mechanics and energetics of terrestrial locomotion: bipeds to polypeds. In:Wieser W, Gnaiger E (eds) Energy transformation in cells and animals. Thieme, Stuttgart, pp 175–182

Full RJ (1991) The concepts of efficiency and economy in land locomotion. In: Blake RW (ed) Efficiency, economy and related concepts in comparative animal physiology. Cambridge Univ Press (in press)

Full RJ, Herreid CF (1983) The aerobic response to exercise of the fastest land crab. Am J Physiol 244: R530–R536
Full RJ, Herreid CF (1984) Fiddler crab exercise: the energetic cost of running sideways. J Exp Biol 109: 141–161
Full RJ, Tu MS (1990) Mechanics of six-legged runners. J Exp Biol 148: 129–146
Full RJ, Anderson BD, Finnerty CM, Feder ME (1988) Exercising with and without lungs: 1. The effects of metabolic cost, maximal oxygen transport and body size on terrestrial locomotion in salamander species. J Exp Biol 138: 471–485
Full RJ, Zuccarello DA, Tullis A (1990) Effect of variation in form on the cost of terrestrial locomotion. J Exp Biol 150: 233–246
Gleeson TT (1979) Foraging and transport cost in the Galapagos marine iguana, *Amblyrhynchus cristatus*. Physiol Zool 52: 549–557
Gold A (1973) Energy cost of animal locomotion. Science 183: 275–276
Goldspink G (1977) Energy cost of locomotion. In: Alexander R McN, Goldspink G (eds) Chapman and Hall, London, pp 153–167
Goller F, Esche H (1990) Comparative study of chill coma temperatures and muscle potentials in insect flight muscles. J Exp Biol 150: 221–231
Graham JB, Dewar H, Lai NC, Lowell WR, Arce SM (1990) Aspects of shark swimming performance determined using a large water tunnel. J Exp Biol 151: 175–192
Heglund NC, Taylor CR (1988) Speed, stride frequency and energy cost per strise: how do they change with body size and gait? J Exp Biol 138: 301–318
Heglund NC, Taylor CR, McMahon TA (1974) Scaling stride frequency and gait to animal size: mice to horses. Science 186: 1112–1113
Heglund NC, Fedak MA, Taylor CR, Cavagna GA (1982) Energetics and mechanics of terrestrial locomotion. IV. Total mechanical energy changes as a function of speed and body size in birds and mammals. J Exp Biol 97: 57–66
Heinrich B (1975) Thermoregulation in bumblebees. II. Energetics of warm-up and free flight. J Comp Physiol B96: 155–166
Heinrich B, Bartholomew GA (1971) An analysis of pre-flight warm-up in the sphinx moth, *Manduca sexta*. J Exp.Biol 55: 223–239
Herreid CF, Full RJ, Prawel DA (1981) Energetics of cockroach locomotion. J Exp Biol 94: 189–202
Hill AV (1950) The dimensions of animals and their muscular dynamics. Sci Prog 38: 209–230
Hoppeler H, Kayar SR, Claassen H, Uhlmann E, Karas RH (1987) Adaptive variation in the mammalian respiratory system in relation to energetic demand. III. Skeletal muscles: setting the demand for oxygen. Respir Physiol 69: 27–46
Hoyt DF, Kenagy GJ (1988) Energy costs of walking and running gaits and their aerobic limits in golden-mantled ground squirrels. Physiol Zool 61: 34–40
Hoyt DR, Taylor CR (1981) Gait and the energetics of locomotion in horses. Nature (Lond) 292: 239–240
Hudson D, Bernstein MH (1983) Gas exchange and energy cost of flight in the white necked raven, *Corvus cryptoleucus*. J Exp Biol 103: 121–130
Hui CA (1988a) Penguin swimming I. Hydrodynamics. Physiol Zool 61: 333–343
Hui Ca (1988b) Penguin swimming II. Energetics and behavior. Physiol Zool 61: 344–350
Jayne BC, Bennett AF, Lauder GV (1990) Muscle recruitment during terrestrial locomotion: how speed and temperature affect fiber type use in a lizand. J Exp Biol 152: 101–128
John-Alder HB, Bennett AF (1981) Thermal dependence of endurance and locomotory energetics in a lizard. Am J Physiol 241: R342–R349
John-Alder HB, Lowe CH, Bennett AF (1982) Thermal dependence of locomotory energetics and aerobic capacity of the Gila Monster (*Heloderma suspectum*). J Comp Physiol 151: 119–126
Johnston IA (1985) Sustained force development: specializations and variation among the vertebrates. J Exp Biol 115: 239–252
Johnston IA, Davison W, Goldspink G (1977) Energy metabolism of carp swimming muscle. J Comp Physiol 114: 203–216
Josephson RK (1981) Temperature and mechanical performance of insect muscle. In: Heinrich B (ed) Insect thermoregulation. Wiley, New York, pp 9–44
Josephson RK (1985a) Mechanical power output from straited muscle during cyclic contraction. J Exp Biol 114: 493–512

Josephson RK (1985b) The mechanical power output of a tettigoniid wing muscle during singing and flight. J Exp Biol 117: 357–368
Josephson RK, Stevenson RD (1991) Mechanical efficiency of an flight muscle of *Schistocerca gregaria*. J Physiol Lond (in press)
Josephson RK, Young D (1981) Synchronous and asynchronous muscles in cicadas. J Exp Biol 91: 219–237
Kayar SR, Hoppeler H, Mermod L, Weibel ER (1988) Mitochondrial size and shape in equine skeletal muscle: a three dimensional reconstruction study. Anat Rec 222: 333–339
Kayar SR, Hoppeler H, Lindstedt SL, Claassen H, Jones JH, Essen-Gustavsson B, Taylor CR (1989) Total muscle mitochondrial volume in relation to aerobic capacity of horses and steers. Pflügers Arch 413: 343–347
Kaufmann R (1990) Respiratory cost of swimming in larval and juvenile cyprinids. J Exp Biol 150: 343–366
Kram R, Taylor CR (1990) Energetics of running: a new perspective. Nature (Lond) 346: 265–267
Lasiewski RC, Galey FR, Vasquez C (1965) Morphology and physiology of the pectoral muscles of hummingbirds. Nature (Lond) 205: 404–405
Lighthill J (1971) Large amplitude elongated body theory of fish locomotion. Proc R Soc Lond B179: 125–138
Lighton JRB (1985) Minimum cost of transport and ventilatory patterns in three African beetles. Physiol Zool 58: 390–399
Machin KE, Pringle JWS (1959) The physiology of insect fibrillar muscle II. Mechanical properties of a beetle flight muscle. Proc R Soc Lond B151: 204–225
Marsh RL, Taigen TL (1987) Properties enhancing aerobic capacity of calling muscles in gray gree frogs *Hyla versicolor*. Am J Physiol 252 (Regulatory Integrative Comp Physiol 21): R786–R793
Pennycuick CJ (1969) The mechanics of bird migration. Ibis 111: 525–556
Pennycuick CJ (1975) On the running of the gnu (*Connochaetes taurinus*) and other animals. J Exp Biol 63: 775–779
Pennycuick CJ, Rezende MA (1984) The specific power output of aerobic muscle, related to the power density of mitochondria. J Exp Biol 108: 377–392
Prange HD (1976) Energetics of swimming in a sea turtle. J Exp Biol 64: 1–12
Prange HD, Schmidt-Nielsen K (1970) The metabolic cost of swimming in ducks. J Exp Biol 53: 763–777
Rall JA (1986) Energetic aspects of skeletal muscle contraction: implication of fiber types. In: Terjung RL (ed) Exer Sports Sci Rev 13 Macmillan, New York, pp 33–74
Rayner JVM (1979) A new approach to animal flight mechanics. J Exp Biol 80: 17–54
Rome LC, Loughna PT, Goldspink G (1984) Muscle fiber activity in carp as a function of swimming speed and muscle temperature. Arm J Physiol (Reg Int Comp Physiol 16): R272–R279
Rothe HJ, Bissel W, Nachtigall W (1987) Pigeon flight in a wind tunnel II. Gas exchange and power requirements. J Comp Physiol B157: 99–109
Schmidt-Nielsen K (1972) Locomotion: energetic cost of swimming, flying and running. Science 177: 222–228
Schwerzmann K, Hoppeler H, Kayar SR, Weibel ER (1989) Oxidative capacity of muscles and mitochondria: correlations of physiological, biochemical, and morphometric characteristics. Proc Natl Acad Sci 86: 1583–1587
Sotavalta O (1954) The essential factor regulating the wingstroke of insects in wing mutilation and loading experiments and in experiments at subatmospheric pressure. Ann Zool Soc Vanamo 15: 1–67
Sotavalta O, Laulajainen E (1961) On the sugar consumption of the drone fly (*Eristalis tenax* L.) in flight experiments. Ann Acad Sci Fenn Ser A4 53: 1
Stainsby WN, Gladden LB, Barclay JK, Wilson BA (1980) Exercise efficiency; validity of base-line subtractions. J Appl Physiol 48: 518–522
Steudel K (1990) The work and energetic cost of locomotion I. The effects of limb mass distribution in quadrupeds. J Exp Biol 154: 273–285
Stevens ED (1988) Effect of pH and stimulation phase on work done by isolated frog muscle during cyclic contraction. J Muscle Res Cell Motil 9: 329–333
Stevens ED, Josephson RK (1977) Metabolic rate and body temperature of singing katydids. Physiol Zool 50: 31–42

Stevenson RD, Josephson RK (1990) Effects of operating frequency and temperature on mechanical power output from moth muscle. J Exp Biol 149: 61–78
Stokes DR, Josephson RK (1988) The mechanical power output of a crab respiratory muscle. J Exp Biol 140: 287–299
Taylor CR (1985) Force development during sustained locomotion: a determinant of gait, speed and metabolic power. J Exp Biol 115: 253–262
Taylor CR, Schmidt-Nielsen K, Raab JL (1970) Scaling of energetic cost to body size in mammals. Am J Physiol 210: 1104–1107
Taylor CR, Shkolnick A, D'meil R, Baharav D, Bonut A (1974) Running in cheetahs, gazelles, and goats: energy cost and limb configuration. Am J Physiol 227: 848–850
Taylor CR, Heglund NC, McMahon TA, Lonney TR (1980) Energetic cost of generating muscular force during running: a comparison of large and small animals. J Exp Biol 86: 9–18
Taylor CR, Heglund NC, Maloiy GMO (1982) Energetics and mechanics of terrestrial locomotion. 1. Metabolic energy consumption as a function of speed and body size in birds and mammals. J Exp Biol 97: 1–21
Taylor CR, Karas RH, Weibel ER, Hoppeler H (1987) Adaptive variation in the mammalian respiratory system in relation to energetic demand. Respir Physiol 69: 1–127
Thomas SP (1975) Metabolism during flight in two species of bats, *Phyllostomus hastatus* and Pteropus gouldii. J Exp Biol 63: 273–293
Torre-Bueno JR, LaRochelle J (1978) The metabolic cost of flight in unrestrained birds. J Exp Biol 75: 223–229
Tucker VA (1968) Respiratory exchange and evaporative water loss in the flying budgerigar. 846. J Exp Biol 48: 67–87
Tucker VA (1970) Energetic cost of locomotion in animals. Comp Biochem Physiol 34: 841–846
Tucker VA (1972) Metabolism during flight in the laughing gull. Am J Physiol 222: 237–245
Tucker VA (1973) Bird metabolism during flight: evaluation of a theory. J Exp Biol 58: 689–709
Videler JJ (1989) Energetic consequences of the interactions between animals and water. In: Wieser W, Gnaiger E (eds) Energy transformation in cells and animals. Thieme, Stuttgart, pp 175–182
Vleck D, Gleeson TT, Bartholomew GA (1981) Oxygen consumption during swimming in Galapagos marine iguanas and its ecological correlates. J Comp Physiol 141: 531–536
Webb PW (1971a) The swimming energetics of trout. I: Thrust and power output at cruising speeds. J Exp Biol 55: 489–520
Webb PW, (1971a) The swimming energetics of trout. II. Oxygen consumption and swimming efficiency. J Exp Biol 55: 521–540
Webb PW (1984) Body form, locomotion and foraging in aquatic vertebrates. Am Zool 24: 107–120
Webb PA, Kostecki PT, Stevens ED (1984) The effect of size on locomotor kinematics of rainbow trout. J Exp Biol 109: 77–95
Weibel ER, Taylor CR (1981) Design of the mammalian respiratory system. Respir Physiol 44: 1–164
Weis-Fogh T (1973) Quick estimates of flight fitness of hovering animals including novel mechanisms for lift production. J Exp Biol 59: 169–230
Weis-Fogh T, Alexander R McN (1977) The sustained power output from striated muscle. In: Pedley TJ (ed) Scale effects in animal locomotion. Academic Press, London, pp 511–525
Williams TM (1983a) Locomotions of the North American mink, a semi-aquatic mammal. Swimming energetics and body drag. J Exp Biol 103: 155–168
Williams TM (1983b) Locomotion of the North American mink, a semi-aquatic mammal. 11. The effect of an elongate body on running energetics. J Exp Biol 48: 153–161
Williams TM, Kooyman GL (1985) Swimming performance and hydrodynamic characteristics of Harbor seals *Phoca vitulina*. Physiol Zool 58(5): 576–589
Woledge RC (1989) Energy transformation in living muscle. In: Wieser W, Gnaiger E, Energy transformation in cells and animals. Thieme, Stuttgart, pp 36–45
Wolf TJ, Schmidt-Hempel P, Ellington CP, Stevenson RD (1989) Physiological correlates of foraging efforts in honeybees: oxygen consumption and nectar load. Funct Ecol 3: 417–424

Chapter 9

Optimization of Locomotion

A. Houston

Contents

1 Introduction

This chapter differs from the others in this book in that it is concerned with the consequences of locomotion for an animal's Darwinian fitness. An animal's fitness depends on its ability to survive and reproduce. By moving to a new habitat, an animal may obtain benefits, such as an increased chance of finding food or a mate. It may also incur costs, such as increased energy expenditure or danger of being killed by a predator. We expect evolution to favour the move if the benefits outweight the costs. While this idea is simple to state, it is usually difficult to

NERC Unit of Behavioural Ecology, Department of Zoology, Oxford University, South Parks Road, Oxford OX1 3PS, UK measure the various costs and benefits in the field and to relate them to fitness. Part of the difficulty lies in the fact that the fitness consequences of an action depend on the animal's subsequent behaviour (McNamara and Houston 1986). Locomotion provides a clear example of this point. We cannot determine the best way of travelling unless we know what the animal is going to do when it arrives. Thus in the context of travel speed there can be various different optimal speeds, depending on context in which travel occurs (e.g. Pyke 1981a).

One way to approach the problem of assessing the fitness consequences of an animal's behaviour is to use some sort of currency as a measure of fitness. Any such currency should be closely related to fitness but easier to measure. For animals that are not reproducing, currencies based on energy spent and energy gained are often used. This can be illustrated in the context of an animal moving between patches of food.

2 When to Leave a Patch: the Marginal Value Theorem

Imagine an environment that contains well-defined patches of food. An average time τ is spent in travelling between patches. By eating in a patch, an animal depletes the available food. As a result its intake rate decreases. When should it leave the current patch to search for a new one? To answer this problem we need to specify how fitness depends on energetic gain. One possibility is that fitness is maximized by maximizing the long-term rate of energetic gain. Given that the time to travel between patches can vary and that the energy obtained from a patch might be a random variable, it is not obvious what the long-term rate might be. A basic result from renewal theory is useful in this context. We consider a renewal point, at which the system is always in the same state. In this case, leaving a patch is such a point. The period between two renewal points is called a cycle. It can be shown (e.g. Johns and Miller 1963) that the long-term rate R(t) is given by the expected reward obtained on a cycle divided by the expected duration of a cycle. To simplify the problem, assume that all patches are the same, that the time required for travel between patches is τ and that the cumulative gain in energy is a deterministic function G(t) of the time t spent on a patch. Then

$$R(t) = G(t)/(t + \tau). \tag{1}$$

The best time to leave a patch is time t^* which results in the highest possible value of the rate of energetic gain. This time can be found by the graphical approach shown in Fig. 1. The cumulative gain $G(t)$ is plotted against t, and the time axis is extended to the left of $t = 0$ by an amount τ. The tangent from τ to the curve $G(t)$ touches the curve at $G(t^*)$. To see why this is so, consider line i from τ to the curve $G(t)$ in Fig. 1. This line corresponds to staying in the patch for a time t_i. It can be seen that the slope of this line is $G(t_i)/(t_i + \tau)$ which is the rate that results from leaving the patch after time t_i. This argument is general – the slope of any line is the rate achieved by the corresponding time spent in the patch. Thus the

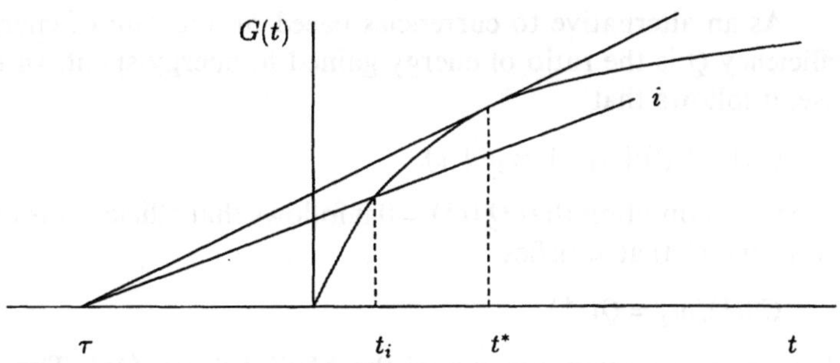

Fig. 1. Graphical illustration of the marginal value theorem. The time in patch t is plotted to the *right* of the y axis. The travel time τ is plotted to the *left*. The cumulative gain as a function of time in patch is the curve $G(t)$. The *slope* of line i is the rate that results for staying in the patch for time t_i. The highest possible slope results when a line from τ is tangent to the gain curve. The resulting optimal time in the patch is t^*

maximum possible rate is achieved by the line from τ that is tangent to the curve $G(t)$.

So far, I have only considered the energy gained by foraging; this means that $R(t)$ is the gross rate of energetic gain. Only a minor modification is required to give us the net rate. Let m_1 be the metabolic rate while travelling and let m_2 be the metabolic rate while in the patch. We can also introduce an energetic cost c incurred on each journey regardless of its duration. This cost seems a reasonable component of a model of flight, where it represents the energy spent in reaching the velocity used for the journey (Houston 1986). The net rate of energetic gain $N(t)$ is given by the equation

$$N(t) = \frac{G(t) - m_1\tau - m_2 t - c}{t + \tau}. \tag{2}$$

Although it is possible to solve this equation graphically (e.g. Kacelnik and Houston 1984), I will use an analytic approach. The time t^* that maximizes $N(t)$ satisfies the equation

$$dN(t^*)/dt = 0. \tag{3}$$

From the rule for differentiating a quotient, it follows that t^* can be found from the equation

$$(t^* + \tau)[G'(t^*) - m_2] = G(t^*) - m_1\tau - m_2 t^* - c, \tag{3}$$

which can be written as

$$G'(t^*) - m_2 = N(t^*). \tag{4}$$

This equation is the marginal value theorem (MVT) (Charnov 1976a). The left-hand side of this equation is the marginal net rate of gain at t^*, and the right-hand side is the long-term net rate of gain if the optimal time in a patch t^* is used. It follows that the MVT requires an animal to leave a patch when the marginal rate falls to the overall rate under the optimal policy.

As an alternative to currencies based on the rate of energetic gain, energetic efficiency Q is the ratio of energy gained to energy spent. In the context of patch use, it follows that

$$Q(t) = G(t)/(m_1\tau + m_2 t + c). \tag{5}$$

From the condition that $Q'(t^*) = 0$ it follows that efficiency is maximized by leaving at a time t^* that satisfies

$$G'(t^*)/m_2 = Q(t^*). \tag{6}$$

This is the efficiency version of the MVT [cf. Eq. (4)]. The left-hand side is the marginal efficiency, so it is optimal to leave when this quantity falls to the efficiency under the optimal policy.

In the context of a parent bird feeding its young, Norberg (1981a) suggested a currency that I will refer to as the overall rate, Y. This rate takes account of the time that the parent must spend in replacing the energy that it uses. If the parent can gain energy at a rate γ, then the time taken to replace the energy used is $t_R = (m_1\tau + m_2 t + c)/\gamma$. The overall rate is then given by

$$Y(t) = \frac{G(t)}{t + \tau + t_R}. \tag{7}$$

Data from parent birds feeding their young (e.g. Carlson and Moreno 1982; Kacelnik 1984) are not predicted by the maximization of efficiency but are consistent with at least one form of rate – see Houston (1987) for further discussion. Welham and Ydenberg (1988) observed ring-billed gulls (*Larus delawarensis*) foraging behind tractors. As a tractor moves round a field, gulls land behind, forage, then fly up to land behind the tractor again. The observed time foraging was between the time predicted by maximizing net rate and maximizing efficiency.

3 Optimal Travel Speed in Two Cases

3.1 Foraging While Travelling

Some animals gather food as they travel. Ware (1975) developed a model for such cases that has two main components. By increasing its velocity, the animal increases the rate at which it consumes items of food, but it also increases the rate at which it uses energy. To model the effect of velocity on feeding rate, Ware used Holling's disc equation (e.g. Holling 1966). The equation can be derived as follows: assume that the encounter rate with food items is λ, so that the mean time between encounters is $1/\lambda$. Items have a mean energy content e and require a mean time h to "handle". It follows from the result of Johns and Miller (1963) that the mean rate of gain R is given by the equation

$$R = \frac{e}{1/\lambda + h}. \tag{8}$$

To link this equation to the velocity of the animal, Ware assumed that $\lambda = kv$, where v is the animal's velocity and k is a positive constant that depends on the density of prey items and the area that the animal can search. The metabolic rate m was assumed to have the following dependence on velocity:

$$m = m_0 + \alpha v^{\beta}, \tag{9}$$

where α and β are empirically determined constants. Combining these equations we have

$$N(v) = ykve/(1 + kvh) - (m_0 - \alpha v^{\beta}), \tag{10}$$

where y is a positive constant that depends on assimilation efficiency. Ware was concerned with growth in fish, and used this equation to find the velocity that maximizes growth. This equation expresses the net rate of energetic gain as a function of velocity, so I have used "N" for the left-hand side, rather that "G" as used by Ware.

Ware (1975) pointed out that there were various currencies that might be appropriate for fish. In addition to N, as defined by Eq. (10), he also considered growth efficiency and growth efficiency per unit unit ration. When data from bleak (Ivlev 1960) were used to estimate the parameters of the equations, N was clearly the best currency of the three, with $v^* = 3.08 \times 10^{-2}\,\mathrm{m\,s^{-1}}$ compared to an observed velocity of $2.97 \times 10^{-2}\,\mathrm{m\,s^{-1}}$. (The optimum velocities for efficiency and efficiency per unit ration are $1.19 \times 10^{-2}\,\mathrm{m\,s^{-1}}$ and $1.11 \times 10^{-2}\,\mathrm{m\,s^{-1}}$ respectively). Pyke (1981a) gives a general analysis of this sort of model. If an animal's fitness increases with the amount of energy obtained, but decreases with the time spent foraging (because this time prevents the animal from doing other things or exposes the animal to predators), then the animal's speed should maximize its net rate of energetic gain. Pyke also shows that if there is a speed at which the rate of energy expenditure is minimized, then it will be optimal to travel at a faster speed.

For a general discussion of body size and velocity of aquatic animals see Gerristen 1984. The flight speed of the black skimmer (*Rynchops nigra*) is discussed by Blake (1985). Searching for cryptic prey is discussed by Gendron and Staddon (1983, 1984). For changes in searching behaviour over the breeding cycle see Robinson (1986).

3.2 No Foraging While Travelling

As a simple example, consider an animal that travels a distance D to a foraging area, where it obtains a net amount of energy e in a time h. If it travels with speed v, using energy at rate $P(v)$, then its net rate of gain $N(v)$ is given by the equation

$$N(v) = \frac{e - P(v)D/v}{h + D/v}.$$

From the condition $dN(v)/dv = 0$ we find that

$$v^* dP(v^*)/dv - P(v^*) = N(v^*). \tag{11}$$

This equation is a special case of Eq. (10) of Pyke (1981a) and Eq. (12) of Houston (1986).

4 Different Modes of Locomotion

Instead of choosing a best velocity from a range of possible velocities, it may sometimes be more relevant to consider the consequences of adopting qualitatively different modes of locomotion, for example walking as opposed to flying. In an exploration of the time and energy budget of a foraging animal, Norberg (1977) pointed out that it can be advantageous for the animal to switch between modes of foraging as a function of prey density. Norberg developed his argument on the basis of minimizing the foraging time required if the animal is to balance its energy budget. We can capture the essence of the argument by working directly with net rate of gain, N. We have

$$N = a\lambda_0 e - m,$$

where a is a measure of the predator's ability to capture items, λ_0 is the density of items, e is the net energetic value of capturing an item (energy gained minus the cost of capture) and m is the metabolic rate while searching (NB there is no handling time in this model).

Now consider two search modes, indicated by subscripts 1 and 2 on a and m. Following Norberg, we assume that method 2 is more efficient (i.e. $a_2 > a_1$) but is also the method with the higher metabolic rate (i.e. $m_2 > m_1$). We can find a critical value $\hat{\lambda}_0$ for a switch between search methods by equating the resulting net rates of gain:

$$a_2\hat{\lambda}_0 e - m_2 = a_1\hat{\lambda}_0 e - m_1.$$

From this equation we find that

$$\hat{\lambda}_0 = (m_2 - m_1)/e(a_2 - a_1),$$

which is equivalent to Eq. (19) of Norberg (1977). (a_i is equivalent to $1/K_i$ and λ_0 is equivalent to N). If λ_0 is below $\hat{\lambda}_0$, then the animal should use foraging method 1, whereas if λ_0 is above, $\hat{\lambda}_0$ then the animal should use foraging method 2.

Masman et al. (1988) discuss Norberg's model in their analysis of the allocation of time to various activities by the kestrel (*Falco tinnunculus*). The kestrel has two foraging methods, flight-hunting and perch-hunting. The respective metabolic rates are 92 W and 10.8 W. Masman et al. estimate that during winter, flight-hunting provides a gain of about 220 W, whereas perch-hunting provides only 22 W. Thus the more expensive method results in the higher net rate, and should be adopted. Masmam et al. reach the same conclusion by a graphical analysis (see Fig. 2). In fact, the kestrels use a mixture of both methods.

Pyke (1981b) noted that although hummingbirds (Trochilidae) and honey-eaters (Meliphagidae) both feed on nectar, hummingbirds tend to hover in front of the flowers whereas honeyeaters tend to pech on a suitable stem. Pyke

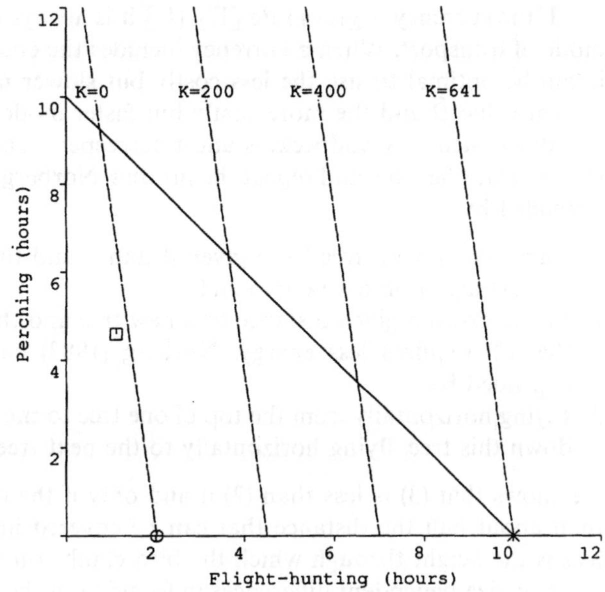

Fig. 2. Energetic consequences of a kestrel's allocation of time between two methods of hunting. The *solid line* indicates the boundary of possible time allocations in a winter's day of 10 h. Each *broken line* corresponds to a given level K of net gain (in K_J) over a day. The gain increases *from left to right*. The line at the *extreme left* is the zero gain case; the *circle at the bottom of this line* is the allocation that minimizes the time to get enough energy to survive the day. It involves no time spent perching. This time-minimizing solution has the same form as the allocation that maximizes gain, which is indicated by the *star*. The observed allocation is indicated by the *square*. (After Masmann et al. 1988)

suggests that this difference may be a result of the time and energy consequences of mode of foraging. A bird that hovers can move between flowers in a shorter time than a bird that perches, but it will use energy at a higher rate. The energetic cost of hovering will be relatively more important as body size increases (see Chap. 3) Honeyeaters tend to be larger than hummingbirds, so this argument suggests that it will be more economical for them to perch than to hover. Pyke checked this idea using data from honeyeaters foraging on *Lambertia formosa.* Assuming that a hypothetical hovering honeyeater would visit the same number of flowers per inflorescence as a perching honeyeater is observed to visit, and would extract the same amount of nectar, Pyke found that hovering gave an advantage in terms of net rate of nearly 2 W. For further discussion of perching versus hovering, see Hainsworth (1991).

Houston (1990) analyses the choice between walking and flying, with special reference to parent birds feeding their young. Norton-Griffiths (1969) found that oystercatchers (*Haematopus ostralegus*) walk when the distance from the young to the feeding area is less than about 50 m, but fly when it is greater. Houston shows that several currencies result in a prediction of such a critical distance.

Two cases are considered: (a) the parent's gain while collecting food is an increasing but decelerating function (as in the MVT), (b) the parent colllects a fixed amount of food.

In case (a) the animal can choose not only its mode of travel but the time that it spends collecting food, and hence the amount of food that it brings back to its young. The analysis is simplified by the fact that the payoff under the optimal time in patch t_W^* if the bird walks is greater than the payoff under the optimal time in patch t_F^* if the bird files if and only if $t_W^* < t_F^*$.

If the currency is gross rate [Eq. (1)] it is always optimal to adopt the quicker mode of transport. When a currency includes the energetic cost of travelling, then it can be optimal to use the less costly but slower mode when D is below some critical value $\tilde{D}$ and the more costly but faster mode when D is above $\tilde{D}$.

Birds such as woodpeckers and treecreepers search a tree for food and then travel to another tree and repeat the process. Norberg (1981b) compares the energy expended by

1. Climbing down a tree for a given distance and then flying up by this amount to start again on a new tree and
2. Flying down a given distance to a new tree and then climbing up it. He shows that (2) requires less energy. Norberg (1983) compares (2) with the energy expended by
3. Flying horizontally from the top of one tree to the top of the next tree, foraging down this tree, flying horizontally to the next tree, and foraging up it.

He shows that (3) is less than (2) if and only if the distance between trees is less than about half the distance that can be covered in a glide in which the height loss is the height through which the bird climbs on a tree.

For size-dependent differences in foraging in the marine iguana, see Trillmich and Trillmich (1986).

5 Moving Versus Staying

Janetos (1982) develops a simple model for the rate of gain R_s of an animal that does not move ("sit-and-wait forager") and the rate of gain R_m of an animal that moves under certain conditions. The site that an animal is in is either good or bad. At the end of each day there is a probability a of a good site turning bad and a probability b of a bad site turning good. If the rate of gain in a good site is G and the rate of gain in a bad site is B then

$$R_s = pG + qB,$$

where p is the equilibrium probability that a site is in the good state $[= b/(a+b)]$ and $q = 1 - p$ is the equilibrium probability that the site is in the bad state.

The active forager stays in a site if it is good but moves to a new site if its site becomes bad. If C is thecost of moving then

$$R_m = p'G + q'(B - C),$$

where

$$p' = p/(a + p)$$

and

$$q' = 1 - p'.$$

It is better to adopt the active behaviour rather to remain in a site if $R_m > R_s$, i.e. if

$$(G - B)(p' - p) > q'c.$$

Hanski (1985) discusses moving versus staying in the context of a shrew trying to survive in a stochastic environment. Once again the environment can be in one of two states, good or bad. Food is readily available when the environment is good but is scarce when it is bad. The shrew's problem is to survive through a bad period that starts at time $t = 0$. Time is discrete (say in hours) and there is a constant probability p of the state being good by $t + 1$, given that it is bad at t. The shrew can either rest or be active. If it rests, it dies if the bad period does not end before T, where T is the number of time units for which the shrew's energy reserves will keep it alive. This occurs with probability $(1 - p)^T$, so that the shrew's probability of survival if it rests is $1 - (1 - p)^T$. Hanski assumes that if the shrew is active then there is probability q of death in each time unit until the transition from bad to good. This leads to a three-state Markov chain with two absorbing states, dead and alive in a good environment. The probability of absorbing in this state is $p/[p + (1 - p)q]$, and this is the shrew's probability of survival if it is active. Thus the shrew should rest if

$$1 - (1 - p)^T > p/[p + (1 - p)q].$$

Across species of shrews, T will increase as body size increases. When T is about 4 h it is optimal to be active for most values of $0 < p < 0.5$ and $0 < q < 0.1$, whereas when T is about 20 h, it is optimal to rest in most of this region. Thus larger shrews should be likely to respond to a bad period by resting, whereas small species should respond by being active. Hanski found this trend over four species of shrews.

The models of Janetos and Hanski are concerned with an animal that faces an environment that is changing but whose parameters are known to the animal. One can also look at moving versus staying in the context of uncertainty about environmental parameters.

For a review of data on search modes see McLaughin (1989). For a model of area-restricted search see Knoppien and Reddingius (1985). For a saltatory search (moving between periods of searching while stationary), see O'Brien et al. (1989).

6 Travelling in Groups

There has been considerable discussion of the energetic consequences of being in a group. For birds see Lissaman and Shollenberger (1970), Heppner (1974), Higdon and Corrsin (1978), Badgerow (1988), Norberg (1990). For fish see Weihs (1973), Pitcher (1986).

7 Avian Flight

Although models of avian flight differ in their details, they agree in predicting a U-shaped relationship between speed v and rate of energy expenditure $P(v)$. The basic reason for this sort of curve is described in Chap. 3. In this chapter I will take the relationship for granted and explore some of its consequences. One obvious point is that there is a speed at which rate of expenditure is minimized. Because rate of expenditure is power, this velocity will be referred to as the minimum power speed v_{mp}. An example is shown in Fig. 3. The figure also shows another speed that might be advantageous. This is the speed at which the energy expended per unit distance flown in minimized. The time to fly unit distance is $1/v$ and the rate of expenditure is $P(v)$, so this velocity minimizes $P(v)/v$. This speed also maximizes

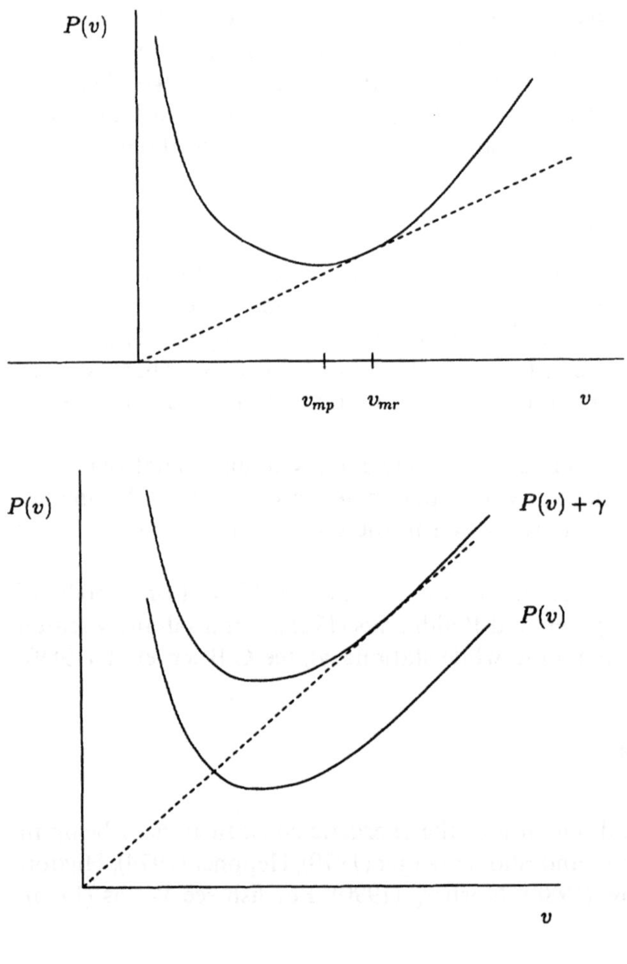

Fig. 3. a The power expended in flight $P(v)$ as a function of speed v. Power is minimized at the minimum power speed v_{mp}. Energy expended per unit distance is minimized at the maximum range speed v_{mr}, which can be found by constructing a tangent to the power curve from the origin. The effect of wind speed can be incorporated by changing the point from which the tangent is constructed as described by Pennycuick (1978). **b** The graphical method suggested by Norberg (1981a) for solving the equation $dP(v^*)/dv = [\gamma + P(v^*)]/v^*$. The lower curve is $P(v)$, the upper curve is $P(v) + \gamma$. By the argument behind **a**, the tangent from the origin to this curve solves the equation

the animal's flight range for a given amount of energy spent, and so it is referred to as the maximum range speed v_{mr}. Figure 3a shows how v_{mr} can be found graphically. The logic is very similar to that used in the graphical solution of the patch use problem (Fig. 1). Any line from the origin to the power curve $P(v)$ has slope $P(v)/v$, so the line with the minimum slope must be tangent to the curve. Pennycuick (1978) describes how the figure can be modified to include the effect of wind speed.

Norberg (1981a) argues that a parent bird should use a flight speed that is greater than v_{mr}. Norberg's argument is based on the overall rate of energy Y delivered to the young. Assume that the parent collects an amount of energy e in time h and that the round-trip distance that it flies is D. If the bird flies at speed v then it spends a time D/v flying and hence its energy consumption is $P(v)D/v$. The time required to replace this amount of energy is $P(v)D/v\gamma$, where γ is the rate at which the parent can gain energy. If we ignore other sources of energy consumption then

$$Y(v) = e/[h + D/v + P(v)D/v\gamma]. \quad (12)$$

It follows that Y is maximized by choosing v so that the time

$$\varphi = D/v + P(v)D/v\gamma \quad (13)$$

is minimized. The optimal speed v^* can be found from the condition

$$d\varphi/dv = 0. \quad (14)$$

This amounts to requiring that at v^* the decrease in time that results from flying slightly faster is equal to the increase in time that results from the need to replace the extra energy spent. From Eqs. (13) and (14) it follows that v^* is given by the equation

$$dP(v^*)/dv = [\gamma + P(v^*)]/v^*. \quad (15)$$

Norberg (1981a) shows how this equation can be solved graphically. The procedure is illustrated in Fig. 3b. The maximum range speed emerges as a special case corresponding to $\gamma = 0$. If we rewrite Eq. (15) as

$$v^* dP(v^*)/dv - P(v^*) = \gamma$$

we see that it has the same form as Eq. (11). It can be seen from Eq. (15) that v^* does not depend on the energy gained e, the time to collect it h or the distance D. McLaughlin and Montgomerie (1985) use data obtained from female Lapland longspurs (*Calcarius lapponicus*) feeding dependent young to test Norberg's model. Unfortunately, they assume that γ is based on the net rate of foraging for the young [c.f. Eq. (11)], rather than the rate at which the parent can gain energy for itself. On the basis of this assumption, they concluded that Norberg's model predicts an increase in flight speed with increasing rate of delivery of energy to the nestlings, as indicated by nestling age. The absence of such a correlation was taken as evidence against Norberg's model. Given that γ is the parent's rate, Norberg's model does not predict a correlation between flight speed and nestling age. Although this removes one problem from Norberg's model, it creates another.

McLaughlin and Montgomerie performed a quantitative test of Norberg's model by finding optimal speed for $\gamma = 2.22$ W (based on the rate of delivery of energy to the young). The observed speed was less than the predicted speed. The female can presumably achieve a higher rate when feeding herself than when feeding her young, so γ will be greater than 2.22 W and the predicted speed will be even further from the observed speed. There is, however, still a possibility that Norberg's general idea can account for the data. Houston (1986) argued that flight was likely to involve an energetic cost $c(v)$ that depends on the flight velocity v. One possibility is that at the start of each flight the energy required is given by the associated kinetic energy (e.g. Oster 1976; for a discussion of the costs of take-off and landing, see Norberg 1990, pp 146–151). As long as dc/dv, d^2c/dv^2 and d^2P/dv^2 are all positive, then the optimal velocity will increase with the round trip distance D (Houston 1986). The parameters concerning how energy is collected for the young still have no effect on the optimal velocity. For low values of D, the increase in v^* with D is quite marked. McLaughlin and Montgomerie (1985) report such an increase with distance for flights of less than 50 m, but McLaughlin and Montgomerie (1990) find that the quantitative predictions of this approach do not agree with the data.

Schnell and Hellack (1979) and Pennycuick (1987) found that flight speeds tend to be between v_{mp} and v_{mr}, whereas Gill (1985) and Blake et al. (1990) observed speeds that were greater than v_{mr}.

7.1 The Effect of Mass

For some animals, the food that they obtain on a foraging trip may significantly change their energetic expenditure while flying. This effect has been investigated in bees (Schmid-Hempel et al. 1985; Kacelnik et al. 1986) and in hummingbirds (De Benedictis et al. 1978; Tamm 1989). Even if the gross gain from foraging is constant, so that the loading curve $G(t)$ is a straight line, the increase in energetic expenditure with load can make the net gain a decelerating function of time spent collecting food. Schemid-Hempel et al. (1985) and Kacelnik et al. (1986) compared the predictions of maximizing net rate of energetic gain and maximizing energetic efficiency concerning how big a load a bee should take before returning to the hive. Under both currencies it can be optimal for the bee to return before its crop is full, but the predictions using energetic efficiency are much closer to the data.

An increase in a bird's mass will increase its metabolic expenditure at any flight speed. It will also increase v_{mp} and v_{mr} (e.g. Pennycuick 1975). Many birds become lighter while feeding their young (see Moreno 1989 for a review). Freed (1981) and Norberg (1981c) argued that instead of being an indication of energetic stress, this reduction in mass may be a strategy for reducing the energetic cost of flight. Norberg (1981c) analyses the parent's energy saving based on the assumption that the bird files at v_{mr}. Houston (1986) looks at the saving when parents use the velocity that maximizes the overall rate to the young. Computations based on a starling, including a cost of take-off based on kinetic energy, showed an increase in velocity as mass decreased, but the effect was small.

Videler et al. (1988) show that kestrels reduce their flight speed when lead weights are attached to their feet. A general analysis of the effect of changes in mass on optimal flight speed has not been attempted, but it is possible that the maximization of net rate of energetic gain in a model that includes a mass-dependent take-off cost will produce the observed effects.

For flight and predation see Pomeroy (1990), for flight and injury see Cuthill and Guildford (1990).

8 Diving

Many birds and mammals hunt for food by diving underwater. The need to return to the surface to breathe puts constraints on foraging (e.g. Dunstone and O'Connor 1979; Houston and McNamara 1985). Kramer (1988) uses a "marginal value" argument to explore the effect of the depth to which an animal dives on the length of time that the animal spends at the surface.

A diving animal may use either aerobic or anaerobic respiration (e.g. Kooyman 1989). Although anaerobic respiration is less efficient in energetic terms, it enables the animal to make longer dives. Ydenberg and Clark (1989) consider the conditions under which anaerobic respiration should be used. They develop a model based on the western grebe (*Aechmorphus occidentalis*) hunting for fish. When fish occur in dense shoals that are hard to find and are likely to be lost if the grebe has to surface, a grebe will catch more fish over a given period by using anaerobic rather than aerobic respiration.

For dividing and foraging in ducks, see Tome (1988) and Ball (1990).

9 Predator–Prey Interactions

A simple model of which types of prey a predator should eat can be constructed from the Holling disc equation. It can be shown that a prey type that has energy content e_i and handling time h_i should be included in the diet if and only if

$$e_i/h_i > N^*,$$

where N^* is the net rate of energetic gain under the optimal rule for choosing prey items (see for example Charnov 1976b). This rule has an obvious resemblance to the MVT – see Eq. (4). Anholt et al. (1987) investigate a modification of the prey choice model in which the predator has to pursue its prey. If the probability of capture declines with the time spent chasing the prey, then it can be advantageous for the predator to abandon the chase and search for a new prey item. In fact, the pursuit of the prey can be thought of as the exploitation of a patch, with the time at which the predator gives up corresponding to leaving the patch. Anholt et al. show in the context of a single prey type that the predator should give up when a suitably defined local rate falls to N^*. This local rate is based on the expectation of capture a very sort time into the future (cf. the reward rate defined by McNamara and Houston 1985). There are several reasons why this expectation might decrease with t: the predator may tire sooner than the prey, or the predator might become aware that this particular prey item is more than usually fast. In the standard model of prey choice, a given prey type is either always included in the optimal diet or always rejected – all encounters with a type have the same outcome. Furthermore, the outcome depends only on items in the optimal diet, because only these items determine N^*. In the model based on pursuit times, the fact that it can be optimal for a predator to give up chasing a prey type means that some outcomes with a prey type result in capture and some do not. For two prey types with $e_1/h_1 > e_2/h_2$, increasing the encounter rate with type 1 items results in a gradual decrease of type 2 items in the predator's diet.

The model of Anholt et al. (1987) looks at the predator–prey interaction from the predator's point of view; Weihs and Webb (1984) investigate how a potential prey item should behave if it wishes to escape. Weihs and Webb consider two aspects of this problem. In avoidance, the prey animal see the predator before the predator sees it. The prey animal might behave so as to minimize the probability that the predator will detect it or to improve its chances of escape should a chase develop. Once the predator has started its pursuit of a prey animal, then the prey's problem is one of evasion.

Weihs and Webb show that the probability of detection by the predator is minimized by the prey animal moving along a straight line at right angles to the line that gives the limit of the predator's angle of sight. This statement of the solution is for a moving coordinated system based on the predator's head. A prey animal that wishes to maximize its distance from the predator at any future time should in general always move directly away from the predator. Weihs and Webb suggest that if all members of a group of fish adopt this rule when a predator is approaching but has not yet fixed upon a target, the fish will spread round the predator, producing the "fountain" effect (e.g. Potts 1970; Nursall 1973; Magurran and Pitcher 1987). Stotting, a peculiar gait used by some antelopes when disturbed by predators, is discussed by Caro (1986) and Fitzgibbon and Fanshawe (1988).

10 The Ideal Free Distribution

Fretwell and Lucas (1970) introduced a model for how animals arriving in an area should choose a habitat, but their idea provides a general principle for how animals should be distributed. They assumed that habitats differed in quality, such that a solitary animal in a good quality habitat would have a higher fitness than a solitary animal in a poor quality habitat. They also assumed that in each habitat an animal's fitness decreased as the number of conspecifies in the habitat increased. Given that animals are aware of habitat quality and are free to move between habitats, animals arriving in an area will start by choosing the best habitat. They will continue to choose this habitat until the number of animals in this habitat reduces its quality to below that of the next best habitat. Animals will choose both these habitats until their quality is reduced below that of the next best habitat, and so on. The resulting pattern is known as the ideal free distribution (IFD). At the IFD, no animal can increase its fitness by moving to another habitat. In the original model of Fretwell and Lucas, all the animals are equal, in the sense that they have the same ability to compete for resources. At the IFD, all animals have the same fitness. Parker and Sutherland (1986) discuss the effects of differences in competitive ability. The IFD is still characterized by the fact that no animal can increase its fitness by moving, but it is not longer the case that all animals have the same fitness. Animals that have a high competitive ability have higher fitness than animals with low competitive ability, but all animals with the same ability have the same fitness. For a different model of competition see Korona (1989).

The IFD has been used to study how a group of animals uses two feeding areas. A typical experiment involves providing food at rate r_1 in patch 1 and at rate r_2 in patch 2. Assume that all animals have the same competitive ability and an individual in a group of n_i animals in patch i achieves an intake rate r_i/n_i. If maximizing intake rate maximizes fitness then the IFD is given by $r_1/n_1 = r_2/n_2$. Milinski (1988) reviews this area. McNamara and Houston (1990a) find the IFD when animals maximize survival probability rather than intake rate.

The reasoning behind the IFD can also be applied to animals searching for mates (Parker 1978). The dungfly *Scatophaga stercoraria* lays its eggs in cow pats. Females tend to approach fresh pats from downwind, flying over the pat to land upwind and walk back to the pat. Once a male finds a female, he mounts her and mates. The male remains on the female while she oviposits. A male can achieve a mating either by finding an unaccompanied female or by displacing another male from a female (see Parker 1978 for a review). Shortly after the cow pat has been deposited, the displacement of other males is rare enough to be ignored in calculating reproductive success. Parker (1974) used the IFD to predict the distribution of males in various zones around the cow pat and found that it provided a good account of the data. Davies and Halliday (1979) obtained comparable results when male toads were searching for females.

For a model that links decisions about when to leave a patch to the distribution of animals, see Bernstein et al. (1988).

11 Maximizing Rate of Reproduction Versus Maximizing Total Reproduction

The dung fly system also provides an example of the MVT. The "patch" is the female, and the question is how long a male should spend in copulation. On the basis of sperm competition experiments it is possible to estimate the proportion of a female's eggs that a male will fertilise as a function of the duration t of copulation. This proportion increases with t but at a decreasing rate, so that eventually it is advantageous for a male to abandon the current female and search for a new one (Parker and Stuart 1976). The travel time in this case is the time for a male to find a female, and has a mean of 156.5 min. With an function $1-\exp(-t/16)$ fitted to the data on the proportion of eggs fertilised, the optimal time for a male to spend in copulation is 41.4 min, which is close to the observed value of 35.5 min (Parker and Stuart 1976).

The model used by Parker and Stuart (1976) maximizes the rate at which the male obtains fertilizations. If the mating season has a fixed duration, then a male's fitness is maximized by maximizing this rate. This assumption may not be realistic for an insect. It is quite likely that a male dung fly will die during the mating season. This possibility can be captured by working directly with the male's expected reward, where the reward is measured in terms of eggs fertilised. In essence, the possibility of death discounts future matings. Houston and McNamara (1986) develop a model in which the rate of mortality while travelling between patches is β and the rate of mortality while on a patch is α. The time to travel between patches is denoted by τ, and the instantaneous reward rate after a time t has been spent on the patch is $r(t)$. Let F^* be the expected future total reward for an animal which has just arrived at a patch and follows the optimal policy. Then it is optimal to stay on a patch if and only if

$$r(t) > \alpha e^{-\beta\tau} F^*.$$

When $\alpha = \beta = \mu$, the time on a patch that is predicted by this condition is greater than the time predicted by rate maximization, but approaches the rate-maximizing solution as μ decreases. It is worth emphasising that this result is based on a constant value of τ. In this model behaviour is not determined by the expected travel time but by the distribution of travel times.

12 The Effects of Predation

The maximization of total reward in the example of copula duration in the dung fly illustrates one way in which differential mortality can be included in models. This section looks at the risk of mortality in animals that are not reproducing. It is clear that the models based solely on energy that were introduced at the start of the chapter cannot be applied in such cases.

In general, the expected contribution to fitness that results from moving to a given habitat depends not just on expected energetic gain but also on the probability of being killed by a predator. There is considerable evidence that habitat choice depends on both these things – see Lima and Dill (1990) for a review. In the context of a growing animal it is possible to obtain a formula for how to choose between habitats that differ in intake rate and predation rate. Assume that an animal must reach some critical size before it can reproduce, and that its fitness does not depend on the time at which this size is reached. (For the effect of a time constraint, see

Ludwig and Rowe 1990). Then the animal's fitness will be maximized by minimizing the mortality associated with reaching this size. If predation is the only source of mortality, then the animal should minimize μ/g, where μ is the mortality rate (i.e. probability of death per unit time) and g is the energetic gain. (For further discussion and generalization see Gilliam 1982; Werner and Gilliam 1984). This approach has been applied successfully to habitat choice by bluegill sunfish (*Lepomis macrochirus*). Werner et al. (1983) studied bluegills of three sizes in an experimentally manipulated pond. In the absence of predators, the size-specific choice of habitat was predicted quite accurately by the maximization of net rate of energetic gain. In the presence of predators (largemouth bass, *Micropterus salmoides*), the two larger size classes still chose the habitat that gave the highest net rate of energetic return, whereas the smallest size class shifted to a habitat with a lower rate of gain. This size class was the only one that incurred significant mortality in the presence of the predators. Werner and Hall (1988) studied habitat use by bluegills in several small lakes with depths between 10 and 15 m. Bluegills hatch in the littoral zone and then move to the pelagic zone, where they feed on zooplankton. Werner and Hall found that in four lakes the bluegills return to the littoral when they are about 12.5 mm long. They stay for several years, feeding on invertebrates, after which they return to the pelagic zone. Across five lakes the size of fish when this move occurred was between 50 and 83 mm, and was correlated with the density of largemouth bass. Werner and Hall showed that all sizes of bluegill have a higher feeding rate in the pelagic zone, but predation on small bluegills is 40 to 80 times higher in open water than it is in vegetation. The two switches between habitats can be understood in terms of the size-dependent mortalities and growth rates shown in Fig. 4.

Gilliam (1990) shows how the μ/g criterion can be used to predict patch-residence time. If the mortality rate while on the patch is α and the mortality rate while travelling is β, then the mortality rate while foraging is

$$\mu = \frac{\alpha t + \beta \tau}{t + \tau}$$

and the gross rate of gain is

$$\frac{G(t)}{t + \tau},$$

so μ/g is minimized by maximizing $G(t)/(\alpha t + \beta \tau)$. This is equivalent to maximizing

$$\frac{G(t)}{t + \tau\alpha/\beta},$$

from which it can be seen that the graphical approach used in Fig. 1 can be modified to include this case by replacing the travel time τ by $\tau\alpha/\beta$. Thus if $\alpha = \beta$, the patch residence time that minimizes μ/g also maximizes gross rate. This is in contrast to the model based on total expected reward (Houston and McNamara 1986) in which it is optimal to stay longer than the rate maximizing solution when $\alpha = \beta$. The effect is small when α and β are small. In fact, Gilliam's result is a

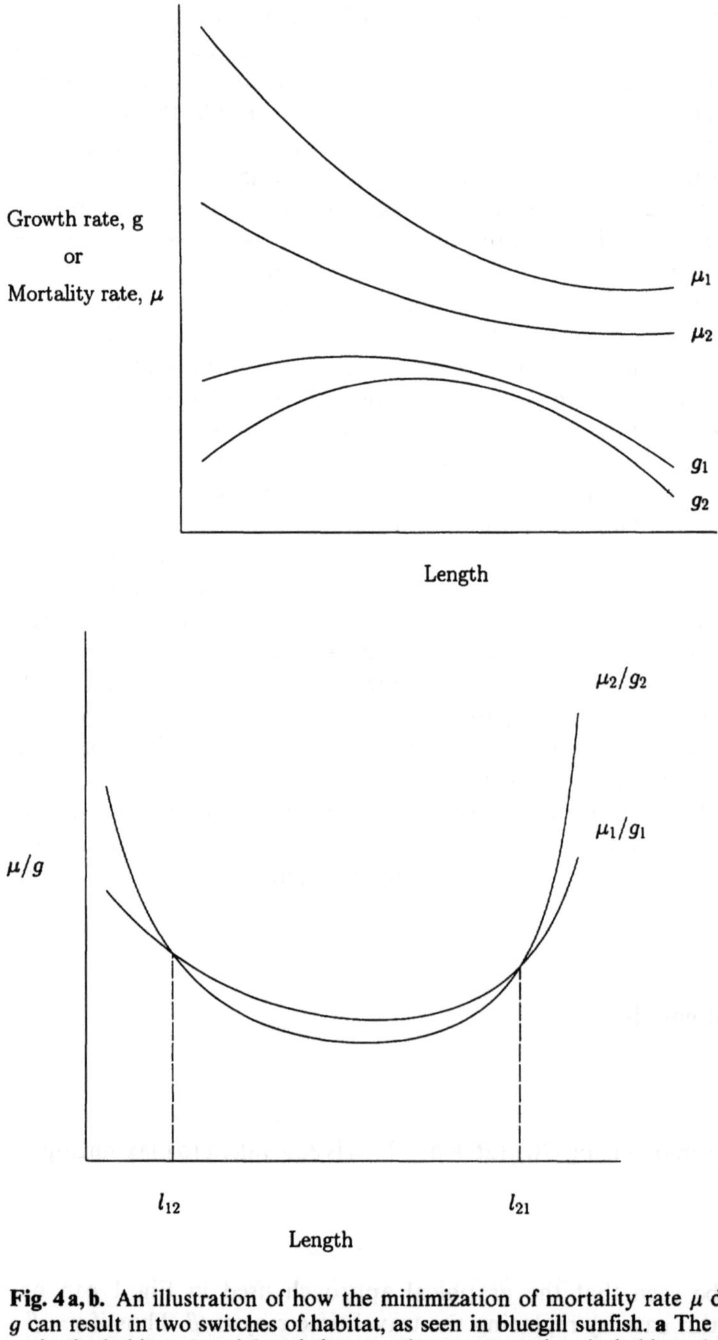

Fig. 4 a, b. An illustration of how the minimization of mortality rate μ divided by growth rate g can result in two switches of habitat, as seen in bluegill sunfish. **a** The mortality rates μ_1 and μ_2 in the habitats 1 and 2 and the growth rates g_1 and g_2 in habitats 1 and 2 as a function of body size, indicated by length. **b** The ratio μ/g in each habitat, obtained from the curves in **a**. The optimal choice of habitat minimizes μ/g. A fish that starts with a length that is less than l_{12} should stay in habitat 1 until its length gets above l_{12}, whereupon it should move to habitat 2 until its length gets above l_{21}, when it should move to habitat 1. (After Werner and Hall 1988)

special case of the results of Houston and McNamara, corresponding to low mortalities – see Houston and McNamara (1986), Appendix 4.

The μ/g currency is based on an animal that is growing but not reproducing. No comparable formula for combining the influence of mortality rate and energetic gain has been obtained for animals that are neither growing nor reproducing, but some general principles have emerged from the dynamic programming approach developed by McNamara and Houston (1986) and Mangel and Clark (1986, 1988). The essence of these models is that the animal is characterised by its level of energy reserves, x. When x falls to zero, the animal dies of starvation. A policy is a rule for choosing between habitats as a function of x. The optimal policy maximizes the animal's long-term survival probability. When reserves are low, the danger of starvation may make it worth choosing a habitat with a high expected gain, even though it also has a high associated risk of predation. As reserves increase, avoiding predation becomes relatively more important (McNamara 1990a, b; McNamara and Houston 1990b).

13 Migration

Baker (1978) provides a comprehensive review of this subject, but this definition of migration as "the act of moving from one spatial unit to another" has been criticised as too broad. Sinclair (1983) defines migration as "a regular round trip within a life-span of the individual", and this is the sense in which the term will be used in this chapter.

13.1 Diadromy

A diadromous fish lives in both freshwater and saltwater during the course of its lifetime. Within this category, anadromous fish migrate from saltwater to spawn in freshwater, whereas catadromous fish migrate from freshwater to spawn in saltwater. There is a tendency for anadromy to be associated with temperate latitudes and catadromy to be associated with tropical latitudes (Baker 1978; Gross et al. 1988). Baker was puzzled by this pattern, but suggested that a possible explanation might be based on the pattern of food availability.

Gross et al. (1988) give such an explanation. In temperate latitudes, the sea tends to be more productive than freshwater, resulting in higher growth rates in saltwater. In tropical latitudes the relationship is reversed, so that the sea is less productive than freshwater.

It is likely that saltwater and freshwater habitats will differ in terms of predation risk as well as food availability, and that predation risk in a given habitat will depend on size (see Ricker 1976 for data on salmon). Some idea of the importance of these effects can be obtained by using the μ/g currency.

For migration in birds, see Biebach et al. (1986), Alerstam and Lindström (1990). For stable migration patterns, see Lundberg and Alerstam (1986).

14 Dispersal

Sinclair (1983) defines dispersal as a one-way movement with no pre-determined direction. Hamilton and May (1977) show that dispersal can be favoured by natural selection even in a stable and saturated environment. They introduce their analysis with a schematic model based on a parthenogenetic species that lives in an environment with a fixed number of sites. In each site just one adult can live. At the end of a fixed season each adult produces a given number of offspring. The mother determines the fraction u of these offspring that will migrate to other sites. The fraction $1-u$ of offspring remain at their mother's site. All these offspring survive, whereas the migrants have a probability p of surviving. The migrants are distributed equally over all sites. Once this process is complete, all offspring at a site have an equal chance of becoming the reproducing adult at the site. Hamilton and May point out that if nearly all members of a population do not send out migrants, then a mutant individual that does send out migrants will have a higher fitness than other members of the population. This is because the mutant has some chance of reproducing in sites other than its own, whereas the other members of the population only have a chance of reproducing in their own site. The evolutionarily stable strategy (ESS) in this case is to disperse a fraction u^* of offspring, where

$$u^* = 1/(2-p).$$

Taylor (1988) generalises this result and discusses it in the context of kin selection. Crespi and Taylor (1990) apply a particular model of this type to data on thrips (*Hoplothhrips karnyi*).

15 Concluding Remarks

The optimization of locomotion is a vast topic and I have not attempted to cover all the issues. Instead, I have tried to concentrate on simple models about when to move and how fast to move. I hope that the examples that I have used illustrate how the application of optimization theory to locomotion can link an animal's physiology, behaviour and ecology.

Acknowledgments. I thank Neill Alexander, Anne Magurran, Ulla Norberg and Wayne Thompson for comments on previous versions of this chapter. This work was supported by the Natural Environment Research Council.

References

Alerstam T, Lindström Å (1990) Optimal bird migration: the relative importance of time, energy and safety. In: Gwinner E (ed) Bird migration. Springer, Berlin Heidelberg New York Tokyo

Anholt BR, Ludwig D, Rasmussen JB (1987) Optimal pursuit times: how long should predators pursue their prey? Theor Popul Biol 31: 453–464

Badgerow JP (1988) An analysis of function in the formation flight of Canada Geese. Auk 105: 749–755

Baker RR (1978) The evolutionary ecology of animal migration. Hodder and Stoughton, London

Ball JP (1990) Active diet selection or passive reflection of changing food availability: the underwater foraging behaviour of canvasback ducks. In: Hughes RN (ed) Bebavioural mechanisms of food selection. Springer, Berlin Heidelberg New York Tokyo, pp 95–109

Bernstein C, Kacelnik A, Krebs JR (1988) Individual decisions and the distribution of predators in a patchy environment. J Anim Ecol 57: 1007–1026

Biebach H, Friedrich W, Heine G (1986) Interaction of bodymass, fat, foraging and stopover period in trans-Sahara migrating passerine birds. Oecologia 69: 370–379

Blake RW, (1985) A model of foraging efficiency and daily energy budget in the black skimmer (*Rynchops nigra.*) Can J Zool 63: 42–48

Blake RW, Kolotylo R, de la Cueva H (1990) Flight speeds of the barn swallow, *Hirundo rustica.* Can J Zool 68: 1–5

Carlson A, Moreno J (1982) The loading effect in central place foraging Wheatears (*Oenanthe oenanthe* L). Behav Ecol Sociobiol 11: 173–185

Caro TM (1986) The functions of stotting: a review of the hypotheses. Anim Behav 34

Charnov EL (1976a) Optimal foraging: the marginal value theorem. Theor Popul Biol 9: 129–136

Charnov EL (1976b) Optimal foraging: attack strategy of a mantid. Am Nat 110: 141–151

Crespi BJ, Taylor PD (1990) Dispersal rates under variable patch density. Am Nat 135: 48–62

Cuthill I, Guildford T (1990) Perceived risk and obstacle avoidance in flying birds. Anim Behav 40: 188–190

Davies NB, Halliday TR (1979) Competitive mate seaching in male common toads. *Bufo bufo.* Anim Behav 27: 1253–1267

DeBenedictis PA, Gill FB, Hainsworth FR, Pyke GH, Wolf LL (1978) Optimal meal size in hummingbirds. Am Nat 112: 301–316

Dunstone N, O'Connor RJ (1979) Optimal foraging in an amphibious mammal. I. The aqualung effect. Anim Behav 27: 1182–1194

FitzGibbon CD, Fanshawe JH (1988) Stotting in Thomson's gazelles: an honest signal of condition. Behav Ecol Sociobiol 23: 69–74

Freed LA (1981) Loss of mass in breeding wrens: stress or adaptation? Ecology 62: 1179–1186

Fretwell DS, Lucas HL (1970) On territorial behaviour and other factors influencing habitat distribution in birds. Acta Biotheor 19: 16–36

Gendron RP, Staddon JER (1983) Searching for cryptic prey: the effect of search rate. Am Nat 121: 172–185

Gendron RP, Staddon JER (1984) A laboratory simulation of foraging behavior: the effect of search rate on the probability of detecting prey. Am Nat 124: 407–415

Gerritsen J (1984) Size efficiency reconsidered: a general foraging model for free-swimming aquatic animals. Am Nat 123: 450–467

Gill FB (1985) Hummingbird flight speeds. Auk 102: 97–101

Gilliam JF (1982) Foraging under mortality risk in size-structured populations. PhD Thesis, State Univ, Michigan

Gilliam JF (1990) Hunting by the hunted: optimal prey selection by foragers under predation hazard. In: Hughes RN (ed) Behavioral mechanisms of food selection. Springer, Berlin Heidelberg New York Tokyo, pp 797–819

Gross MR, McDowall RM, Coleman RM (1988) Aquatic productivity and the evolution of diadromous fish migration. Science 239: 1291–1293

Hainsworth FR (1991) Discrimination between foraging rules and why hummingbirds hover. Anim Behav 41: 902–903

Hamilton WD, May RM (1977) Dispersal in stable habitats. Nature (Lond) 269: 578–581

Hanski I (1985) What does a shrew do in an energy crisis? In: Sibly RM, Smith RH (eds) Behavioural ecology: ecological consequences of adaptive behaviour. Blackwell, Oxford, pp 247–252
Heppner F (1974) Avian flight formations. Bird-Banding 45: 160–169
Higdon JJL, Corrsin S (1978) Induced drag of a bird flock. Am Nat 112: 724–744
Holling CS (1966) The functional response of invertebrate predators to prey density. Mem Entomol Soc Can 48: 1–86
Houston AI (1986) The optimal flight velocity for a bird exploiting patches of food. J Theor Biol 119: 345–362
Houston AI (1987) Optimal foraging by parent birds feeding dependent young. J Theor Biol 124: 251–274
Houston AI (1990) Parent birds feeding young – to walk or to fly. J Theor Biol 142: 141–147
Houston AI, McNamara JM (1985) A general theory of central-place foraging for single-prey loaders. Theor Popul Biol 28: 233–262
Houston AI, McNamara JM (1986) The influence of mortality on the behaviour that maximizes reproductive success in a patchy environment. Oikos 47: 267–274
Ivlev VS (1960) On the utilization of food by planktophage fishes. Bull Math Biophys 22: 371–389
Janetos AC (1982) Active foragers versus sit-and-wait predators: a simple model. J Theor Biol 95: 381–385
Johns M, Miller RG (1963) Average renewal loss rate. Ann Math Stat 34: 396–401
Kacelnik A (1984) Central place foraging in starlings (*Sturnus vulgaris*) I: Patch residence time. J Anim Ecol 53: 283–300
Kacelnik A, Houston A (1984) Some effects of energetic costs on foraging strategies. Anim Behav 32: 609–614
Kacelnik A, Houston A, Schmid-Hempel P (1986) Central-place foraging in honey-bees: the effects of travel time and nectar flow on crop-filling. Behav Ecol Sociobiol 19: 19–24
Knoppien P, Reddingius J (1985) Predators with two modes of searching: A mathematical model. J Theor Biol 114: 273–301
Kooyman GL (1989) Diverse divers. Springer, Berlin Heidelberg New York Tokyo
Korona F (1989) Ideal free distribution of unequal competitors can be determined by the form of competition. J Theor Biol 138: 347–352
Kramer DL (1988) The behavioral ecology of air breathing by aquatic animals. Can J Zool 66: 89–94
Lima SL, Dill LM (1990) Behavioral decisions made under the risk of predation: a review and prospectus. Can J Zool 68: 619–640
Lissaman PBS, Shollenberger CA (1970) Formation flight of birds. Science 168: 1003–1005
Ludwig D, Rowe L (1990) Life-history strategies for energy gain and predator avoidance under time constraints. Am Nat 135: 686–707
Lundberg S, Alerstam T (1986) Bird migration patterns: conditions for stable geographical population segregation. J Theor Biol 123: 403–414
Magurran AE, Pitcher RJ (1987) Provenance, shoal size and the sociobiology of predator-evasion behaviour in minnow shoals. Proc R Soc Lond Ser B229: 439–465
Mangel M, Clark CW (1986) Towards a unified foraging theory. Ecology 67: 1127–1138
Mangel M, Clark CW (1988) Dynamic modeling in behavioral ecology. Princeton Univ Press, Princeton, NJ
Masman D, Daan S, Dijkstra C (1988) Time allocation in the kestrel (*Falco tinnunculus*) and the principle of energy maximization. J Anim Ecol 57: 411–432
McLaughlin RL (1989) Search modes of birds and lizards: evidence for alternative movement patterns. Am Nat 133: 654–670
McLaughlin RL, Montgomerie RD (1985) Flight speeds of central place foragers: female Lapland longspurs feeding nestlings. Anim Behav 33: 810–816
McLaughlin RL, Montgomerie RD (1990) Flight speeds of parent birds feeding nestlings: maximization of foraging efficiency or food delivery rate? Can J Zool 68: 2269–2274
McNamara JM (1990a) The policy which maximizes long-term survival of an animal faced with the risks of starvation and predation. Adv Appl Probab 22: 295–308
McNamara JM (1990b) The starvation-predation trade-off and some behavioural and ecological consequences. In: Hughes RN (ed) Behavioural mechanisms of food selection. Springer, Berlin Heidelberg New York Tokyo, pp 39–59

McNamara JM, Houston A (1985) A simple model of information use in the exploitation of patchily distributed food. Anim Behav 33: 553–560
McNamara JM, Houston AI (1986) The common currency for behavioral decisions. Am Nat 127: 358–378
McNamara JM, Houston AI (1990a) The state-dependent ideal-free distribution. Evol Ecol 4: 298–311
McNamara JM, Houston AI (1990b) Stavation and predation in a patchy environment. In: Swingland I, Shorrocks B (eds) Living in a patchy, environment. Oxford Univ Press, Oxford, pp 23–43
Milinski M (1988) Games fish play: making decisions as a social forager. Trends in Ecol Evol 3: 325–330
Moreno J (1989) Strategies of mass change in breeding birds. Biol J Linn Soc 37: 297–310
Norberg RÅ (1977) An ecological theory on foraging time and energetics and choice of optimal food searching method. J Anim Ecol 46: 511–529
Norberg RÅ (1981a) Optimal flight speed in birds when feeding young. J Anim Ecol 50: 473–477
Norberg RÅ (1981b) Why foraging birds should climb and hop upwards rather than downwards. Ibis 123: 281–288
Norberg RÅ (1981c) Temporary weight decrease in breeding birds may result in more fledged young. Am Nat 118: 838–850
Norberg RÅ (1983) Optimal locomotion modes of foraging birds in trees. Ibis 125: 172–180
Norberg UM (1990) Vertebrate flight. Springer, Berlin Heidelberg New York Tokyo
Norton-Griffiths M (1969) The organisation, control, and development of parent feeding in the oystercatcher (*Haematopus ostralegus*). Behaviour 34: 55–114
Nursall JR (1973) Some behavioural interactions of spottail shiners (*Notropis hudsonius*) yellow perch (*Perca flavescens*) and northern pike (*Esox lucius*). J Fish Res Board Can 30: 1161–1178
O'Brien WJ, Evans BI, Browman HI (1989) Flexible search tactics and efficient foraging in saltatory searching animals. Oecologia 80: 100–110
Oster G (1976) Modeling social insect populations. I. Economics of foraging and population growth in bumblebees. Am Nat 110: 215–245
Parker GA (1974) The reproductive behavior and the nature of sexual selection in *Scatophaga sterroraria* L. IX. Spatial distribution of fertilization rates and evolution of male search strategy within the reproductive area. Evolution 28: 93–108
Parker GA (1978) Searching for mates. In: Krebs JR, Davies NB (eds) Behavioural ecology. Blackwells, Oxford, pp 214–244
Parker GA, Stuart RA (1976) Animal behavior as a strategy optimizer: evolution of resource assessment strategies and optimal emigration thresholds. Am Nat 110: 1055–1076
Parker GA, Sutherland WJ (1986) Ideal free distributions when individuals differ in competitive ability: phenotype-limited ideal free models. Anim Behav 34: 1222–1242
Pennycuick CJ (1975) Mechanics of flight. In: Farner DS, King JR (eds) Avian biology. Academic Press, New York, pp 1–75
Pennycuick CJ (1978) Fifteen testable predictions about bird flight. Oikos 30: 165–176
Pennycuick CJ (1987) Flight of Auks (Alcidae) and other northern seabirds compared with southern Procellariiformes: ornithodolite observations. J Exp Biol 128: 335–347
Pitcher TJ (1986) Functions of shoaling behaviour in teleosts. In: Pitcher TJ (eds) The behaviour of teleost fishes. Croom Helm, London, pp 294–337
Pomeroy D (1990) Why fly? The possible benefits for lower mortality. Biol J Linn Soc 40: 53–65
Potts GW (1970) The schooling ethology of *Lutjianus monostigma* in the shallow reef environment of Aldabra. J Zool (Lond) 161: 223–235
Pyke GH (1981a) Optimal travel speeds of animals. Am Nat 118: 475
Pyke GH (1981b) Why hummingbirds hover and honeyeaters perch. Anim Behav 29: 861–867
Ricker WE (1976) Review of the rate of growth and mortality of Pacific salmon in salt water and noncatch mortality caused by fishing. J Fish Res Board Can 33: 1483–1524
Robinson SK (1986) Three-speed foraging during the breeding cycle of yellow- rumped caciques (Icterinae: *Cacicus cela*). Ecology 67: 394–405
Schmid-Hempel P, Kacelnik A, Houston A (1985) Honeybees maximise efficiency by not filling their crop. Behav Ecol Sociobiol 17: 61–66
Schnell GD, Hellack JJ (1979) Bird flight speeds in nature: optimized or a compromise? Am Nat 113: 53–66

Sinclair ARE (1983) The function of distance movement in vertebrates. In: Swingland IR, Greenwood PJ (eds) The ecology of animal movement. Clarendon Press, Oxford, pp 240–258
Tamm S (1989) Importance of energy costs in central place foraging by hummingbirds. Ecology 70: 195–205
Taylor PD (1988) An inclusive fitness model for dispersal of offspring. J Theor Biol 130: 363–378
Tome MW (1988) Optimal foraging: food patch depletion by ruddy ducks. Oecologia 76: 27–36
Trillmich KGK, Trillmich F (1986) Foraging strategies of the marine iguana, *Amblyrhynchus cristatus*. Behav Ecol Sociobiol 18: 259–266
Videler JJ, Vossebelt G, Gnodde M, Groenewegen A (1988) Indoor flight experiments with trained kestrels. I. Flight strategies in still air with and without added weight. J Exp Biol 134: 173–183
Ware DM (1975) Growth, metabolism and optimal swimming speed of a pelagic fish. J Fish Res Board Can 32: 33–41
Weihs D (1973) Hydromechanics of fish schooling. Nature (Lond) 241: 290–291
Weihs D, Webb WP (1984) Optimal avoidance and evasion tactics in predator-prey interactions. J Theor Biol 106: 189–206
Welham CVJ, Ydenberg RC (1988) Net energy versus efficiency maximizing by foraging ring-billed gulls. Behav Ecol Sociobiol 23: 75–82
Werner EE, Gilliam JF (1984) The ontogenetic niche and species interactions in size-structured populations. Annual Rev Ecol Syst 15: 393–425
Werner EE, Gilliam JF, Hall DJ, Mittelbach GG (1983) An experimental test of the effects of predation risk on habitat use in fish. Ecology 64: 1540–1548
Werner EE, Hall DJ (1988) Ontogenetic habitat shifts in bluegill: the foraging rate – predation risk trade-off. Ecology 69: 1352–1366
Ydenberg RC, Clark CW (1989) Aerobiosis and anaerobiosis during diving by western grebes: an optimal foraging approach. J Theor Biol 139: 437–449

Subject Index

MIX
Papier aus verantwortungsvollen Quellen
Paper from responsible sources
FSC® C105338

If you have any concerns about our products,
you can contact us on
ProductSafety@springernature.com

In case Publisher is established outside the EU,
the EU authorized representative is:
Springer Nature Customer Service Center GmbH
Europaplatz 3, 69115 Heidelberg, Germany

Printed by Libri Plureos GmbH
in Hamburg, Germany